Cambridge Studies in Advanced Mathematics 53

GROUPS AS GALOIS GROUPS

1 W.M.L. Holcombe *Algebraic automata theory*
2 K. Petersen *Ergodic theory*
3 P.T. Johnstone *Stone spaces*
4 W.H. Schikhof *Ultrametric calculus*
5 J.-P. Kahane *Some random series of functions, 2nd edition*
6 H. Cohn *Introduction to the construction of class fields*
7 J. Lambek & P.J. Scott *Introduction to higher-order categorical logic*
8 H. Matsumura *Commutative ring theory*
9 C.B. Thomas *Characteristic classes and the cohomology of finite groups*
10 M. Aschbacher *Finite group theory*
11 J.L. Alperin *Local representation theory*
12 P. Koosis *The logarithmic integral I*
13 A. Pietsch *Eigenvalues and s-numbers*
14 S.J. Patterson *An introduction to the theory of the Riemann zeta-function*
15 H.J. Baues *Algebraic homotopy*
16 V.S. Varadarajan *Introduction to harmonic analysis on semisimple Lie groups*
17 W. Dicks and M. Dunwoody *Groups acting on graphs*
18 L.J. Corwin and F.P. Greenleaf *Representations of nilpotent Lie groups and their applications*
19 R. Fritsch and R. Piccinini *Cellular structures in topology*
20 H. Klingen *Introductory lectures on Siegel modular forms*
21 P. Koosis *The logarithmic integral II*
22 M.J. Collins *Representations and characters of finite groups*
24 H. Kunita *Stochastic flows and stochastic differential equations*
25 P. Wojtaszczyk *Banach spaces for analysts*
26 J.E. Gilbert and M.A.M. Murray *Clifford algebras and Dirac operators in harmonic analysis*
27 A. Frohlich and M.J. Taylor *Algebraic number theory*
28 K. Goebel and W.A. Kirk *Topics in metric fixed point theory*
29 J.F. Humphreys *Reflection groups and Coxeter groups*
30 D.J. Benson *Representations and cohomology I*
31 D.J. Benson *Representations and cohomology II*
32 C. Allday and V. Puppe *Cohomological methods in transformation groups*
33 C. Soule et al. *Lectures on Arakelov geometry*
34 A. Ambrosetti and G. Prodi *A primer of nonlinear analysis*
35 J. Palis and F. Takens *Hyperbolicity, stability and chaos at homoclinic bifurcations*
37 Y. Meyer *Wavelets and operators 1*
38 C. Weibel *An Introduction to homological algebra*
39 W. Bruns and J. Herzog *Cohen–Macaulay rings*
40 V. Snaith *Explicit Brauer induction*
41 G. Laumon *Cohomology of Drinfeld modular varieties: Part I*
42 E.B. Davies *Spectral theory of differential operators*
43 J. Diestel, H. Jarchow, and A. Tonge *Absolutely summing operators*
44 P. Mattila *Geometry of sets and measures in Euclidean spaces*
45 R. Pinsky *Positive harmonic functions and diffusion*
46 G. Tenenbaum *Introduction to analytic and probabilistic number theory*
47 C. Peskine *Complex projective geometry*
48 Y. Meyer and R. Coifman *Wavelets and operators II*
49 R. Stanley *Enumerative combinatorics*
50 I. Porteous *Clifford algebras and the classical groups*

GROUPS AS GALOIS GROUPS

An Introduction

HELMUT VÖLKLEIN
University of Florida
and
Universität Erlangen

CAMBRIDGE UNIVERSITY PRESS
Cambridge, New York, Melbourne, Madrid, Cape Town, Singapore, São Paulo

Cambridge University Press
The Edinburgh Building, Cambridge CB2 8RU, UK

Published in the United States of America by Cambridge University Press, New York

www.cambridge.org
Information on this title: www.cambridge.org/9780521562805

First published 1996
This digitally printed version 2008

A catalogue record for this publication is available from the British Library

Library of Congress Cataloguing in Publication data
Völklein, Helmut.
Groups as Galois groups: an introduction / Helmut Völklein.
p. cm. – (Cambridge studies in advanced mathematics; 53)
Includes index.
ISBN 0-521-56280-5 (hardcover)
1. Inverse Galois theory. I. Title. II. Series.
QA247.V65 1996
512′.3 – dc20 95-46746
CIP

ISBN 978-0-521-56280-5 hardback
ISBN 978-0-521-06503-0 paperback

MEINEN ELTERN
Max und Edeltraut Völklein
UND MEINER FRAU
Sommai
IN LIEBE GEWIDMET

Contents

Preface

The goal of the book is to lead the reader to an understanding of recent results on the Inverse Galois Problem: The construction of Galois extensions of the rational field $\mathbb{Q}$ with certain prescribed Galois groups. Assuming only a knowledge of elementary algebra and complex analysis, we develop the necessary background from topology (Chapter 4: covering space theory), Riemann surface theory (Chapters 5 and 6), and number theory (Chapter 1: Hilbert's irreducibility theorem). Classical results like Riemann's existence theorem and Hilbert's irreducibility theorem are proved in full, and applied in our context. The idea of **rigidity** is the basic underlying principle for the described construction methods for Galois extensions of $\mathbb{Q}$.

From the work of Galois it emerged that an algebraic equation $f(x) = 0$, say over the rationals, is solvable by radicals if and only if the associated Galois group G_f is a solvable group. As a consequence, the general equation of degree $n \geq 5$ cannot be solved by radicals because the group S_n is not solvable.

This idea of encoding algebraic–arithmetic information in terms of group theory was the beginning of both Galois theory and group theory. Nowadays we learn basic Galois theory in every first-year algebra course. It has become one of the guiding principles of algebra. One aspect of the theory that remains unsatisfactory is the fact that it is very hard to compute the Galois group of a given polynomial. Therefore, the full correspondence between equations of degree n and subgroups of S_n can only be worked out for very small values of n. Since it is probably impossible to get a full understanding of this correspondence for general n, one is naturally led to the following more reasonable question: Do at least all subgroups of S_n occur in this correspondence, that is, does every subgroup of S_n correspond to some equation of degree n? The most important case is that of irreducible equations, which correspond to the transitive subgroups of S_n.

This question is one formulation of the Inverse Problem of Galois Theory. It is often just called the Inverse Galois Problem. Hilbert was the first to study this problem. His irreducibility theorem shows that it suffices to realize groups as Galois groups over the function field $\mathbb{Q}(x)$. This allows us to use methods from Riemann surface theory and algebraic geometry. Hilbert applied his method to obtain Galois realizations of the symmetric and alternating groups. The next milestone was Shafarevich's realization of all solvable groups over $\mathbb{Q}$ (in the 1950s). His approach is purely number-theoretic, and does not extend to nonsolvable groups.

The classification of finite simple groups, completed around 1980, gave a new direction to the work on the Inverse Galois Problem. It now seemed natural to concentrate first on the simple groups, and get the composite groups later by some kind of inductive procedure. It is not yet clear how this inductive procedure – or embedding problem, in technical terms – would work in general. There are, however, quite a few results in this direction, for example, Serre's obstruction theory for central extensions and Matzat's notion of GAR-realization for extensions with centerless kernel. The latter works best if one wants to realize Galois groups over the full cyclotomic field $\mathbb{Q}_{ab}$, instead of over $\mathbb{Q}$ (because all embedding problems over $\mathbb{Q}_{ab}$ with abelian kernel are solvable). If every nonabelian finite simple group has a GAR-realization over $\mathbb{Q}_{ab}(x)$, then the Inverse Galois Problem has a positive solution over $\mathbb{Q}_{ab}$. Moreover, the lattice of all algebraic extensions of $\mathbb{Q}_{ab}$ would then be known. In technical terms, the absolute Galois group of $\mathbb{Q}_{ab}$ would be a free profinite group of countable rank. The latter is known as Shafarevich's conjecture. We will describe the notion of GAR-realization – a Galois realization with particular extra properties – and the related notions of GAL-realization and GAP-realization in Chapter 8.

The above justifies focusing on the simple groups, more generally, on almost simple groups (i.e., groups between a simple group and its automorphism group). That is what the main body of this book is about. It uses the geometric approach of Hilbert, coupled with the idea of **rigidity** (as Thompson called it). The rigidity criterion (in its various versions) gives purely group-theoretic conditions that force a finite group to occur as a Galois group over $\mathbb{Q}$ (actually over every hilbertian field of characteristic 0). It is generally believed to have been found independently by Belyi, Matzat, and Thompson in the early 1980s. But it should be remarked that it is contained implicitly as a special case in earlier work of Fried ([Fr1] 1977).

The elementary level of our approach is the main difference between this and existing books on the subject by Matzat [Ma1] and Serre [Se1], and the forthcoming book [MM] by Malle and Matzat, which give a much higher level presentation. It has not been my goal to state each result in its greatest generality; rather I have tried to give an introduction to the various ideas involved in the subject. Accordingly, there is no claim for completeness. Omissions include the theory of nonsplit abelian embedding problems and the construction of rigid triples in Lie type groups. For a quite complete description of the known results on the Inverse Galois Problem we refer the reader to [MM]. The same holds true for tracing the origin of results – I have tried to attribute proper credit where it seemed appropriate, but again there is no claim for completeness.

Another related topic that is not touched here is the problem of constructing *explicit* polynomials with a given Galois group. There are quite a few results on this, notably polynomials over $\mathbb{Q}$ found by Malle, Matzat, and others, often with the aid of a computer, see [Ma1], [Malle2]. More recently, Abhyankar [Abh2] has found infinite series of polynomials in positive characteristic with various classical groups as Galois groups.

One particular simplification in the first part of the book is that we avoid the descent from $\mathbb{C}$ to $\bar{\mathbb{Q}}$ (usually done by Weil's descent theory), by a simple trick involving Hilbert's irreducibility theorem. This descent is needed in the second part of the book, however, and we present it in Chapter 7, using the Bertini–Noether Lemma. Further, we avoid the technicalities necessary to introduce profinite groups, and phrase everything in terms of finite Galois extensions. Thus it is hoped that now celebrated results – like Thompson's realization of the monster group – become accessible to a wider mathematical audience.

More recent approaches, based on the earlier work of Fried, try to replace the rigidity conditions by the use of moduli spaces and the braid group action. An introduction to this is given in Chapters 9 and 10. We cannot give a full treatment of this theory because it requires deeper methods of algebraic geometry and several complex variables. More important, this theory is still very much in the process of being shaped, connecting, for example, to recent work of Drinfeld, Ihara, and others on the Grothendieck–Teichmüller group, work of Fried on modular towers, and other topics. In addition, the extension to the generalized braid groups introduced by Brieskorn (and possibly other fundamental groups) is yet to be developed.

To keep in line with the main theme of this book – the idea of rigidity – Chapters 9 and 10 show how the braid group action on generating systems naturally arises from the study of weak rigidity. We derive the resulting Outer Rigidity Criterion using the higher-dimensional version of Riemann's existence theorem (which we cannot prove here).

Finally, Chapter 11 gives an introduction to Harbater's patching method. It is essentially independent of the rest of the book. The idea is to imitate the analytic theory of Chapters 4 to 6 for base fields other than the complex numbers. Complex analysis is replaced by ultrametric analysis, which works over any field that is complete with respect to an ultrametric absolute value.

Actually, for our approach very little is required from ultrametric analysis, and we develop it in the first two sections. Riemann's existence theorem does not generalize in its full strength, but certain substitutes are obtained (that also hold over fields of positive characteristic). Results include the solution of the Inverse Galois Problem over the fields $\mathbb{Q}_p(x)$ (where $\mathbb{Q}_p$ is the p-adic field) and a proof of the "geometric case of Shafarevich's conjecture."

The first part of the book (Chapters 1 to 6) gives a full proof of the basic rigidity criteria for the realization of groups as Galois groups. Chapter 1 (Hilbert's irreducibility theorem) is essentially independent of the rest (except for some very basic definitions and lemmas), in the methods as well as in the results. Chapter 1 gives the logical foundation for the other chapters, however; they are concerned with realizing groups G over $\mathbb{Q}(x_1, \ldots, x_n)$, whereas Chapter 1 shows that then G also occurs as a Galois group over $\mathbb{Q}$. The first two sections of Chapter 1 suffice for this conclusion. On a first reading, it may be advisable to skip Chapter 1 and take Hilbert's irreducibility theorem for granted.

Chapter 2 formulates the algebraic version of Riemann's existence theorem and draws some corollaries. Chapter 3 derives the rigidity criterion and gives applications. Chapter 4 is purely topological. It is applied in Chapter 5 to reduce the algebraic version of Riemann's existence theorem to the analytic version. The analytic version is proved in Chapter 6.

The exposition in the second part of the book (especially Chapters 9 and 10) proceeds at a faster pace, whereas I have taken care to keep the first part quite elementary. The first part could be used for a one-semester course for second year graduate students.

This book grew out of notes taken by Ralph Frisch during a course I gave at Erlangen in the summer of 1991. I thank Ralph for his enthusiasm and diligence. Thanks for long years of encouragement, beginning with my first steps in mathematics, are due to Karl Strambach, my teacher and friend. I thank M. Jarden and H. Matzat for long-term invitations to the Institute for Advanced Studies in Jerusalem and to the IWR at the University of Heidelberg, respectively. Further, I thank G. Malle and P. Müller for a critical reading of parts of the manuscript. I acknowledge various remarks and discussions from several colleagues, in particular the above-mentioned and W.-D. Geyer, D. Haran, K. Magaard, J.-P. Serre, J. Thompson, and M. van der Put. Above all, I want to express my deep indebtedness to Mike Fried who introduced me to this exciting area, and in countless conversations and e-mail messages has shared his profound knowledge freely with me.

Helmut Völklein Gainesville

Notation

We let $\mathbb{N}$, $\mathbb{Z}$, $\mathbb{Q}$, $\mathbb{R}$, and $\mathbb{C}$ denote the set of natural numbers and integers, and the field of rational, real, and complex numbers, respectively.

For G a group, $\mathrm{Aut}(G)$ (resp., $\mathrm{Inn}(G)$) denotes its automorphism group (resp., group of inner automorphisms). $Z(G)$ denotes the center of G, and $C_G(H)$ the centralizer of H in G. The direct resp., semi-direct, product of groups is denoted by $G \times H$ resp., $G \cdot H$. All group actions are from the left, unless otherwise stated. A conjugacy class of a group is called nontrivial if it is different from $\{1\}$ (the class consisting of the neutral element).

The symbol $:=$ means "defined to be equal to." (Thus $x := 2$ means x is defined to be 2.) If K/k is a field extension, we let $\mathrm{Aut}(K/k)$ denote the group of automorphisms of K that are the identity on k. If K/k is Galois, $G(K/k)$ (=$\mathrm{Aut}(K/k)$) denotes the Galois group; for any subfield L of K invariant under $G(K/k)$ we let $\mathrm{res}_{K/L}$ denote the restriction homomorphism $G(K/k) \to G(L/k \cap L)$. If U is a subgroup of $G(K/k)$, then K^U denotes the fixed field of U. If K/k and L/k are field extensions, a k-isomorphism from K to L is an isomorphism that is the identity on k. We let $\bar{k}$ denote an algebraic closure of k.

We use the abbreviation "FG-extension" for "finite Galois extension."

Part One

The Basic Rigidity Criteria

1
Hilbert's Irreducibility Theorem

The definition and basic properties of hilbertian fields are given in Section 1.1. Section 1.2 contains the proof of Hilbert's irreducibility theorem (which says that the field $\mathbb{Q}$ is hilbertian). We give the elementary proof due to Dörge [Do] (see also [La]).

Section 1.3 is not necessary for someone interested only in Galois realizations over $\mathbb{Q}$. It centers around Weissauer's theorem, which shows that many infinite algebraic extensions of a hilbertian field are hilbertian. As our main application we deduce that the field $\mathbb{Q}_{ab}$ generated by all roots of unity is hilbertian. Next to $\mathbb{Q}$ itself, this field is the one that has attracted the most attention in the recent work on the Inverse Galois Problem. This is due to Shafarevich's conjecture (see Chapter 8).

In this chapter, k denotes a field of characteristic 0. (Most results remain true in positive characteristic, with suitable modifications; see [FJ], Chs. 11 and 12.) We let $x, y, x_1, x_2, \ldots$ denote independent transcendentals over k. Thus $k[x_1, \ldots, x_m]$ is the polynomial ring, and $k(x_1, \ldots, x_m)$ the field of rational functions over k in $x_1, \ldots, x_m$.

1.1 Hilbertian Fields

1.1.1 Preliminaries

We will use elementary Galois theory, as developed in most introductory algebra books, without further reference. (See, e.g., [Jac], I, Ch. 4). The most useful single result will be Artin's theorem (saying that if G is a finite group of automorphisms of a field K then K is Galois over the fixed field F of G and $G(K/F) = G$).

If K is a field with subfield k, we say K is *regular over* k if k is algebraically closed in K.

Lemma 1.1 *Suppose $x_1, \dots, x_m$ are algebraically independent over k, and set $\mathbf{x} = (x_1, \dots, x_m)$. Let $\bar{k}$ be an algebraic closure of k.*

(i) If k'/k is finite Galois, then $k'(\mathbf{x})/k(\mathbf{x})$ is finite Galois, and the restriction map $G(k'(\mathbf{x})/k(\mathbf{x})) \to G(k'/k)$ is an isomorphism. In particular, every field between $k(\mathbf{x})$ and $k'(\mathbf{x})$ is of the form $k''(\mathbf{x})$, and $[k''(\mathbf{x}) : k(\mathbf{x})] = [k'' : k]$.

(ii) Let $f(\mathbf{x}, y) \in k(\mathbf{x})[y]$ be irreducible over $k(\mathbf{x})$, and let $K = k(\mathbf{x})[y]/(f)$ be the corresponding field extension of $k(\mathbf{x})$. Then K is regular over k if and only if f is irreducible over $\bar{k}(\mathbf{x})$. If this holds, then $f(\mathbf{x}, y)$ is irreducible over $k_1(\mathbf{x})$ for every extension field k_1 of k such that $x_1, \dots, x_m, y$ are independent transcendentals over k_1.

Proof. (i) The group $G = G(k'/k)$ acts naturally on $k'(\mathbf{x})$ (fixing $x_1, \dots, x_m$), with fixed field $k(\mathbf{x})$. By Artin's theorem, $k'(\mathbf{x})/k(\mathbf{x})$ is Galois with group G. The last part of (i) follows now by using the Galois correspondence.

(ii) Let $\hat{k}$ be the algebraic closure of k in K, and let α be the image of y in K (thus $f(\alpha) = 0$). Then α satisfies a polynomial $\hat{f}(y) \in \hat{k}(\mathbf{x})[y]$ of degree $[K : \hat{k}(\mathbf{x})]$, and $\hat{f}$ divides f. It follows that if $\hat{k} \neq k$ then f is not irreducible in $\hat{k}(\mathbf{x})[y]$, hence not in $\bar{k}(\mathbf{x})[y]$.

Conversely, assume $\hat{k} = k$, and let k' be any finite Galois extension of k. Let K' be the composite of K and $k'(\mathbf{x})$ inside some algebraic closure of $k(\mathbf{x})$. By (i) we have $K \cap k'(\mathbf{x}) = k''(\mathbf{x})$ for some k'' between k and k'. Then $k'' \subset \hat{k}$, hence $k'' = k$. Thus $K \cap k'(\mathbf{x}) = k(\mathbf{x})$. Since $k'(\mathbf{x})/k(\mathbf{x})$ is Galois (by (i)), it follows that $[K' : k'(\mathbf{x})] = [K : k(\mathbf{x})]$. But $K' = k'(\mathbf{x})[\alpha]$, hence f is irreducible over $k'(\mathbf{x})$. Since k' was an arbitrary finite Galois extension of k, it follows that f is irreducible over $\bar{k}(\mathbf{x})$.

For the last claim, suppose f decomposes as $f = gh$ for $g, h \in k_1(\mathbf{x})[y]$, of degree ≥ 1 in y. Without loss, g is monic in y. We may assume that k_1 is generated over k by the coefficients of g (where g is viewed as a rational function in $x_1, \dots, x_m, y$), and that one such coefficient, call it t, is transcendental over k. By Remark 1.2 below, k_1 is finite over a field $k_2 = k(t_1, \dots, t_s)$, where $t_1, \dots, t_s$ are independent transcendentals over k, and $t = t_1$. There is an infinite subset $A \subset \mathrm{Aut}(k_2/k)$ such that all $\alpha \in A$ take distinct values on t (e.g., $\alpha(t) = t + c, c \in k$, and $\alpha(t_i) = t_i$ for $i > 1$). These α can be extended to embeddings of k_1 into $\bar{k}_2$, and further to embeddings of $k_1(\mathbf{x})[y]$ into $\bar{k}_2(\mathbf{x})[y]$ (fixing $x_1, \dots, x_m, y$). Applying these embeddings to g we obtain infinitely many (distinct) divisors of f in $\bar{k}_2(\mathbf{x})[y]$, all of them monic in y. This contradiction completes the proof. □

Remark 1.2 Suppose $k_1 = k(a_1, \dots, a_r)$ is a finitely generated extension of k. If $t_1, \dots, t_s$ is a collection of elements among $a_1, \dots, a_r$, maximal with respect to being algebraically independent over k, then k_1 is finite over the purely transcendental extension $k(t_1, \ldots, t_s)$ of k. (Indeed, k_1 is finitely generated and algebraic, hence finite over $k(t_1, \ldots, t_s)$.)

Lemma 1.3 *Let α be algebraic over the field L. Let $f(y) = \sum_{i=0}^{n} a_i y^i$ be a polynomial over L of degree $n > 0$ with $f(\alpha) = 0$. Then*

$$g(Y) = Y^n + \sum_{i=0}^{n-1} a_i a_n^{n-i-1} Y^i$$

is a monic polynomial of degree n with $g(a_n\alpha) = 0$. Clearly, $L(\alpha) = L(a_n\alpha)$.

Proof. Clear. □

Let $f(y) \in D[y]$ be a polynomial over the factorial domain D of degree ≥ 1. Recall that $f(y)$ is irreducible in $D[y]$ if and only if it is irreducible in $F[y]$, where F is the field of fractions of D. Further, $f(y)$ is called primitive if it is nonzero, and the g.c.d. of its nonzero coefficients is 1. If $g(y)$ is a nonzero polynomial over F, then there is $d \in F$, unique up to multiplication by units of D, such that $d \cdot g(y)$ is primitive. Further, a polynomial ring in any (finite) number of variables over a field is factorial. (For all this, see, e.g., [Jac], I, Ch. 2.)

Lemma 1.4 *Let $f(x_1, \ldots, x_s)$ be a polynomial in $s \geq 2$ variables over k, of degree ≥ 1 in x_s. Then f is irreducible as polynomial in s variables if and only if f is irreducible and primitive when viewed as polynomial in x_s over the ring $D = k[x_1, \ldots, x_{s-1}]$. Note that f is irreducible over D if and only if f is irreducible over $F = k(x_1, \ldots, x_{s-1})$.*

Proof. First assume f is irreducible and primitive when viewed as polynomial in x_s over D. If then $f = gh$ for polynomials g, h in $x_1, \ldots, x_s$ then one of these polynomials, say g, must actually be a polynomial in $x_1, \ldots, x_{s-1}$. Since f is primitive, it follows that g is a unit in D, hence $g \in k$. This proves that f is irreducible as a polynomial in s variables. The converse is clear. For the last statement in the Lemma, see above. □

1.1.2 Specializing the Coefficients of a Polynomial

First a basic lemma about specializing a Galois extension. This lemma will be used several times, in particular in Chapter 10 for a problem in positive characteristic. Therefore we allow fields of any characteristic (just in this Lemma 1.5).

Recall that a polynomial (in one variable) is called separable if it has no multiple roots. The discriminant of a monic polynomial $p(y)$ is a polynomial function (over $\mathbb{Z}$) in the coefficients of p. It is nonzero if and only if p is separable.

Lemma 1.5 *Let K/F be a finite Galois extension with Galois group G. Let R be a subring of F, having F as a field of fractions. Let α be a generator for K over F, satisfying $f(\alpha) = 0$ for some monic polynomial $f(y) \in R[y]$ of degree $n = [K : F]$. Finally, let A be a finite subset of K containing α, and invariant under G. Let $S = R[A]$ (the subring of K generated by R and A). Then there is $u \neq 0$ in R such that for each (ring-) homomorphism ω from R to a field F' satisfying $\omega(u) \neq 0$ the following holds:*

1. *ω extends to a homomorphism $\tilde{\omega} : S \to K'$, where K' is a finite field extension of F'. We may assume that K' is generated over F' by $\tilde{\omega}(S)$.*
2. *For each such $\tilde{\omega}$, the field K' is Galois over F', and is generated over F' by $\alpha' = \tilde{\omega}(\alpha)$. We have $f'(\alpha') = 0$, where $f'(y) \in F'[y]$ is the polynomial obtained by applying ω to the coefficients of f. Thus $[K' : F'] = [K : F]$ if and only if f' is irreducible. In this case, K' is F'-isomorphic to $F'[y]/(f')$.*
3. *Now suppose f' is irreducible. Then for each $\tilde{\omega}$ as in (1), there is a unique isomorphism $G \to G' = G(K'/F')$, $\sigma \mapsto \sigma'$, such that $\tilde{\omega}(\sigma(s)) = \sigma'(\tilde{\omega}(s))$ for all $\sigma \in G$, $s \in S$.*

Proof. Since K/F is Galois, the polynomial $f(y)$ is separable, hence its discriminant D_f is a nonzero element of R. Further, $\omega(D_f)$ is the discriminant of the polynomial $f'(y)$ obtained by applying ω to the coefficients of f. We will only consider such ω with $\omega(D_f) \neq 0$. Then $f'(y)$ is separable.

The ideal I of $R[y]$ generated by f is the kernel of the natural map $R[y] \to R[\alpha]$, $h \mapsto h(\alpha)$. Indeed, if $h \in R[y]$ with $h(\alpha) = 0$ then by elementary field theory we have $h = gf$ for some $g \in F[y]$. Write $f = \sum_{i=0}^{n} a_i y^i$, $g = \sum_{j=0}^{m} b_j y^j$ with $a_i \in R$, $b_j \in F$. Since f is monic in y, it follows that $b_m \in R$ (because it equals the highest y-coefficient of h). The second highest y-coefficient of h equals $b_{m-1} + b_m a_{n-1}$, hence $b_{m-1} \in R$. Continuing like this, we see that all $b_j \in R$. Hence $g \in R[y]$, and thus $h \in I$. This yields a natural isomorphism

$$\phi : R[y]/I \to R[\alpha].$$

Step 1 We first consider the special case that $R[A] = R[\alpha]$. We show that (1)–(3) hold for each homomorphism $\omega : R \to F'$ with $\omega(D_f) \neq 0$.

Extend ω to a map $R[y] \to F'[y]$ (fixing y). This map sends f to f', hence induces a homomorphism

$$\psi : R[y]/I = R[y]/fR[y] \to F'[y]/f'F'[y] = F'[y]/(f').$$

Let

$$\chi = \psi \circ \phi^{-1} : R[\alpha] \to F'[y]/(f').$$

(1). Set $K' = F'[y]/(g')$, where g' is an irreducible factor of f'. Then K' is a finite field extension of F'. Composing χ with the natural map $F'[y]/(f') \to F'[y]/(g') = K'$ we obtain a homomorphism $S = R[\alpha] \to K'$ that extends ω. This proves (1).

(2). We have $K' = F'[\tilde{\omega}(S)] = F'[\tilde{\omega}(\alpha)] = F'[\alpha']$ (because $S = R[\alpha]$ by hypothesis in Step 1).

The conjugates $\alpha_1, \ldots, \alpha_n$ of α over F all lie in $A \subset S$ (by hypothesis). Let $\alpha'_1, \ldots, \alpha'_n$ be their $\tilde{\omega}$-images. Applying $\tilde{\omega}$ to $f(y) = (y - \alpha_1) \cdots (y - \alpha_n)$ we get $f'(y) = (y - \alpha'_1) \cdots (y - \alpha'_n)$. Hence K' contains all conjugates of α' over F', and therefore is normal over F'. Also, K'/F' is separable (since f' is), hence K'/F' is Galois. The rest of (2) is clear.

(3). Assume f' is irreducible. Then $\alpha'_1, \ldots, \alpha'_n$ are all conjugate over F' (and are pairwise distinct since f' is separable). Thus for each $i = 1, \ldots, n$ there is a unique $\sigma'_i \in G' = G(K'/F')$ mapping α' to α'_i. Also, there is a unique $\sigma_i \in G = G(K/F)$ mapping α to α_i. Thus $\sigma_i \mapsto \sigma'_i$ is a bijection from G to G'.

Now fix some s in $S = R[\alpha]$. We can write it in the form $s = h(\alpha)$ with $h(y) \in R[y]$. Let $h'(y) \in F'[y]$ be obtained by applying ω to the coefficients of h. Then $\sigma'_i(\tilde{\omega}(s)) = \sigma'_i(\tilde{\omega}(h(\alpha))) = \sigma'_i(h'(\alpha')) = h'(\alpha'_i) = \tilde{\omega}(h(\alpha_i)) = \tilde{\omega}(\sigma_i(h(\alpha))) = \tilde{\omega}(\sigma_i(s))$. This proves that $\sigma'(\tilde{\omega}(s)) = \tilde{\omega}(\sigma(s))$ for all $s \in S$ and $\sigma \in G$. In particular, $(\sigma\tau)'(\alpha') = (\sigma\tau)'(\tilde{\omega}(\alpha)) = \tilde{\omega}(\sigma\tau(\alpha)) = \sigma'(\tilde{\omega}(\tau(\alpha))) = \sigma'\tau'(\alpha')$. Thus the map $\sigma \mapsto \sigma'$ is homomorphic, hence isomorphic. This proves (3).

Step 2 The general case.

Each $a \in A$ can be written as

$$a = \sum_{i=0}^{n-1} b_i \alpha^i$$

with $b_i \in F$. Choose $v \neq 0$ in R such that $vb_i \in R$ for all occurring b_i (as a ranges over A). This is possible because F is the field of fractions of R. Set $u = vD_f$ and $\tilde{R} = R[u^{-1}]$. Then all $b_i \in \tilde{R}$, hence $A \subset \tilde{R}[\alpha]$ and so $\tilde{R}[A] = \tilde{R}[\alpha]$.

If $\omega : R \to F'$ is a homomorphism with $\omega(u) \neq 0$, then ω extends uniquely to a homomorphism $\tilde{R} \to F'$. Now apply Step 1 to $\tilde{R}$, and we are done. □

The next Lemma can be viewed as a very weak analogue of Hilbert's irreducibility theorem (noting that an irreducible polynomial in characteristic 0 is separable). We use the phrase "for almost all" to mean "for all but finitely many."

Lemma 1.6 *Let L be a field, and $f(x, y) \in L[x, y]$ separable as polynomial in y over $L(x)$. Then the specialized polynomial $f(b, y) \in L[y]$ is separable for almost all $b \in L$.*

Proof. By Lemma 1.3 we may assume f is monic as polynomial in y. Its discriminant is an element $D(x) \in L[x]$, nonzero because f is separable (in y). For each $b \in L$, the polynomial $f(b, y) \in L[y]$ has discriminant $D(b)$. Thus $f(b, y)$ is separable for all $b \in L$ different from the roots of $D(x)$. □

Proposition 1.7 *Let K be a Galois extension of $k(x)$ of finite degree $n > 1$. Then there is a polynomial $f(x, y) \in k[x, y]$, monic and of degree n in y, and a generator α of K over $k(x)$ with $f(x, \alpha) = 0$. Further:*

(i) For almost all $b \in k$ the following holds: If the specialized polynomial $f_b(y) := f(b, y)$ is irreducible in $k[y]$, then the field $k[y]/(f_b)$ is Galois over k, with Galois group isomorphic to $G = G(K/k(x))$.

(ii) Suppose ℓ is a finite extension of k contained in K. Let $h(x, y) \in \ell[x, y]$ be irreducible as polynomial in y over $\ell(x)$, and assume the roots of this polynomial are contained in K. Then for almost all $b \in k$ the following holds: If $f(b, y)$ is irreducible in $k[y]$, then $h(b, y)$ is irreducible in $\ell[y]$.

(iii) There is a finite collection of polynomials $p_I(x, y) \in k[x][y]$, irreducible and of degree >1 when viewed as polynomial in y over $k(x)$, such that for almost all $b \in k$ the following holds: If none of the specialized polynomials $p_I(b, y) \in k[y]$ has a root in k, then $f(b, y)$ is irreducible in $k[y]$.

Proof. Each generator α of K over $k(x)$ satisfies some polynomial $f(y)$ of degree n over $k(x)$. Multiplying f by some element of $k[x]$ we may view $f = f(x, y)$ as polynomial in two variables over k. By Lemma 1.3 we may assume that f is monic in y. Thus $f(y) = (y - \alpha_1) \cdots (y - \alpha_n)$, where $\alpha_1, \ldots, \alpha_n$ are the conjugates of α over $k(x)$.

For $b \in k$ let $\omega_b : k[x] \to k$ be the evaluation homomorphism $h(x) \mapsto h(b)$. We apply Lemma 1.5 with $F = k(x)$, and with $\omega = \omega_b : R = k[x] \to F' = k$. Then $f'(y)$ (obtained by applying ω to the coefficients of $f(y)$) equals the polynomial $f_b(y) = f(b, y)$. Let $u = u(x) \in R = k[x]$ be as in Lemma 1.5. Then $\omega_b(u) = u(b)$, hence assertions (1) to (3) in Lemma 1.5 hold for all $b \in k$ different from the finitely many roots of u. Assertions (2) and (3) imply claim (i).

Assume from now on that $b \in k$ is not a root of u. Then ω_b extends to $\tilde{\omega} : S \to K'$ where S is a subring of K containing $k[x][\alpha_1, \ldots, \alpha_n]$, and K' a finite Galois extension of k generated by $\tilde{\omega}(S)$. Let $\alpha'_1, \ldots, \alpha'_n$ be the $\tilde{\omega}$-images of $\alpha_1, \ldots, \alpha_n$. Then $f_b(y) = (y - \alpha'_1) \cdots (y - \alpha'_n)$.

(iii). Let I be a proper, nonempty subset of $\{1, \ldots, n\}$. Since f is irreducible as polynomial in y over $k(x)$, the partial product $\prod_{i \in I}(y - \alpha_i)$ cannot lie in $k(x)[y]$. Thus it has some coefficient d_I with $d_I \notin k(x)$. This d_I lies in S (since the α_i are in S), and it satisfies some irreducible polynomial p_I over $k(x)$ of degree >1. We may choose p_I to have coefficients in $k[x]$.

Now assume that f_b is not irreducible. Then there is some I as above such that the polynomial $\prod_{i \in I}(y - \alpha'_i)$ lies in $k[y]$. It follows that $c := \tilde{\omega}(d_I)$ lies in k (since it is a coefficient of this polynomial). Applying $\tilde{\omega}$ to the equation $p_I(x, d_I) = 0$ we obtain $p_I(b, c) = 0$. This proves (iii).

(ii). Assume f_b is irreducible, and write h as

$$h(x, y) = h_0(x) \prod_{i=1}^{t} (y - \beta_i) \tag{1.1}$$

with $h_0(x) \in \ell[x]$ and $\beta_i \in K$. We may assume that the β_i lie in the finite set A from Lemma 1.5, hence in S. Set $\beta'_i = \tilde{\omega}(\beta_i)$.

We may further assume that A contains a generator of ℓ over k. Then $\ell \subset S$. Thus $\tilde{\omega}$ maps the field ℓ isomorphically to a subfield of K' that we identify with ℓ (via $\tilde{\omega}$). Under this identification we get

$$h(b, y) = h_0(b) \prod_{i=1}^{t} (y - \beta'_i)$$

(applying $\tilde{\omega}$ to (1.1)). Further, the map in assertion (3) of Lemma 1.5 maps the subgroup $H = G(K/\ell(x))$ of G onto a subgroup H' of $G(K'/\ell)$.

Since h is irreducible as polynomial in y over $\ell(x)$, it is separable (since $\mathrm{char}(k) = 0$) and the group $H = G(K/\ell(x))$ permutes its roots β_i transitively. Then H' permutes the β'_i transitively. Exclude those finitely many b with $h_0(b) = 0$, and those for which $h(b, y)$ is not separable (see Lemma 1.6). Then the polynomial $h(b, y)$ is separable, and the group $H' \subset G(K'/\ell)$ permutes its roots β'_i transitively. Hence $h(b, y)$ is irreducible over ℓ. □

Corollary 1.8 *The following conditions on k are equivalent:*

(1) For each irreducible polynomial $f(x, y)$ in two variables over k, of degree ≥ 1 in y, there are infinitely many $b \in k$ such that the specialized polynomial $f(b, y)$ (in one variable) is irreducible.
(2) Given a finite extension ℓ/k, and $h_1(x, y), \ldots, h_m(x, y) \in \ell[x][y]$ that are irreducible as polynomials in y over the field $\ell(x)$, there are infinitely many $b \in k$ such that the specialized polynomials $h_1(b, y), \ldots, h_m(b, y)$ are irreducible in $\ell[y]$.
(3) For any $p_1(x, y), \ldots, p_t(x, y) \in k[x][y]$ that are irreducible and of degree >1 when viewed as polynomial in y over $k(x)$, there are infinitely many $b \in k$ such that none of the specialized polynomials $p_1(b, y), \ldots, p_t(b, y)$ has a root in k.

Proof. Clearly, (2) implies (1) and (3) (cf. Lemma 1.4). It remains to prove that each of (1) and (3) implies (2).

Let $h_1(x, y), \ldots, h_m(x, y) \in \ell[x][y]$ be as in (2). Let S_0 be the set of all roots of these polynomials in some algebraic closure of $\ell(x)$. Choose a finite extension K of $\ell(x)$ that contains S_0, and is Galois over $k(x)$.

Now apply the above Proposition: Part (ii) shows the implication (1) $\Rightarrow$ (2). (Note that the polynomial $f(x, y)$ from the Proposition (defining the extension $K/k(x)$) is irreducible as polynomial in two variables by Lemma 1.4.) For the implication (3) $\Rightarrow$ (2), use additionally part (iii). □

Definition 1.9 *A field k is called* **hilbertian** *if it satisfies (one of) the 3 equivalent conditions (1), (2), (3).*

Using (1) and (2) we see that every finite extension of a hilbertian field is hilbertian. In the next section we prove that the field $\mathbb{Q}$ is hilbertian. Thus every algebraic number field (of finite degree over $\mathbb{Q}$) is hilbertian.

1.1.3 Basic Properties of Hilbertian Fields

Lemma 1.10 *Suppose k is hilbertian, and $f(x_1, \ldots, x_s)$ is an irreducible polynomial in $s \geq 2$ variables over k, of degree ≥ 1 in x_s.*

(i) Then there are infinitely many $b \in k$ such that the polynomial $f(b, x_2, \ldots, x_s)$ (in $s - 1$ variables) is irreducible over k.
(ii) For any nonzero $p \in k[x_1, \ldots, x_{s-1}]$ there are $b_1, \ldots, b_{s-1} \in k$ such that $p(b_1, \ldots, b_{s-1}) \neq 0$ and $f(b_1, \ldots, b_{s-1}, x_s)$ is irreducible (as polynomial in one variable).

Proof. First we derive (ii) from (i). We use induction on s. The case $s = 2$ is just (i). Now assume $s > 2$, and the claim holds for $s - 1$. Write p as a polynomial in $x_2, \ldots, x_{s-1}$, with certain coefficients $c_j(x_1) \in k[x_1]$. By (i) there is $b_1 \in k$ such that $f'(x_2, \ldots, x_s) := f(b_1, x_2, \ldots, x_s)$ is irreducible, and $c_j(b_1) \neq 0$ for some j. Then $p'(x_2, \ldots, x_{s-1}) := p(b_1, x_2, \ldots, x_{s-1})$ is nonzero. Now the induction hypothesis yields $b_2, \ldots, b_{s-1} \in k$ such that $p'(b_2, \ldots, b_{s-1}) \neq 0$ and $f'(b_2, \ldots, b_{s-1}, x_s)$ is irreducible. Thus $(b_1, \ldots, b_{s-1})$ is as desired.

It remains to prove (i). Let d be an integer bigger than the highest power of any variable occurring in f. Kronecker's specialization of f is defined as $S_d f(x, y) = f(x, y, y^d, \ldots, y^{(d^{s-2})})$ (a polynomial in two variables). Write

$$S_d f(x, y) = g(x) \prod_i g_i(x, y)$$

a product of irreducible polynomials $g_i(x, y)$, of degree ≥ 1 in y, and $g(x) \in k[x]$. Since k is hilbertian, there are infinitely many $b \in k$ such that all $g_i(b, y)$ are irreducible. (Use condition (2) and Lemma 1.4.) Consider only such b from now on. We may additionally assume that $g(b) \neq 0$.

Now assume that $f(b, x_2, \ldots, x_s)$ is reducible, say $f(b, x_2, \ldots, x_s) = h(x_2, \ldots, x_s)h'(x_2, \ldots, x_s)$, where h and h' are both not constant. The Kronecker specializations $S_d h(y)$ and $S_d h'(y)$ are defined similarly as above. We have $S_d f(b, y) = S_d h(y) S_d h'(y)$, hence $S_d h(y)$ and $S_d h'(y)$ are each a product of certain $g_i(b, y)$ (up to factors from k). Let $H(x, y)$ and $H'(x, y)$ be the product of the corresponding $g_i(x, y)$. Then $S_d f(x, y) = g(x)H(x, y)H'(x, y)$.

Because of the uniqueness of the d-adic expansion of an integer, there are unique polynomials $\tilde{h}(x_1, \ldots, x_s), \tilde{h}'(x_1, \ldots, x_s)$ with $S_d\tilde{h} = gH$, $S_d\tilde{h}' = H'$, such that the highest power of $x_2, \ldots, x_s$ occurring in $\tilde{h}, \tilde{h}'$ is less than d. If the latter would also hold for $\tilde{f} := \tilde{h}\tilde{h}'$ then we would have $\tilde{f} = f$ because of the uniqueness of the d-adic expansion. This contradicts the irreducibility of f because $\tilde{f} = \tilde{h}\tilde{h}'$ with $\tilde{h}, \tilde{h}'$ not constant.

Thus $\tilde{f}$, when written as polynomial in $x_2, \ldots, x_s$, contains a monomial $\kappa(x_1)x_2^{i_2} \ldots x_s^{i_s}$ where some $i_\nu \geq d$, and $\kappa \neq 0$. Note that $\tilde{h}(b, x_2, \ldots, x_s)$ is a scalar multiple of $h(x_2, \ldots, x_s)$. (Compare their Kronecker specializations.) Similarly for h'. It follows that $\tilde{f}(b, x_2, \ldots, x_s)$ is a (nonzero) scalar multiple of $f(b, x_2, \ldots, x_s)$. This implies that $\kappa(b) = 0$.

There are only finitely many possibilities for κ (up to multiplication with elements of k), corresponding to all decompositions $S_d f = gHH'$. If we choose b distinct from the (finitely many) zeroes of all these κ, then $f(b, x_2, \ldots, x_s)$ is irreducible. This proves (i). □

Corollary 1.11 *If k is hilbertian then so is every finitely generated extension field of k.*

Proof. First we consider a purely transcendental extension $F = k(x_1, \ldots, x_m)$. Set $D = k[x_1, \ldots, x_m]$. Let $f(x, y) \in F[x, y]$ be irreducible, of degree ≥ 1 in y. Then f is irreducible as polynomial in y over $F(x)$ (Lemma 1.4). We may assume that $f \in D[x, y]$. By Lemma 1.4 we may further assume that f is irreducible as polynomial in $x_1, \ldots, x_m, x, y$. Thus by the preceding lemma, there are infinitely many $b \in k$ such that $f(x_1, \ldots, x_m, b, y)$ is irreducible. Then $f(x_1, \ldots, x_m, b, y)$ is irreducible when viewed as polynomial in y over D, hence over F. Thus F is hilbertian.

Now consider an arbitrary finitely generated extension K of k. Such K is finite over k or over a purely transcendental extension F of k (Remark 1.2). Then F is hilbertian by the preceding paragraph. Hence K is hilbertian since the hilbertian property is preserved under finite extensions (see the remarks after Definition 1.9). □

Remark 1.12 Actually, $k(x_1, \ldots, x_m)$ is hilbertian for *any* field k, and $m \geq 1$ (see [FJ], Th. 12.10).

Theorem 1.13 *Suppose k is hilbertian. If a finite group G occurs as Galois group over $k(x_1, \ldots, x_m)$ then G also occurs as Galois group over k.*

Proof. If $m > 1$ then $k(x_1, \ldots, x_m) = k(x_1, \ldots, x_{m-1})(x_m)$. Since $k(x_1, \ldots, x_{m-1})$ is hilbertian by the preceding result, we can reduce inductively to the case $m = 1$. The case $m = 1$ follows from Proposition 1.7(i). □

Definition 1.14 *Let G be a finite group. We say G* **occurs regularly over** *k if for some $m \geq 1$ there is a Galois extension of $k(x_1, \ldots, x_m)$, regular over k, with Galois group isomorphic to G.*

One of the main properties of regular realizations is their invariance under extension of the base field k.

Corollary 1.15 *Suppose G occurs regularly over k. Then G occurs regularly over every extension field k_1 of k. Thus G is a Galois group over k_1 if k_1 is hilbertian.*

Proof. Suppose $x_1, \ldots, x_m$ are independent transcendentals over k_1, and set $\mathbf{x} = (x_1, \ldots, x_m)$. We can assume $G = G(K/k(\mathbf{x}))$, with K regular over k. Set $n = |G| = [K : k(\mathbf{x})]$. Write K in the form $K = k(\mathbf{x})[y]/(f)$, for some

$f(\mathbf{x}, y) \in k(\mathbf{x})[y]$. Then f is irreducible over $\bar{k}_1(\mathbf{x})$ by Lemma 1.1(ii). Hence $K_1 = k_1(\mathbf{x})[y]/(f)$ is a field extension of $k_1(\mathbf{x})$ of degree n, regular over k_1. Clearly, K_1 is Galois over $k_1(\mathbf{x})$ (because all the roots of f over $k_1(\mathbf{x})$ are already contained in K). Now $G(K_1/k_1(\mathbf{x}))$ and $G(K/k(\mathbf{x}))$ have the same order, and the former group embeds into the latter via restriction. Hence they are isomorphic. This proves the first claim. The second follows by Theorem 1.13. □

Remark 1.16 (a) If k is hilbertian and the (nontrivial) group G occurs regularly over k then there are actually infinitely many linearly disjoint Galois extensions of k with group G. (Exercise! Use Lemma 1.1(ii) and Proposition 1.7(i), (ii).)

(b) In Definition 1.14, one might wonder which integers m work for a given group G. The answer is, either all or none: If G occurs regularly over k, then for each $m \geq 1$ there is a Galois extension of $k(x_1, \dots, x_m)$, regular over k, with Galois group isomorphic to G. At the end of this chapter we give a proof for hilbertian k (which works for any k if we assume Remark 1.12).

Example 1.17 (The symmetric group S_n occurs regularly over every field k).

Consider the polynomial with indeterminate coefficients

$$f(y) = y^n + x_1 y^{n-1} + \cdots + x_n = \prod_{i=1}^{n} (y - t_i)$$

as polynomial in y over $k(x_1, \dots, x_n)$. Then its roots $t_1, \dots, t_n$ are also independent transcendentals over k. Thus the natural action of S_n on $t_1, \dots, t_n$ (permuting the subscripts) extends to an action of S_n on the field $k(t_1, \dots, t_n)$ (through field automorphisms). The fixed field F of S_n in $k(t_1, \dots, t_n)$ contains $x_1, \dots, x_n$, and $[k(t_1, \dots, t_n) : F] = |S_n| = n!$. On the other hand, since $t_1, \dots, t_n$ are the roots of a polynomial of degree n over $k(x_1, \dots, x_n)$, we have $[k(t_1, \dots, t_n) : k(x_1, \dots, x_n)] \leq n!$. It follows that $F = k(x_1, \dots, x_n)$. Hence by Artin's theorem,

$$G(k(t_1, \dots, t_n)/k(x_1, \dots, x_n)) = S_n.$$

Thus S_n occurs regularly over k.

1.2 The Rational Field Is Hilbertian

We are going to show that $\mathbb{Q}$ is hilbertian, using the condition (3) in Definition 1.9. Thus we have to test whether certain auxiliary polynomials $p_i(b, y) \in \mathbb{Q}[y]$ have rational roots. Using the analyticity of roots (i.e., that simple roots are locally analytic functions of the coefficients), this can be transformed into the question whether certain meromorphic functions take integral values. The latter is answered by a beautiful theorem whose proof uses only elementary calculus.

1.2.1 Analyticity of Roots

Theorem 1.18 *Let $f(x, y) \in \mathbb{C}[x, y]$ be of degree $n \geq 1$ in y. Let $c_0 \in \mathbb{C}$ such that the polynomial $f(c_0, y) \in \mathbb{C}[y]$ is separable of degree n. Then there exist holomorphic functions $\psi_1, \ldots, \psi_n$ defined in a neighborhood U of c_0 such that for each $c \in U$ the polynomial $f(c, y)$ has the n distinct roots $\psi_1(c), \ldots, \psi_n(c)$.*

The theorem follows from the complex-analytic version of the usual theorem on implicit functions: Indeed, the separability of $f(c_0, y)$ means that $(\partial f/\partial y)(c_0, \gamma_i) \neq 0$ for all roots γ_i of $f(c_0, y)$. Thus there is a holomorphic function ψ_i defined around c_0 with $\psi_i(c_0) = \gamma_i$, such that for $(c, d) \in \mathbb{C}^2$ close to (c_0, γ_i) we have $f(c, d) = 0$ if and only if $d = \psi_i(c)$. Those ψ_i are as desired.

Since Theorem 1.18 is basic for our development (especially for the construction of the Riemann surface of an algebraic function, see Section 5.4), however, we give a more algebraic proof, closer to the realm of our methods. The argument is due to Cauchy (see [E], Ch. III, §1). It also yields the analogous result in ultrametric analysis (see Theorem 11.2).

Proof. It suffices to show that for each root γ of $f(c_0, y)$ there is a holomorphic function ψ defined around c_0 with $\psi(c_0) = \gamma$ and $f(c, \psi(c)) = 0$ for all c close to c_0. Indeed, for c close enough to c_0 the values of these n functions ψ on c will be distinct, hence comprise all roots of $f(c, y)$.

Replacing x by $x - c_0$ and y by $y - \gamma$ we may assume that $c_0 = \gamma = 0$. Thus we have to find holomorphic ψ with $\psi(0) = 0$ and

$$f(t, \psi(t)) = 0 \tag{1.2}$$

for all t close to 0. Such ψ has a Taylor expansion of the form $\psi(t) = \sum_{i=1}^{\infty} a_i t^i$ around 0.

By hypothesis we have $f(0, 0) = 0$, hence

$$f(x, y) = ax + by + \text{higher order terms.}$$

Here $b = (\partial f/\partial y)(0, 0) \neq 0$ (since 0 is a simple root of $f(0, y)$). Dividing by b we may assume $b = 1$. Then

$$g(x, y) := y - f(x, y)$$

has no constant term and no y-term. Condition (1.2) is equivalent to

$$\psi(t) = g(t, \psi(t)).$$

This equation allows us to compute the coefficients a_i of ψ recursively, if we develop the right-hand side into a power series around 0. Indeed, the t^i-coefficient on the right-hand side involves only such a_j with $j < i$ (and the coefficients of g). After completing the recursion, a_i appears as a polynomial, with non-negative integer coefficients, in the coefficients of g. Note that only the $x^\mu y^\nu$ coefficients (of g) with $\mu + \nu \leq i$ occur.

The coefficients a_i computed above yield the (unique) power series $\psi(t) = \sum_{i=1}^{\infty} a_i t^i$ that solves (1.2) in the formal sense (i.e., in the ring of formal power series $\mathbf{C}[[t]]$; see 2.1.1 below). It remains to see that this power series has a positive radius of convergence.

Let C be a positive constant bounding the absolute value of the coefficients of g. Consider the function

$$\tilde{g}(t, u) = C\left(-1 - u + \frac{1}{(1-t)(1-u)}\right).$$

Solving the quadratic equation $u = \tilde{g}(t, u)$ for u in terms of t we see that

$$\tilde{\psi}(t) = \frac{1}{2(C+1)}\left(1 - \frac{\sqrt{1 - t(1 + (1+2C)^2) + t^2(1 + (1+2C)^2)}}{1 - t}\right)$$

is the unique holomorphic function defined around 0 with $\tilde{\psi}(0) = 0$ and

$$\tilde{\psi}(t) = \tilde{g}(t, \tilde{\psi}(t)). \tag{1.3}$$

(Here we choose that branch of the square root function around 1 with $\sqrt{1} = 1$.) The geometric series formula yields

$$\tilde{g}(t, u) = C(t + t^2 + tu + u^2 + \cdots) = C\left(-1 - u + \sum_{\mu,\nu=0}^{\infty} t^\mu u^\nu\right)$$

for $|t| < 1$, $|u| < 1$. From this and (1.3) we see that the Taylor coefficients $\tilde{a}_i$ of $\tilde{\psi}$ are obtained using the same polynomials as for the a_i, now applied to the coefficients of $\tilde{g}$, which are all equal to C. Since C bounds the absolute value of the coefficients of g, it follows that $|a_i| \leq \tilde{a}_i$ for all i. But the Taylor series

$\sum_{i=1}^{\infty} \tilde{a}_i t^i$ of $\tilde{\psi}$ has a positive radius of convergence, hence the same holds for $\psi(t) = \sum_{i=1}^{\infty} a_i t^i$. □

1.2.2 The Rational Field Is Hilbertian

Definition 1.19 *Let $M \subset \mathbb{N}$. We say M is* **sparse** *if there is a real number κ with $0 < \kappa < 1$ such that*

$$|M \cap \{1, \ldots, N\}| \leq N^{\kappa}$$

for almost all N.

Remark 1.20 Clearly, each finite set is sparse. More generally, a finite union of sparse sets is sparse.

Theorem 1.21 *Let $i_0 \in \mathbb{Z}$, and let*

$$\phi(t) = \sum_{i=i_0}^{\infty} a_i t^i$$

be a Laurent series with complex coefficients, converging for all $t \neq 0$ in a neighborhood of 0 in $\mathbb{C}$. Let $B(\phi)$ be the set of all $b \in \mathbb{N}$ for which $\phi(1/b)$ (is defined and) is an integer. Then $B(\phi)$ is a sparse set unless ϕ is a Laurent polynomial (i.e., unless almost all a_i vanish).

The proof of the Theorem is just elementary calculus. We postpone it until the end of this section. First we derive that $\mathbb{Q}$ is hilbertian.

Lemma 1.22 *Let $p(x, y) \in \mathbb{Q}[x][y]$ be irreducible over $\mathbb{Q}(x)$ and of degree $r > 1$ (in y). Then for almost all $x_0 \in \mathbb{Z}$ the following holds:*

(a) There are $\epsilon > 0$ and holomorphic functions $\psi_1(t), \ldots, \psi_r(t)$ defined for complex t with $|t| < \epsilon$, such that $\psi_1(t), \ldots, \psi_r(t)$ are the roots of the polynomial $p(x_0 + t, y) \in \mathbb{Q}[y]$.
(b) If some $\psi_i(t)$ is a rational function of t (with complex coefficients), then there are only finitely many $q \in \mathbb{Q}$ with $\psi_i(q) \in \mathbb{Q}$.
(c) Let $B(p, x_0)$ be the set of all $b \in \mathbb{N}$ such that $p(x_0 + \frac{1}{b}, c) = 0$ for some $c \in \mathbb{Q}$. Then $B(p, x_0)$ is a sparse set.

Proof. The polynomial p is irreducible, hence separable (over $\mathbb{Q}(x)$). Thus $p(x_0, y)$ is separable for almost all $x_0 \in \mathbb{Z}$ (Lemma 1.6). Consider only such x_0 in the following.

(a) Follows from Theorem 1.18.

(b) Suppose $\psi := \psi_i$ is a rational function of t (i.e., a quotient of two polynomials). Then $p(x_0 + t, \psi(t))$ is identically zero (as a rational function in t). Thus $p(x_0 + x, \psi(x)) = 0$ in $\mathbb{C}(x)$, for x a transcendental element over $\mathbb{C}$. Then $\psi(x) \in \mathbb{C}(x)$ is algebraic over $\mathbb{Q}(x)$, hence over $\bar{\mathbb{Q}}(x)$. But $\bar{\mathbb{Q}}(x)$ is algebraically closed in $\mathbb{C}(x)$. (Indeed, any irreducible polynomial over $\bar{\mathbb{Q}}(x)$ remains irreducible over $\mathbb{C}(x)$ by Lemma 1.1(ii).) It follows that $\psi(x) \in \bar{\mathbb{Q}}(x)$.

For each $\beta \in G(\bar{\mathbb{Q}}/\mathbb{Q})$ we can consider the rational function ψ^β obtained by applying β to the coefficients of ψ. Then $\psi^\beta(q) = \psi(q)$ for all $q \in \mathbb{Q}$ with $\psi(q) \in \mathbb{Q}$. If there are infinitely many such q, it follows that $\psi^\beta = \psi$ for all β, hence ψ has rational coefficients. Then $\psi(x - x_0) \in \mathbb{Q}(x)$ is a zero of $p(x, y)$ over $\mathbb{Q}(x)$. This contradicts the fact that p is irreducible over $\mathbb{Q}(x)$ (of degree >1). Hence (b).

(c) We may assume $p(x, y) \in \mathbb{Z}[x, y]$. Write

$$p(x, y) = \sum_{i=0}^{r} p_i(x) y^i$$

with $p_i(x) \in \mathbb{Z}[x]$. For suitably large R the expression

$$x^R p\left(x_0 + \frac{1}{x}, y\right) = \sum_{i=0}^{r} x^R p_i\left(x_0 + \frac{1}{x}\right) y^i$$

is an element of $\mathbb{Z}[x, y]$. Denote its y^i-coefficient by $p_i'(x)$. Then $h(x) := p_r'(x)$ is a nonzero element of $\mathbb{Z}[x]$. Following the trick from Lemma 1.3, we define

$$p'(x, Z) = Z^r + \sum_{i=0}^{r-1} p_i'(x) h(x)^{r-i-1} Z^i,$$

an element of $\mathbb{Z}[x, Z]$, monic in Z.

Suppose now that $p(x_0 + \frac{1}{b}, c) = 0$ for $c \in \mathbb{Q}$, $b \in \mathbb{Z}$. Then $p'(b, h(b)c) = 0$. Since $p'(b, Z) \in \mathbb{Z}[Z]$ is monic, it follows that $h(b)c$ is integral over $\mathbb{Z}$. Thus $h(b)c \in \mathbb{Z}$ (because it lies in $\mathbb{Q}$).

If additionally $|1/b| < \epsilon$, then $c = \psi_i(1/b)$ for some $i = 1, \dots, r$ (by (a)). Thus $h(b)\psi_i(1/b) = h(b)c \in \mathbb{Z}$.

Set $\phi_i(t) = h(t^{-1})\psi_i(t)$ (for $0 < |t| < \epsilon$, $i = 1, \dots, r$). The above shows that if $b \in B(p, x_0)$ and $1/b < \epsilon$ then $\phi_i(1/b) = h(b)\psi_i(1/b) \in \mathbb{Z}$ for some $i = 1, \dots, r$. Thus, up to a finite set, $B(p, x_0)$ lies in the union of the $B(\phi_i)$. By

Theorem 1.21 the set $B(\phi_i)$ is sparse if ϕ_i is not a rational function. However, if ϕ_i (hence ψ_i) is rational, then $B(\phi_i)$ is finite by (b). By Remark 1.20, it follows that $B(p, x_0)$ is sparse. Hence (c). □

Actually, the set $B(p, x_0)$ is sparse for all $x_0 \in \mathbb{Q}$, but for the sake of simplicity we restrict ourselves to the unramified case. Now we are ready to prove:

Theorem 1.23 (Hilbert's irreducibility theorem) *The field $\mathbb{Q}$ is hilbertian.*

Proof. Given polynomials $p_j(x, y) \in \mathbb{Q}[x][y]$ as in condition (3) from Definition 1.9, we can choose an $x_0 \in \mathbb{Z}$ as in Lemma 1.22 that works for all p_j. Let C be the set of $b \in \mathbb{N}$ such that none of the specialized polynomials $p_j(x_0 + \frac{1}{b}, y)$ has a root in $\mathbb{Q}$. Set $B = \mathbb{N} \setminus C$. Then B is the union of the $B(p_j, x_0)$. The latter sets are sparse by Lemma 1.22(c). Hence B is sparse (Remark 1.20). Therefore, its complement C is infinite. This proves that $\mathbb{Q}$ is hilbertian. □

1.2.3 Integral Values of Meromorphic Functions

It remains to prove Theorem 1.21. This is based on a generalized mean value theorem due to H.A. Schwarz:

Lemma 1.24 *Let $s_0 < s_1 < \cdots < s_m$ be real numbers, where $m \geq 1$. Let $\chi(s)$ be a real-valued function defined for $s_0 \leq s \leq s_m$, and m times continuously differentiable. Let V_m be the Vandermonde determinant*

$$V_m = \begin{vmatrix} 1 & s_0 & s_0^2 & \cdots & s_0^m \\ \vdots & \vdots & \vdots & \cdots & \vdots \\ 1 & s_m & s_m^2 & \cdots & s_m^m \end{vmatrix} = \prod_{i>j}(s_i - s_j).$$

Then there exists a number σ with $s_0 < \sigma < s_m$ such that

$$\frac{\chi^{(m)}(\sigma)}{m!} = \frac{1}{V_m} \begin{vmatrix} 1 & s_0 & \cdots & s_0^{m-1} & \chi(s_0) \\ \vdots & \vdots & \cdots & \vdots & \vdots \\ 1 & s_m & \cdots & s_m^{m-1} & \chi(s_m) \end{vmatrix}.$$

Proof. Let $F(s)$ be the function

$$F(s) = \begin{vmatrix} 1 & s_0 & \cdots & s_0^{m-1} & \chi(s_0) \\ \vdots & \vdots & \cdots & \vdots & \vdots \\ 1 & s_{m-1} & \cdots & s_{m-1}^{m-1} & \chi(s_{m-1}) \\ 1 & s & \cdots & s^{m-1} & \chi(s) \end{vmatrix}.$$

Set

$$c = \frac{F(s_m)}{(s_m - s_0)\cdots(s_m - s_{m-1})}$$

and

$$G(s) = F(s) - c(s - s_0)\cdots(s - s_{m-1}).$$

The function $G(s)$ vanishes at the $(m+1)$ points $s = s_0, \ldots, s_m$. Hence $G^{(m)}(s)$ vanishes at least at one point σ between s_0 and s_m. Since $G^{(m)}(s) = F^{(m)}(s) - m!c$ we get

$$F^{(m)}(\sigma) = m!c.$$

On the other hand, expanding the determinant defining $F(s)$ we get

$$F(s) = \sum_{i=0}^{m-1} c_i s^i + V_{m-1}\chi(s)$$

where the c_i are constants (depending on $s_0, \ldots, s_m$), and V_{m-1} is the Vandermonde determinant of $s_0, \ldots, s_{m-1}$. Hence

$$F^{(m)}(\sigma) = V_{m-1}\chi^{(m)}(\sigma).$$

Comparing the two expressions for $F^{(m)}(\sigma)$ we obtain

$$\frac{\chi^{(m)}(\sigma)}{m!} = \frac{c}{V_{m-1}} = \frac{F(s_m)}{(s_m - s_0)\cdots(s_m - s_{m-1})V_{m-1}} = \frac{F(s_m)}{V_m}.$$

By the definition of $F(s_m)$, we are done. □

Proof of Theorem 1.21. Let $\phi(t)$ be as in Theorem 1.21, and assume it is not a Laurent polynomial. We may further assume that $B(\phi)$ is infinite. Then we have:

Claim 1 The coefficients a_i of ϕ are all real.

Proof. The series

$$\bar{\phi}(t) = \sum_{i=i_0}^{\infty} \bar{a}_i t^i$$

with complex conjugate coefficients has the same radius of convergence as ϕ. We have $\bar{\phi}(1/b) = \phi(1/b)$ for all $b \in B(\phi)$. Since $B(\phi)$ is infinite, it follows that $\bar{\phi} = \phi$. This proves Claim 1.

It follows that $\chi(s) := \phi(s^{-1})$ is a real-valued function, defined (and analytic) for large real values of s. We have

$$\chi(s) = \sum_{i=i_0}^{\infty} a_i s^{-i}.$$

Claim 2 There is $\lambda > 0$, and $m, S \in \mathbb{N}$ such that the following holds: Whenever $s_0, \ldots, s_m \in \mathbb{Z}$ with $\chi(s_0), \ldots, \chi(s_m) \in \mathbb{Z}$ and $S < s_0 < \cdots < s_m$ then

$$s_m - s_0 \geq s_0^{\lambda}.$$

Proof. For large enough m the series

$$\chi^{(m)}(s) = \sum_{i=\mu}^{\infty} d_i s^{-i}$$

has only terms with negative powers of s, that is, $\mu > 0$. Here the d_i are real numbers, and we can assume $d_\mu \neq 0$ (since ϕ is not a Laurent polynomial). Then $s^\mu \chi^{(m)}(s)$ tends to d_μ as s goes to infinity. Hence there is $S > 0$ such that $0 < |s^\mu \chi^{(m)}(s)| < |2d_\mu|$ for $s \geq S$.

Now assume $s_0, \ldots, s_m$ are as in Claim 2, and choose σ according to the previous Lemma. Then $\frac{V_m \chi^{(m)}(\sigma)}{m!}$ is a nonzero integer, hence has absolute value ≥ 1. Thus $V_m \geq \frac{1}{|\chi^{(m)}(\sigma)|}$, and so

$$(s_m - s_0)^{(m+1)(m+2)/2} \geq V_m \geq \frac{1}{|\chi^{(m)}(\sigma)|} \geq \frac{1}{|2d_\mu|}\sigma^\mu \geq \frac{1}{|2d_\mu|}s_0^\mu.$$

Hence

$$s_m - s_0 \geq \left(\frac{1}{|2d_\mu|}\right)^{2/(m+1)(m+2)} s_0^{2\mu/(m+1)(m+2)}.$$

Thus any positive $\lambda < 2\mu/(m+1)(m+2)$ satisfies Claim 2.

Claim 3 Let $b_1 < b_2 < \cdots$ be an infinite sequence of positive integers with $b_{i+1} - b_i \geq b_i^\lambda$, for some $\lambda > 0$. Then the set $B = \{b_1, b_2, \ldots\}$ is sparse.
Proof. For each positive integer N, let N' be the number of $b \in B$ with $\sqrt{N} < b \leq N$. Then $(N' - 1)\sqrt{N}^\lambda \leq N$. Hence

$$N' - 1 \leq N^{1-\frac{\lambda}{2}}.$$

Thus

$$|B \cap \{1, \ldots, N\}| \leq \sqrt{N} + N' \leq \sqrt{N} + N^{1-\frac{\lambda}{2}} + 1.$$

This implies that B is sparse.
Claim 4 $B(\phi)$ is sparse.
Proof. Recall that $B(\phi)$ consists of all integers b such that $\chi(b)$ $(= \phi(1/b))$ is an integer. Delete from $B(\phi)$ all integers $\leq S$, where S is as in Claim 2. By Claim 2, the remaining set can be written as a union of m subsets B each of which satisfies the condition from Claim 3. These sets B are sparse by Claim 3, hence $B(\phi)$ is sparse. □

1.3 Algebraic Extensions of Hilbertian Fields

Every finitely generated extension of a hilbertian field is hilbertian (Corollary 1.11). This is nicely complemented by Weissauer's theorem about infinite algebraic extensions of a hilbertian field (which are not finitely generated!). Weissauer's original proof used nonstandard methods [W]; later Fried [Fr5] gave an algebraic proof whose idea we follow here.

1.3.1 Weissauer's Theorem

Lemma 1.25 *Let k be hilbertian, and ℓ/k finite. Let $\pi, \tilde{\pi} \in \ell[x_1, x_2][y]$, monic in y, where x_1, x_2 are algebraically independent over ℓ. View π and $\tilde{\pi}$ as polynomials in y over $\ell(x_1, x_2)$, and suppose π has no root in a splitting field of $\tilde{\pi}$. Then for any nonzero $v \in \ell[x_1, x_2]$ there are $b_1, b_2 \in k$ with $v(b_1, b_2) \neq 0$ such that $\pi(b_1, b_2, y)$ has no root in a splitting field of $\tilde{\pi}(b_1, b_2, y)$. Here we view $\pi(b_1, b_2, y)$ and $\tilde{\pi}(b_1, b_2, y)$ as polynomials in y over ℓ.*

Proof. We proceed similarly as in the proof of Proposition 1.7. Let $K/\ell(x_1, x_2)$ be a finite Galois extension containing all roots of π and $\tilde{\pi}$. Pick a generator α of $K/\ell(x_1, x_2)$ with $f(\alpha) = 0$ for monic and irreducible $f(y) = f(x_1, x_2, y) \in \ell[x_1, x_2][y]$. By Lemmas 1.4 and 1.10(ii) there are $b_1, b_2 \in \ell$ with $v(b_1, b_2) \neq 0$ such that $f(b_1, b_2, y)$ is irreducible over ℓ. Using the property of the hilbertian field k from Corollary 1.8(2), the proof of Lemma 1.10(ii) shows that we can actually find such b_1, b_2 in k. Consider only such b_1, b_2 from now on.

Let $\omega_{b_1 b_2} : \ell[x_1, x_2] \to \ell$ be the evaluation homomorphism $h(x_1, x_2) \mapsto h(b_1, b_2)$. By Lemma 1.5 we may assume (after replacing v by vu for suitable $u \in \ell[x_1, x_2]$) that $\omega_{b_1 b_2}$ extends to a homomorphism $\omega : S \to K'$, where S is a subring of K containing all roots of π and $\tilde{\pi}$, and K' is a finite Galois extension of ℓ. Since $f(b_1, b_2, y)$ is irreducible, we may further assume there is an isomorphism $G = G(K/\ell(x_1, x_2)) \to G' = G(K'/\ell)$, $\sigma \mapsto \sigma'$, such that $\omega\sigma(s) = \sigma'\omega(s)$ for all $s \in S, \sigma \in G$.

We may certainly assume that π is separable. (Replace it by the product of its distinct irreducible monic factors.) Then its discriminant D is a nonzero element of $\ell[x_1, x_2]$. Replacing v by Dv we may assume that $\pi(b_1, b_2, y)$ is also separable (see the proof of Lemma 1.6).

Write $\pi(y) = (y-\beta_1)\cdots(y-\beta_s)$. Then $\pi(b_1, b_2, y) = (y-\beta_1')\cdots(y-\beta_s')$, where $\beta_i' = \omega(\beta_i)$. Similarly, $\tilde{\pi}(y) = (y-\gamma_1)\cdots(y-\gamma_t)$, and $\tilde{\pi}(b_1, b_2, y) = (y-\gamma_1')\cdots(y-\gamma_t')$. The hypothesis of the Lemma means that for each β_i there is $\sigma \in G$ fixing all γ_ν, but not β_i; say $\sigma(\beta_i) = \beta_j$ with $\beta_i \neq \beta_j$.

Then σ' fixes all γ_ν'. Since $\pi(b_1, b_2, y)$ is separable we have $\beta_i' \neq \beta_j'$, hence σ' does not fix β_i'. This implies the claim. □

Theorem 1.26 (Weissauer) *Let k be hilbertian, let N be a (possibly infinite) Galois extension of k, and let M be a finite extension of N with $M \neq N$. Then M is hilbertian.*

Proof. We use criterion (3) for hilbertianity (from Definition 1.9). Thus consider a finite collection of polynomials $p_j(x, y) \in M[x][y]$, irreducible and of degree >1 as polynomial in y over $M(x)$. We have to show that there are infinitely many $b \in M$ such that none of the specialized polynomials $p_j(b, y)$ has a root in M.

By Lemma 1.3 we can assume that the p_j are monic in y. (Just exclude the finitely many b which are zeroes of the highest y-coefficient $a_j(x)$ of some p_j.) Clearly we may further assume that the p_j are all distinct. Then their product $p(x, y)$ is separable as polynomial in y. We have to show that there are infinitely many $b \in M$ such that $p(b, y)$ has no root in M.

By the following Claim 1, we may assume there is no $\gamma \in \bar{M}(x)$ with $p(x, \gamma) = 0$.

Claim 1 Suppose some $q = p_j$ satisfies $q(x, g(x)) = 0$ for some $g(x) \in \bar{M}(x)$. Then there are only finitely many $b \in M$ such that $q(b, y)$ has a root in M.

Proof. Since $\bar{M}(x)$ is Galois over $M(x)$, all roots of q over $M(x)$ lie in $\bar{M}(x)$. Thus $q(x, y) = \prod_i (y - g_i(x))$ with $g_i(x) \in \bar{M}(x)$, but $g_i(x) \notin M(x)$ (since q is irreducible of degree >1 in y). Then there are only finitely many $b \in M$ with $g_i(b) \in M$ (same argument as in the proof of Lemma 1.22(b)). But the $g_i(b)$ are the roots of $q(b, y) = \prod_i (y - g_i(b))$ (for almost all $b \in k$). Hence Claim 1.

Embed M into some algebraic closure $\bar{k}$ of k. Let $M = N(\theta)$. Then $\theta \notin N$. Let ℓ be a finite Galois extension of k (in $\bar{k}$) containing θ and the coefficients of $p(x, y)$. Pick $\tilde{\theta} \in \ell$ that is conjugate to θ over $\hat{k} := N \cap \ell$, but $\tilde{\theta} \neq \theta$. (Such $\tilde{\theta}$ exists because $\theta \notin \hat{k}$.)

Now introduce new variables x_1, x_2, and consider $\pi(y) := p(x_1 + \theta x_2, y)$ and $\tilde{\pi}(y) := p(x_1 + \tilde{\theta} x_2, y)$ as polynomials in y over $\ell(x_1, x_2)$.

Claim 2 π has no root in a splitting field of $\tilde{\pi}$.

Proof. Set $t = x_1 + \theta x_2$, $\tilde{t} = x_1 + \tilde{\theta} x_2$. Then $\ell(x_1, x_2) = \ell(t, \tilde{t})$, hence t is transcendental over $\ell(\tilde{t})$. We have $\pi(y) = p(t, y) \in \ell[t, y]$. Let $\bar{\pi}(y)$ be an irreducible factor of $\pi(y)$ in $\bar{\ell}(t)[y]$. Then $\bar{\pi}(y)$ has degree >1 in y because $\pi(y)$ has no root in $\bar{\ell}(t)$ (by the assumption that $p(x, y)$ has no y-root in $\bar{M}(x) = \bar{\ell}(x)$).

Let $\tilde{L}$ be a splitting field of $\tilde{\pi}(y)$ $(= p(\tilde{t}, y))$ over $\bar{\ell}(\tilde{t})$. Since t is transcendental over $\bar{\ell}(\tilde{t})$, hence over $\tilde{L}$, the polynomial $\bar{\pi}(y)$ remains irreducible over $\tilde{L}(t)$ (Lemma 1.1(ii)). But $\tilde{L}(t)$ contains all roots of $\tilde{\pi}(y)$ over $\ell(x_1, x_2)$ and contains $\ell(x_1, x_2)$, hence contains a splitting field $\tilde{S}$ of $\tilde{\pi}(y)$ over $\ell(x_1, x_2)$. Thus $\bar{\pi}$ has no root in $\tilde{S}$. Then π also has no root in $\tilde{S}$ (since $\bar{\pi}$ was any irreducible factor of π). This proves Claim 2.

Let $\hat{M}$ be the composite of N and ℓ (inside $\bar{k}$). Then $\hat{M}$ is a (possibly infinite) Galois extension of k.

Claim 3 If F is a finite Galois extension of ℓ inside $\hat{M}$, then F is Galois over $\hat{k}$.

Proof. This is just elementary Galois theory. There is a finite Galois extension N_0 of $\hat{k}$ inside N such that F lies in the composite $\hat{M}_0$ of N_0 and ℓ. We have $N_0 \cap \ell = \hat{k}$, hence $G(\hat{M}_0/\hat{k}) = G(\hat{M}_0/\ell) \times G(\hat{M}_0/N_0)$. Hence each normal subgroup of $G(\hat{M}_0/\ell)$ is normal in $G(\hat{M}_0/\hat{k})$. This implies Claim 3.

Claim 4 For $b_1, b_2 \in k$ consider the polynomials $\pi_{b_1 b_2}(y) := p(b_1 + \theta b_2, y)$ and $\tilde{\pi}_{b_1 b_2}(y) := p(b_1 + \tilde{\theta} b_2, y)$ in $\ell[y]$. If $\pi_{b_1 b_2}$ has a root in $\hat{M}$, then this root lies in a splitting field of $\tilde{\pi}_{b_1 b_2}$ (over ℓ).

Proof. Let $F = \ell(\beta_1, \dots, \beta_m)$, where the β_i are the roots of $\pi_{b_1 b_2}$ in $\hat{M}$. Then F is Galois over ℓ, hence over $\hat{k}$ (by Claim 3). Thus each $\sigma \in G(\ell/\hat{k})$ extends to an automorphism σ_F of F. There is such σ with $\sigma(\theta) = \tilde{\theta}$. This σ maps $\pi_{b_1 b_2}$ to $\tilde{\pi}_{b_1 b_2}$, hence the $\tilde{\beta}_i = \sigma_F(\beta_i)$ are the roots of $\tilde{\pi}_{b_1 b_2}$ in $\hat{M}$. Thus $F = \ell(\tilde{\beta}_1, \dots, \tilde{\beta}_m)$ lies in a splitting field of $\tilde{\pi}_{b_1 b_2}$ over ℓ. Hence Claim 4.

Conclusion of the Proof. By the preceding Lemma and Claim 2, there are infinitely many $b_1, b_2 \in k$ such that $\pi_{b_1 b_2}$ has no root in a splitting field of $\tilde{\pi}_{b_1 b_2}$. Then $\pi_{b_1 b_2}$ has no root in $\hat{M}$ by Claim 4. Thus $\pi_{b_1 b_2}$ has no root in M (since $M \subset \hat{M}$). Hence the $b = b_1 + \theta b_2$ yield infinitely many elements $b \in M$ such that $p(b, y)$ $(=\pi_{b_1 b_2}(y))$ has no root in M. This completes the proof. □

Remark 1.27 The proof shows that the desired elements $b \in M$ (with the property that none of the $p_j(b, y)$ has a root in M) can be chosen in $k(\theta)$, where θ is any generator of M over N. Thus by Proposition 1.7(ii), (iii), for any $h(x, y) \in M[x][y]$, irreducible over $M(x)$, there are infinitely many $b \in k(\theta)$ such that $h(b, y)$ is irreducible in $M[y]$. However, k itself may not contain such b (see Example 1.29 below).

1.3.2 Applications

Weissauer's theorem yields a host of hilbertian subfields of $\bar{\mathbb{Q}}$ that are infinite over $\mathbb{Q}$. Note, however, that the hypothesis $M \neq N$ is crucial, that is, not every (infinite) Galois extension of $\mathbb{Q}$ is hilbertian. A trivial example for this is $\bar{\mathbb{Q}}$, which is certainly not hilbertian. However:

Corollary 1.28 *The field $\mathbb{Q}_{ab}$ obtained by adjoining all roots of unity to $\mathbb{Q}$ is hilbertian.*

Proof. Take $k = \mathbb{Q}$, $N = \mathbb{Q}_{ab} \cap \mathbb{R}$ (where $\mathbb{R}$ is the real field), and $M = \mathbb{Q}_{ab}$. We have $M = N(\sqrt{-1})$ (because complex conjugation leaves $\mathbb{Q}_{ab}$ invariant, hence for $z = a + \sqrt{-1}b \in \mathbb{Q}_{ab}$ we have $a = 1/2(z + \bar{z}) \in N$ and similarly $b \in N$). Since $\mathbb{Q}$ is hilbertian, we need only check that N is Galois over $\mathbb{Q}$. This reduces to the question whether $\mathbb{Q}_n \cap \mathbb{R}$ is Galois over $\mathbb{Q}$, where $\mathbb{Q}_n$ is the field of nth roots of unity (for any n). It is well known, however, that $G(\mathbb{Q}_n/\mathbb{Q})$ is abelian, hence any field between $\mathbb{Q}$ and $\mathbb{Q}_n$ is Galois over $\mathbb{Q}$. □

By the theorem of Kronecker and Weber, $\mathbb{Q}_{ab}$ is the composite of all finite abelian extensions of $\mathbb{Q}$ (i.e., Galois extensions of $\mathbb{Q}$ with abelian Galois group).

Example 1.29 Consider the polynomial $f(x, y) = y^2 - x \in \mathbb{Q}[x, y]$, which is irreducible over $\bar{\mathbb{Q}}(x)$.

(i) Let $\mathbb{Q}_{solv}$ denote the composite of all finite solvable extensions of $\mathbb{Q}$. The field $\mathbb{Q}_{solv}$ is closed under taking square roots (because it consists of all algebraic numbers that are expressible by iterated radicals – Galois' theorem). Thus $f(b, y)$ is reducible over $\mathbb{Q}_{solv}$ for each $b \in \mathbb{Q}_{solv}$. Thus $\mathbb{Q}_{solv}$ is not hilbertian.

(ii) Since the square root of any rational number lies in $\mathbb{Q}_{ab}$ (by the theorem of Kronecker and Weber), it follows that $f(b, y)$ is reducible in $\mathbb{Q}_{ab}[y]$ for each $b \in \mathbb{Q}$. This shows that in the situation of Weissauer's theorem, the elements $b \in M$ required by the hilbertian property can not always be chosen in k.

A theorem similar to Weissauer's has recently been proved by Haran and Jarden [HJ]: If N and L are (possibly infinite) Galois extensions of a hilbertian field k such that none of the two is contained in the other, then the composite of N and L over k is hilbertian.

We conclude this chapter by giving a proof of the claim in Remark 1.16(b) in the case that k is hilbertian. (The same proof works for general k if one substitutes Corollary 1.11 by Remark 1.12.) The idea (taken from [Ma1], p. 226) is to use the proof of Weissauer's theorem.

Application *Suppose k is hilbertian. If G occurs regularly over k, then for each $m \geq 1$ there is a Galois extension of $k(x_1, \ldots, x_m)$, regular over k, with Galois group isomorphic to G.*

Proof. Using Lemma 1.1(i), we can increase m by just adding new variables. It remains to decrease m.

Let $G = G(K/k(\mathbf{x}))$, with K regular over k, and $\mathbf{x} = (x_1, \ldots, x_m)$. Write K in the form $K = k(\mathbf{x})[y]/(f)$, for some $f(\mathbf{x}, y) \in k(\mathbf{x})[y]$. Then f is irreducible over $\bar{k}(\mathbf{x})$ by Lemma 1.1(ii). Using Lemma 1.3, we can assume that f is monic in y, and lies in $k[\mathbf{x}][y]$.

Assume $m \geq 2$, and set $x = x_m$. We are going to use the set-up of Weissauer's theorem, with $\kappa = k(x_1^2, x_2, \ldots, x_{m-1})$ in place of k, and $N = \bar{k}(x_1^2, x_2, \ldots, x_{m-1})$, $M = \bar{k}(x_1, \ldots, x_{m-1})$. Then $M = N(\theta)$ for $\theta = x_1$. Further, κ is hilbertian by Corollary 1.11. By Remark 1.27, for any $h(x, y) \in M[x][y]$, irreducible over $M(x)$, there are infinitely many $b \in \kappa(\theta)$ such that $h(b, y)$ is irreducible over M.

Set $k' = k(x_1, \ldots, x_{m-1})$. Then $k(\mathbf{x}) = k'(x)$. Thus $f \in k[\mathbf{x}][y] = k'[x][y] \subset M[x][y]$. We know that f is irreducible over $\bar{k}(\mathbf{x}) = M(x)$. Hence by the preceding paragraph, there are infinitely many $b \in \kappa(\theta) = k'$ such that $h_b(y) := f(x_1, \ldots, x_{m-1}, b, y)$ is irreducible over $M = \bar{k}(x_1, \ldots, x_{m-1})$. Then $K' := k'[y]/(h_b)$ is a finite extension of k', regular over k (by Lemma 1.1(ii)). Excluding finitely many b's, we have K' Galois over k' with group G (Proposition 1.7(i)). Thus G occurs as Galois group of an extension of $k' = k(x_1, \ldots, x_{m-1})$, regular over k. By induction, this proves the claim. □

2
Finite Galois Extensions of $\mathbb{C}(x)$

In this chapter we study finite Galois extensions $L/\mathbb{C}(x)$ (and more generally, $L/k(x)$ where k is algebraically closed of characteristic 0). A generating element y for $L/\mathbb{C}(x)$ satisfies an algebraic equation $F(x, y) = 0$. As suggested by analysis, we try to find a solution y that is a power series in x, and more generally, in $x - p$ for any $p \in \mathbb{C}$. This is not always possible, but one can find a solution that is a Laurent series in $(x - p)^{1/e}$ for some integer $e \geq 1$. This is Newton's theorem, proved in Section 2.1.

For each p this yields an integer $e = e_{L,p}$ – the minimal value of e satisfying the above. This integer is called the ramification index of L at p. (We define it also for $p = \infty$, where one considers expansions in $x^{-1/e}$.) A finer invariant is obtained as follows. If $y = \sum_{i=N}^{\infty} a_i((x - p)^{1/e})^i$ solves $F(x, y) = 0$, then so does $y' = \sum_{i=N}^{\infty} a_i(\zeta_e(x - p)^{1/e})^i$, where $\zeta_e = \exp(2\pi\sqrt{-1}/e)$. The element of $G(L/\mathbb{C}(x))$ mapping y to y' is not invariantly defined, but its conjugacy class C_p is. It consists of elements of order $e_{L,p}$.

The above defines a set of invariants of a finite Galois extension of $\mathbb{C}(x)$, which we call the ramification type. The existence and uniqueness problem for extensions of prescribed ramification type is the main theme of the first part of this book. The existence problem is solved by Riemann's existence theorem (RET). We state a first version of RET at the end of this chapter.

Uniqueness for extensions of fixed ramification type holds under the group-theoretic condition of weak rigidity. This is also a consequence of RET. Actually, RET allows us to parametrize the extensions of $\mathbb{C}(x)$ of any given ramification type (using certain tuples of group elements). This parametrization is not canonical, however, and this fact gives rise to the monodromy action of the braid group (see Chapters 9 and 10).

2.1 Extensions of Laurent Series Fields

2.1.1 The Field of Formal Laurent Series over k

Let k be a field. Let Λ be the set of sequences $(a_i)_{i\in\mathbb{Z}}$ of elements $a_i \in k$, indexed by the integers, such that there is $N \in \mathbb{Z}$ with $a_i = 0$ for $i < N$. Define addition by

$$(a_i) + (b_i) = (a_i + b_i)$$

and multiplication by

$$(a_i) \cdot (b_j) = (c_n), \qquad \text{where} \qquad c_n = \sum_{i+j=n} a_i b_j.$$

Note the latter sum is finite. It is routine to check that this makes Λ into a commutative ring, whose 0-element is the all-zeroes sequence, and whose 1-element is the sequence (a_i) with $a_0 = 1$ and $a_i = 0$ otherwise.

Let us check that Λ is even a field. For nonzero (a_i) in Λ there is $N \in \mathbb{Z}$ with $a_i = 0$ for $i < N$, and $a_N \neq 0$. Define $b_j = 0$ for $j < -N$, and $b_{-N} = a_N^{-1}$. Now the equations

$$\sum_{i+j=n} a_i b_j = 0, \qquad n = 1, 2, \ldots$$

can be solved inductively for b_j, $j = -N+1, -N+2, \ldots$. The resulting sequence (b_j) is the inverse of (a_i), and we have shown that Λ is a field.

Clearly k embeds as a subfield of Λ, via the map sending $a \in k$ to the sequence (a_i) with $a_0 = a$ and $a_i = 0$ otherwise. View k as a subfield of Λ via this embedding. Further, let t be the sequence (a_i) with $a_1 = 1$ and $a_i = 0$ otherwise. Then the subring $k[t]$ of Λ generated by k and t is the polynomial ring in one variable over k:

$$\sum_{i=0}^{M} a_i t^i = (a_i)$$

where $a_i = 0$ if $i < 0$ or $i > M$. In general we write

$$\sum_{i=N}^{\infty} a_i t^i = (a_i)$$

where $a_i = 0$ if $i < N$. (This notation is purely formal, with no actual summation involved; it is possible, however, to give Λ a topology such that the above actually becomes a convergent series.) The field operations in Λ correspond to the (formal) addition and multiplication of Laurent series. Thus Λ is called the "field of formal Laurent series over" k, and denoted by $k((t))$. The subring of Λ consisting of all

$$\sum_{i=0}^{\infty} a_i t^i$$

is called the "ring of formal power series over" k, and denoted by $k[[t]]$. Clearly Λ is the field of fractions of $k[[t]]$. In particular, Λ contains $k(t)$, the field of fractions of $k[t]$.

2.1.2 Factoring Polynomials over $k[[t]]$

The map $k[[t]] \to k$ sending $\sum_{i=0}^{\infty} a_i t^i$ to its constant term a_0 is a ring homomorphism ("evaluation at $t = 0$"). If $F(y) \in k[[t]][y]$ is a polynomial in y with coefficients in $k[[t]]$, we let $F_0(y) \in k[y]$ denote the polynomial obtained by applying the above homomorphism to the coefficients of F.

The following result is a special case of Hensel's Lemma. It tells us that coprime factorizations of F_0 lift to factorizations of F.

Lemma 2.1 *Let F be a monic polynomial in y with coefficients in $k[[t]]$. Suppose the associated polynomial $F_0 \in k[y]$ factors as*

$$F_0 = g \cdot h$$

for monic polynomials $g, h \in k[y]$ that are coprime (i.e., g.c.d.$(g, h) = 1$). Then F factors as

$$F = G \cdot H$$

where G, H are monic polynomials in y with coefficients in $k[[t]]$ such that $G_0 = g$ and $H_0 = h$.

Proof. Extending the above notation, we write

$$F = \sum_{i=0}^{\infty} F_i t^i$$

with $F_i \in k[y]$. Let $m := \deg(F) = \deg(F_0)$. Then $\deg(F_i) < m$ for $i > 0$. Let $r = \deg(g)$, $s = \deg(h)$. We want to find

$$G = \sum_{i=0}^{\infty} G_i t^i \quad \text{and} \quad H = \sum_{i=0}^{\infty} H_i t^i$$

with $G_0 = g$, $H_0 = h$, and $G_i, H_i \in k[y]$ of degree $<r$ (resp., $<s$) for $i > 0$. Further, $F = GH$, of course.

The condition $F = GH$ is equivalent to the system of equations

$$F_n = \sum_{i+j=n} G_i H_j \tag{2.1$_n$}$$

for $n = 0, 1, \ldots$. These equations can be solved inductively as follows. For $n = 0$ we just get the hypothesis $F_0 = gh = G_0 H_0$. Now let $n > 0$, and assume G_i and H_j as above have already been found for all $i, j < n$ satisfying the equations $(2.1)_\nu$ for $\nu = 1, \ldots, n-1$. The nth equation can be written as

$$G_0 H_n + H_0 G_n = U_n \tag{2.2}$$

where $U_n = F_n - \sum_{i=1}^{n-1} G_i H_{n-i}$ has degree $<m$. To complete the induction, it only remains to show that this equation (2.2) can be solved for $G_n, H_n \in k[y]$ of degree $<r$ (resp., $<s$).

Since G_0 and H_0 are coprime in $k[y]$ the ideal they generate is all of $k[y]$. Thus there are $P, Q \in k[y]$ with $G_0 P + H_0 Q = U_n$. By the division algorithm, we can write $P = H_0 S + R$ for $R, S \in k[y]$ with $\deg(R) < s$. Set $H_n = R$ and $G_n = Q + G_0 S$. Then (2.2) holds, and $\deg(H_n) < s$ by construction. Since $H_0 G_n = U_n - G_0 H_n$ we have $H_0 G_n$ of degree $<m$, hence G_n has degree $<r$. Thus G_n, H_n are as desired. □

Corollary 2.2 *Let k be an algebraically closed field of characteristic* 0. *Let F be a monic polynomial in y of degree $n \geq 2$, with coefficients in $k[[t]]$. Suppose the y^{n-1}-coefficient of F_0 is zero, and $F_0(y) \neq y^n$. Then F factors as $F = GH$ with G, H monic nonconstant polynomials in y with coefficients in $k[[t]]$.*

Proof. Since k is algebraically closed, the polynomial $F_0 \in k[y]$ factors as a product of monic linear polynomials. If these linear factors are not all equal, then we can write $F_0 = gh$ for nonconstant g, h that are coprime in $k[y]$. Then the claim follows from the Lemma.

It remains to exclude the case that $F_0 = (y-a)^n (a \in k)$. In this case, the y^{n-1}-coefficient of F_0 is $-na$. Hence $a = 0$ from the hypothesis (since $\operatorname{char}(k) = 0$). But then $F_0 = y^n$, contradicting the hypothesis. □

2.1.3 The Finite Extensions of $\bar{k}((t))$

For each positive integer e, let $\mathbb{Z}e^{-1}$ be the set of all rational numbers of the form i/e, with $i \in \mathbb{Z}$. Then $\mathbb{Z}e^{-1}$ is a (additive) group, isomorphic to $\mathbb{Z}$ via the map $i \mapsto i/e$. It contains $\mathbb{Z}$ as a subgroup of index e.

Let Λ_e be the set of sequences $(a_j)_{j\in\mathbb{Z}e^{-1}}$ of elements $a_j \in k$, such that $a_j = 0$ for almost all $j < 0$. Define addition and multiplication as in 2.1.1. This makes Λ_e into a field, isomorphic to Λ via the map sending $(a_j)_{j\in\mathbb{Z}e^{-1}}$ to $(b_i)_{i\in\mathbb{Z}}$, where $b_i = a_{i/e}$. Under this isomorphism, the element t of Λ corresponds to the element $\tau = (a_j)$ in Λ_e with $a_{1/e} = 1$ and $a_j = 0$ otherwise.

We identify Λ with the subfield of Λ_e consisting of all sequences $(a_j)_{j\in\mathbb{Z}e^{-1}}$ with $a_j = 0$ if $j \notin \mathbb{Z}$. Then $\tau^e = t$. In the symbolic notation from 2.1.1, we write the general element (a_j) of Λ_e as $\sum_{j\in\mathbb{Z}e^{-1}} a_j t^j = \sum_{i\in\mathbb{Z}} a_{i/e}\tau^i = \sum_{i\in\mathbb{Z}} b_i\tau^i$, identifying Λ_e with $k((t^{1/e})) = k((\tau))$.

Lemma 2.3 *Suppose k contains a primitive eth root of unity ζ_e. Then Λ_e is Galois over Λ of degree e. The Galois group is cyclic, generated by the element $\omega : \sum_{i\in\mathbb{Z}} b_i\tau^i \mapsto \sum_{i\in\mathbb{Z}}(b_i\zeta_e^i)\tau^i$. Further $\Lambda_e = \Lambda(\tau)$ with $\tau^e = t$.*

Proof. One checks that ω is an automorphism of Λ_e. Its fixed field consists of those $\sum_{i\in\mathbb{Z}} b_i\tau^i$ with $b_i = 0$ unless $\zeta_e^i = 1$, that is, unless e divides i. Thus the fixed field of ω equals Λ. It follows by Artin's theorem that Λ_e is Galois over Λ with Galois group $\langle\omega\rangle$. Further it is clear that ω has order e. Thus $[\Lambda_e : \Lambda] = |G(\Lambda_e/\Lambda)| = e$.

The automorphism ω maps τ to $\zeta_e\tau$. Thus ω^μ maps τ to $\zeta_e^\mu\tau$, and it follows that no nontrivial element of $G(\Lambda_e/\Lambda)$ fixes τ. Hence $\Lambda_e = \Lambda(\tau)$. □

Theorem 2.4 *Let k be an algebraically closed field of characteristic 0. Let Δ be a field extension of $\Lambda = k((t))$ of finite degree e. Then $\Delta = \Lambda(\delta)$ with $\delta^e = t$.*

Thus each finite extension of Λ is isomorphic to some Λ_e. In other words, each polynomial over Λ (in one variable) has a root in some Λ_e. The latter was already known to Newton (see [Abh1], where this is called Newton's theorem; we follow the elementary proof given there). The topological counterpart is the fact that each connected covering of the punctured disc D^* of degree e is equivalent to the covering $D^* \to D^*$, $z \mapsto z^e$ (see Chapters 4 and 5). In terms of infinite Galois theory, the theorem implies that the absolute Galois group of $k((t))$ is isomorphic to $\hat{\mathbb{Z}}$, the profinite completion of $\mathbb{Z}$.

The theorem is equivalent to the following.

Lemma 2.5 *With k as above, suppose F is a monic nonconstant polynomial in y with coefficients in $k[[t]]$. Then F has a root in some Λ_e.*

Proof. Suppose F is of minimal degree violating the claim. Then certainly $n = \deg(F) \geq 2$. Write $F(y) = y^n + \lambda_{n-1}y^{n-1} + \cdots + \lambda_0$ with $\lambda_\nu \in k[[t]]$. Then the polynomial

$$\tilde{F}(y) = F\left(y - \frac{\lambda_{n-1}}{n}\right)$$

has zero y^{n-1}-coefficient. Replace F by $\tilde{F}$ to assume that F has zero y^{n-1}-coefficient. If further $F_0(y) \neq y^n$ then F factors as in Corollary 2.2, which contradicts the minimality assumption on F. Thus $F_0(y) = y^n$, which means that all λ_ν have zero constant term.

There is some $\nu = 0, \ldots, n-2$ with $\lambda_\nu \neq 0$ (otherwise $F = y^n$ has root 0). Consider only those ν for the remainder of this proof. Let m_ν be the lowest t-power occurring with nonzero coefficient in λ_ν:

$$\lambda_\nu = a\, t^{m_\nu} + \text{higher order terms}$$

where $a \in k$ is nonzero. Then $m_\nu > 0$. Let u be the minimum of the numbers $m_\nu/(n-\nu)$. Then u is a positive rational number. Write $u = d/e$ with positive integers d and e.

Now embed Λ into $\Lambda_e = k((\tau))$ as above. Consider the polynomial

$$F^*(y) = \tau^{-dn}F(\tau^d y) = y^n + \sum_{\nu=0}^{n-2} \lambda_\nu \tau^{d(\nu-n)} y^\nu \qquad \in \Lambda_e[y]$$

The y^ν-coefficient of this polynomial – if nonzero – is a Laurent series in τ of the form

$$\begin{aligned} \lambda_\nu \tau^{d(\nu-n)} &= a\, t^{m_\nu} \tau^{d(\nu-n)} + \text{higher order terms} \\ &= a\, \tau^{E_\nu} + \text{higher order terms} \end{aligned}$$

where

$$E_\nu = e(n-\nu)\left(\frac{m_\nu}{n-\nu} - u\right) \geq 0,$$

and $E_\nu = 0$ for at least one ν. Hence each coefficient of F^* is actually a power series in τ, and for at least one ν this power series has nonzero constant term. Thus F^* satisfies the conditions in Corollary 2.2 (with t replaced by τ). Hence $F^* = GH$ factors over $k[[\tau]]$ as in Corollary 2.2. Then H has degree strictly smaller than n, hence has a root in some $\Lambda_e(\tau^{1/e'})$ by minimality of n. Thus F^* has a root in $\Lambda_e(\tau^{1/e'}) = \Lambda_{ee'}$. Then also F has a root in $\Lambda_{ee'}$. □

Now we can deduce the Theorem as follows. Let Δ be as in the Theorem. Write $\Delta = \Lambda(\theta)$, and let $F \in \Lambda[y]$ be an irreducible polynomial with root θ. By Lemma 1.3 we may assume that F is as in Lemma 2.5. Thus F has a root θ' in some $\Lambda_{e'}$. Hence we may assume $\Delta \subset \Lambda_{e'}$.

Since $G(\Lambda_{e'}/\Lambda)$ is cyclic of order e', for each divisor e of e' there is a unique field between Λ and $\Lambda_{e'}$ of degree e over Λ. Hence $\Delta = \Lambda_e = \Lambda(t^{1/e})$ (see Lemma 2.3). Hence the Theorem.

2.2 Extensions of $\bar{k}(x)$

In this section k is an algebraically closed field of characteristic 0, and $k(x)$ the field of rational functions in the indeterminate x over k.

We use the abbreviation **FG-extension** for "finite Galois extension."

2.2.1 Branch Points and the Associated Conjugacy Classes

We fix a compatible system $(\zeta_e)_{e \in \mathbb{N}}$ of primitive eth roots of unity in k. That is, ζ_e is a primitive eth root of unity, and whenever $e = e'e''$ then $\zeta_e^{e''} = \zeta_{e'}$. In the case $k = \mathbb{C}$, the canonical choice is $\zeta_e = \exp(2\pi\sqrt{-1}/e)$.

Let $\Lambda = k((t))$ as in the previous section. Let Δ/Λ be an FG-extension of degree e. Then $\Delta = \Lambda(\delta)$ with $\delta^e = t$ (Theorem 2.4). Hence there is a unique element $\omega \in G(\Delta/\Lambda)$ with $\omega(\delta) = \zeta_e\delta$ (for each such δ). We call ω the "distinguished generator" of $G(\Delta/\Lambda)$. (Note ω generates $G(\Delta/\Lambda)$ because ζ_e is a *primitive* eth root of 1.) For each $\delta' \in \Delta$ with $(\delta')^{e'} = t$ for some integer $e' \geq 1$ we have $\omega(\delta') = \zeta_{e'}\delta'$. It suffices to prove this for $\delta' = \delta^{e/e'}$. (All other δ' differ only by a root of 1; note e/e' is an integer.) Indeed, $\omega(\delta') = \zeta_e^{e/e'}\delta' = \zeta_{e'}\delta'$ by compatibility of the ζ_e. In particular, if $\Lambda \subset \Delta' \subset \Delta$ then $\omega|_{\Delta'}$ is the distinguished generator of $G(\Delta'/\Lambda)$.

Set $\mathbb{P}^1_k = k \cup \{\infty\}$. In this chapter this is a purely formal notation, with ∞ being an element not in k. We extend each $\alpha \in \mathrm{Aut}(k)$ to $\mathbb{P}^1_k$ by setting $\alpha(\infty) = \infty$.

For $p \in \mathbb{P}^1_k$, define an isomorphism $\vartheta_p : k(x) \to k(t)$ as follows: It is the identity on k, and sends x to $t + p$ (resp., to $1/t$) if $p \neq \infty$ (resp., if $p = \infty$).

Proposition 2.6 *Let $L/k(x)$ be an FG-extension, and set $G = G(L/k(x))$. Let $p \in \mathbb{P}^1_k$.*

(a) We can extend $\vartheta_p : k(x) \to k(t)$ to an isomorphism $\vartheta : L \to L_\vartheta$, where L_ϑ is a subfield of some FG-extension Δ of Λ. Then the group $G(\Delta/\Lambda)$ leaves L_ϑ invariant. Define $g_\vartheta \in G$ as

$$g_\vartheta = \vartheta^{-1} \circ \omega \circ \vartheta$$

where ω is the distinguished generator of $G(\Delta/\Lambda)$. If $\tilde{\Delta}$ is another FG-extension of Λ, with subfield $L_{\tilde{\vartheta}}$, and $\tilde{\vartheta} : L \to L_{\tilde{\vartheta}}$ an isomorphism extending ϑ_p, then $g_{\tilde{\vartheta}}$ and g_ϑ lie in the same conjugacy class of G. This conjugacy class C_p depends only on p (and L, of course). We call it **the class of** $G = G(L/k(x))$ **associated with** p.

(b) *Let e be the common order of the elements of the class C_p. This number $e = e_{L,p}$ is called the* **ramification index** *of L at p. It has the following property: Let γ be a primitive element for the extension $L/k(x)$. Then γ satisfies an irreducible polynomial $F(y) \in k(x)[y]$. Let $\vartheta_p F \in k(t)[y]$ be the polynomial obtained by applying ϑ_p to the coefficients of F. Then all irreducible factors H of $\vartheta_p F$ in $\Lambda[y]$ have degree $e = e_{L,p}$. Further, we can take $\Delta = \Lambda_e$ in (a) (where Λ_e is as in 2.1.3).*

(c) *We can choose γ in (b) such that $F(y) = F(x, y) \in k[x, y]$ is monic in y. Then the discriminant $D(x)$ of $F(y)$ over $k(x)$ is a nonzero element of $k[x]$. If $p \in k$ and $D(p) \neq 0$ then $e_{L,p} = 1$.*

(d) *If $L'/k(x)$ is an FG-extension with $L' \subset L$, then the restriction map from G onto $G' = G(L'/k(x))$ maps the class C_p to the class C'_p of G' associated with p.*

Proof. (a) The field L_ϑ is generated over $k(t)$ by the roots of $\vartheta_p F$ (in the notations from (b)). Since $G(\Delta/\Lambda)$ permutes these roots, it leaves L_ϑ invariant.

For the claim about $\tilde{\vartheta}$, we may assume that Δ and $\tilde{\Delta}$ lie in a common FG-extension Δ_0 of Λ. Then both L_ϑ and $L_{\tilde{\vartheta}}$ are subfields of Δ_0 generated over $k(t)$ by the roots of $\vartheta_p F$ in Δ_0. Thus $L_\vartheta = L_{\tilde{\vartheta}}$. Now set $h = \vartheta^{-1}\tilde{\vartheta}$, an element of G. Since the distinguished generator ω_0 of $G(\Delta_0/\Lambda)$ restricted to Δ is the distinguished generator of $G(\Delta/\Lambda)$ (and similarly for $\tilde{\Delta}$), we get $g_{\tilde{\vartheta}} = \tilde{\vartheta}^{-1}\omega_0\tilde{\vartheta} = h^{-1}\vartheta^{-1}\omega_0\vartheta h = h^{-1}g_\vartheta h$. This proves (a) (except for the existence of ϑ, which is proved in (b)).

(b) The field $\Delta = \Lambda[y]/(H)$ is a finite extension of Λ, and H (hence $\vartheta_p F$) has a root γ' in Δ. Thus ϑ_p extends to an isomorphism ϑ from $L = k(x)[\gamma]$ to $L_\vartheta := k(t)[\gamma'] \subset \Delta$. Thus ϑ is as in (a). The restriction homomorphism $G(\Delta/\Lambda) \to G(L_\vartheta/k(t))$ is now injective, because $\Delta = \Lambda(\gamma')$. The degree $[\Delta : \Lambda]$ equals the order of ω, hence the order of $\omega|_{L_\vartheta}$, hence the order of g_ϑ. The latter is $e_{L,p}$ (by definition). Thus $e_{L,p} = [\Delta : \Lambda] = \deg(H)$. For the last claim in (b) note that since $[\Delta : \Lambda] = e$ we have Δ isomorphic to Λ_e over Λ by Theorem 2.4.

(c) By Lemma 1.3 we can assume F is in $k[x, y]$ and is monic in y. View it as polynomial in y over $k(x)$. Then its discriminant $D(x) \in k[x]$ is nonzero because F is irreducible, hence separable. (Note we are in characteristic 0.) If $p \in k$ with $D(p) \neq 0$ then the polynomial $F(p, y) \in k[y]$ is separable (see the proof of Lemma 1.6).

We have $(\vartheta_p F)(y) = F(t+p, y)$. Thus $\vartheta_p F$ is a monic polynomial in y with coefficients in $k[t]$. Further $(\vartheta_p F)_0(y) = F(p, y)$ in the notation of 2.1.2. If $F(p, y)$ is separable, Hensel's Lemma (Lemma 2.1) yields that $\vartheta_p F$ factors as a product of linear factors in $\Lambda[y]$. Thus $e_{L,p} = \deg(H) = 1$.

(d) Let $\vartheta : L \to \Delta$ be as in (a). Then $\vartheta' = \vartheta|_{L'}$ is an isomorphism from L' to $\vartheta'(L') \subset \Delta$ (and it extends ϑ_p). Thus $g_{\vartheta'} = (\vartheta')^{-1}\omega\vartheta' = \vartheta^{-1}\omega\vartheta|_{L'} = (g_\vartheta)_{L'}$. This proves (d). □

Definition 2.7 *Let $L/k(x)$ be an FG-extension, and $p \in \mathbb{P}^1_k$. We say p is a branch point (or ramified place) of $L/k(x)$ if $e_{L,p} > 1$ (equivalently, if the class C_p of $G(L/k(x))$ is nontrivial).*

It follows from (c) above that the number of branch points is finite. We have defined the following invariants of an FG-extension $L/k(x)$: the branch points p, and the associated conjugacy classes C_p. Now we study how these invariants change if we twist L by an automorphism α of k. This is Fried's branch cycle argument.

If C is a conjugacy class in a group, and m an integer, we let C^m denote the class of g^m, $g \in C$.

Lemma 2.8 (Branch cycle argument) *Let $L/k(x)$ and $L'/k(x)$ be FG-extensions, of degree n. For each $p \in \mathbb{P}^1_k$, let C_p (resp., C'_p) be the class of $G = G(L/k(x)$ (resp., $G' = G(L'/k(x))$) associated with p. Let $\alpha \in Aut(k)$, and let m be an integer such that $\alpha^{-1}(\zeta_n) = \zeta_n^m$. Suppose that α extends to an isomorphism $\lambda : L \to L'$ with $\lambda(x) = x$. Let λ^* be the induced group isomorphism from G onto G', mapping $g \in G$ to $\lambda g \lambda^{-1}$. Then*

$$C'_{\alpha(p)} = \lambda^*(C_p)^m.$$

Proof. Fix $p \in \mathbb{P}^1_k$, and set $e = e_{L,p}$. Then e divides $n = |G|$, hence $\alpha^{-1}(\zeta_e) = \zeta_e^m$ (by compatibility of the ζ_e's).

Clearly, α extends to an automorphism $\tilde{\alpha}$ of $\Lambda_e = k((\tau))$, sending $\sum b_i \tau^i$ to $\sum \alpha(b_i)\tau^i$. View again $\Lambda = k((t))$ as a subfield of Λ_e, where $t = \tau^e$, and let ω be the generator of $G(\Lambda_e/\Lambda)$ with $\omega(\tau) = \zeta_e \tau$. Then $\tilde{\alpha}^{-1}\omega\tilde{\alpha}(\tau) = \tilde{\alpha}^{-1}(\omega(\tau)) = \alpha^{-1}(\zeta_e)\tau = \zeta_e^m \tau = \omega^m(\tau)$, hence

$$\tilde{\alpha}^{-1}\omega\tilde{\alpha} = \omega^m.$$

There is an isomorphism ϑ, extending ϑ_p, from L to a subfield of Λ_e

(Proposition 2.6(b)). Then

$$\vartheta' := \tilde{\alpha} \circ \vartheta \circ \lambda^{-1}$$

is an isomorphism, extending $\vartheta_{\alpha(p)}$, from L' to a subfield of Λ_e. Indeed, ϑ' is the identity on k since both λ and $\tilde{\alpha}$ induce the automorphism α of k. Further, $\vartheta'(x) = \tilde{\alpha}(\vartheta(x)) = \tilde{\alpha}(t + p) = t + \alpha(p)$ for $p \neq \infty$, and $\vartheta'(x) = \tilde{\alpha}(\vartheta(x)) = \tilde{\alpha}(1/t) = 1/t$ for $p = \infty$.

Finally, $g_{\vartheta'} = (\vartheta')^{-1}\omega\vartheta' = \lambda\vartheta^{-1}\tilde{\alpha}^{-1}\omega\tilde{\alpha}\vartheta\lambda^{-1} = \lambda\vartheta^{-1}\omega^m\vartheta\lambda^{-1} = \lambda g_\vartheta^m\lambda^{-1}$ $= \lambda^*(g_\vartheta)^m$. This proves the claim. □

As a preparation for the following example, note the following consequence of Hensel's Lemma: Any $g \in k[[t]]$ with nonzero constant term a_0 is an nth power in $k[[t]]$, for each $n \in \mathbb{N}$. Indeed, the polynomial $F(y) = y^n - g$ has $F_0(y) = y^n - a_0$, a separable polynomial (notation from 2.1.2). Hence $F(y)$ has a root in $k[[t]]$ by Hensel's Lemma.

Example 2.9 $L = k(x)(f^{1/n})$, where $f \in k(x)$ and $n \geq 2$. Adjusting by nth powers we may assume $f(x) = \prod_j (x - p_j)^{m_j}$ with $p_j \in k$ mutually distinct, and $1 \leq m_j \leq n - 1$. Then $f(t+p_i) = t^{m_i} \prod_{j \neq i}(t+p_i-p_j)^{m_j}$, hence $f(t+p_i)$ equals t^{m_i} up to a factor that is an nth power in Λ (by the remarks preceding this example). Thus for $e_i = e_{L,p_i}$ we get

$$\Lambda_{e_i} = \Lambda(f(t + p_i)^{1/n}) = \Lambda(t^{m_i/n});$$

hence

$$e_i = \frac{n}{\text{g.c.d.}(n, m_i)}.$$

Similarly one sees that if $p \in k$ is different from all p_i then $e_{L,p} = 1$. Thus the p_i are exactly the finite branch points of L.

To study $p = \infty$, write $f(x) = a_m x^m + \cdots + a_0$ with $a_m \neq 0$. Let $e_\infty = e_{L,\infty}$. Then

$$\Lambda_{e_\infty} = \Lambda(f(1/t)^{1/n}) = \Lambda(t^{-m/n}(a_m + \cdots + a_0 t^m)^{1/n}) = \Lambda(t^{m/n});$$

hence

$$e_\infty = \frac{n}{\text{g.c.d.}(n, m)}$$

where $m = \deg(f)$. Thus $p = \infty$ is a branch point if and only if n does not divide $\deg(f)$.

Exercise Continuing the above example, embed $G = G(L/k(x))$ into the group $\langle \zeta_n \rangle$ of nth roots of 1 by identifying $g \in G$ with $g(f^{1/n})f^{-1/n}$. Show $C_{p_j} = \{\zeta_n^{m_j}\}$, and $C_\infty = \{\zeta_n^{-m}\}$. Show that these elements generate G, and their product is 1. (This is implied by RET, see the next section, but can be checked directly in this simple case.) Deduce that $[L : k(x)] = n$ (equivalently, $y^n - f$ is irreducible in $k(x)[y]$) if and only if the g.c.d. of n and all m_j is 1.

Example 2.10 Quadratic extensions: $[L/k(x)] = 2$. We have $L = k(x)(\sqrt{f})$ for some $f(x) \in k[x]$ that we may take of the form $f(x) = \prod_i (x - p_i)$ for mutually distinct $p_i \in k$. From the above, the finite branch points of L are exactly the p_i. And ∞ is a branch point if and only if the number of p_i's is odd; that is, the total number of branch points is always even. Conversely, for each finite subset P of $\mathbb{P}^1_k$ of even cardinality there is exactly one quadratic extension of $k(x)$ (up to $k(x)$-isomorphism) with P as a set of branch points.

Exercise Galois extensions $k(y)/k(x)$. The simplest case of an extension $L/k(x)$ is that where $L = k(y)$ is itself a rational function field. If this extension is Galois then its group G embeds into the group $\mathrm{Aut}(k(y)/k)$.

(a) Show that for $y' \in k(y)$ we have $k(y') = k(y)$ if and only if y' is of the form $y' = (ay+b)/(cy+d)$ with $a, b, c, d \in k$ satisfying $ad - bc \neq 0$. Deduce that the group $\mathrm{Aut}(k(y)/k)$ is isomorphic to $\mathrm{PGL}_2(k)$ (the two-dimensional projective linear group over k).

(b) Show that each finite cyclic (resp., dihedral) group embeds into $\mathrm{PGL}_2(k)$ (since k is algebraically closed of characteristic 0). The other finite subgroups of $\mathrm{PGL}_2(k)$ are only A_4, S_4 and A_5; and any two finite subgroups of $\mathrm{PGL}_2(k)$ that are isomorphic as abstract groups are conjugate in $\mathrm{PGL}_2(k)$. This was proved by Klein in 1876 (see [Klein] and for a modern presentation, see Lamotke [Lam]).

(c) Deduce from (a) and (b) that if $k(y)/k(x)$ is a Galois extension then its group is either cyclic, dihedral or isomorphic to A_4, S_4 and A_5. By Lüroth's theorem, all those cases actually occur. If $k(y')/k(x')$ is a Galois extension with the same group, then there is a k-isomorphism $k(y) \to k(y')$ mapping $k(x)$ onto $k(x')$. Find the number of branch points and the ramification type of such an extension (at least in the cyclic and dihedral case!). This can be done in the present purely algebraic set-up, but things become clearer in the geometric translation of Chapter 5.

Remark 2.11 We have seen that there are not a lot of Galois extensions of the form $k(y)/k(x)$. The next step is to look at arbitrary (non-Galois) extensions $k(y)/k(x)$, and see what extensions of $k(x)$ we get as their Galois closure. This can be formulated as a purely group-theoretic problem, via RET and the Riemann–Hurwitz formula. The conjecture of Guralnick and Thompson (see [GT]) expects that only finitely many nonabelian simple groups other than A_n can occur as a composition factor of the Galois group of such an extension.

2.2.2 Riemann's Existence Theorem and Rigidity

In this section k is an algebraically closed subfield of $\mathbb{C}$. We write $\mathbb{P}^1 = \mathbb{P}^1_{\mathbb{C}} = \mathbb{C} \cup \{\infty\}$. Further, we take $\zeta_e = \exp(2\pi\sqrt{-1}/e)$ (our compatible system of roots of 1).

We have the following invariants of an FG-extension $L/k(x)$: the Galois group $G = G(L/k(x))$; the finite set $P \subset \mathbb{P}^1$ of branch points; and for each $p \in P$ the conjugacy class C_p of G. This suggests the following.

Definition 2.12 *Consider triples $(G, P, \mathbf{C})$ where G is a finite group, P is a finite subset of $\mathbb{P}^1$, and $\mathbf{C} = (C_p)_{p\in P}$ is a family of nontrivial conjugacy classes of G, indexed by P. Call two such triples $(G, P, \mathbf{C})$ and $(G', P', \mathbf{C}')$* equivalent *if $P = P'$ and there is an isomorphism $G \to G'$ mapping C_p to C'_p for each $p \in P$. Clearly this defines an equivalence relation on triples. Denote the equivalence class of the triple $(G, P, \mathbf{C})$ by $\mathcal{T} = [G, P, \mathbf{C}]$. We call such $\mathcal{T}$ a* **ramification type**, *or just* **type**.

It is clear what we mean by the ramification type of an FG-extension $L/k(x)$. It is an invariant of the $k(x)$-isomorphism class of L (by Lemma 2.8, applied for $\alpha = \mathrm{id}$). Now we can state

Theorem 2.13 (Riemann's Existence Theorem – RET – The Algebraic Version) *Let $\mathcal{T} = [G, P, (C_p)_{p\in P}]$ be a ramification type. Let $r = |P|$, and label the elements of P as $p_1, \ldots, p_r$. Then there exists an FG-extension of $\mathbb{C}(x)$ of type $\mathcal{T}$ if and only if there exist generators $g_1, \ldots, g_r$ of G with $g_1 \cdot \ldots \cdot g_r = 1$ and $g_i \in C_{p_i}$ for $i = 1, \ldots, r$.*

No purely algebraic proof of this theorem is known (neither of the "only if" nor of the "if"-part). We give a full proof in Chapters 4 to 6, using topology and Riemann surfaces.

Remark 2.14 (a) We see in particular that each finite group occurs as Galois group over $\mathbb{C}(x)$ – the Inverse Galois Problem has a positive solution over this field.

(b) A finer version of RET says that there are only finitely many $L/\mathbb{C}(x)$ (up to $\mathbb{C}(x)$-isomorphism) of a given type. They can be parametrized by the Aut(G)-classes of the tuples $(g_1, \ldots, g_r)$ – however, *not canonically*. See Section 5.4 and Proposition 10.14 for the topological interpretation. Alternatively, this can be deduced purely algebraically from the above version of RET. See Remark 7.4.

(c) The condition about the existence of $g_1, \ldots, g_r$ does not depend on the labeling of the branch points. For example, $g_2, g_2^{-1}g_1g_2, g_3, \ldots, g_r$ is another system of generators of G with product 1, but with the classes C_1 and C_2 switched. Here we encounter the *braiding action* for the first time. It accounts for the nonuniqueness of the $g_1, \ldots, g_r$. See Chapters 9 and 10.

(d) It follows immediately from RET that there are no nontrivial FG-extensions of $\mathbb{C}(x)$ with $r < 2$ branch points. And if $r = 2$ then G must be cyclic. (This corresponds to the fact that the Riemann sphere and the complex plane are both simply connected, while the complex plane minus one point has (infinite) cyclic fundamental group; see Chapters 4 and 5.) Thus the first interesting case is that of $r = 3$ branch points.

(e) Actually, RET remains true if we replace $\mathbb{C}$ by an algebraically closed subfield k. See Corollary 7.10.

The fact that there are usually several nonisomorphic FG-extensions of a given ramification type is the major obstacle in the present approach to the Inverse Galois Problem. This obstacle disappears only under a rather strong group-theoretic condition – rigidity. It says that the generators $g_1, \ldots, g_r$ given by RET are as unique as they can be.

Definition 2.15 *Let $(C_1, \ldots, C_r)$ be a tuple of conjugacy classes in a group G. We say it is* **rigid** *(resp.,* **weakly rigid***) in G if the following hold:*

(a) There exist generators $g_1, \ldots, g_r$ of G with $g_1 \cdot \ldots \cdot g_r = 1$ and $g_i \in C_i$ for $i = 1, \ldots, r$.

(b) If $g_1', \ldots, g_r'$ is another system of generators of G with the same properties, then there exists a unique element $g \in G$ (resp., an automorphism γ of G) with $gg_ig^{-1} = g_i'$ (resp., $\gamma(g_i) = g_i'$) for $i = 1, \ldots, r$.

The uniqueness requirement in (b) (in the rigid case) is equivalent to saying that G has trivial center. The automorphism γ is necessarily unique (since prescribed on a generating system). Thus in the rigid case, any automorphism of G fixing each C_i is inner.

Definition 2.16 *A type $\mathcal{T} = [G, P, \mathbf{C}]$ is called* **rigid** *(resp.,* **weakly rigid***) if the elements of P can be labeled as $p_1, \ldots, p_r$ (with $r = |P|$) such that the classes $C_i = C_{p_i}$ form a rigid (resp., weakly rigid) tuple in G.*

Using the braiding action as in Remark 2.14 (c) one sees that if $(C_1, \ldots, C_r)$ is (weakly) rigid then it remains so after any permutation of the C_i's. Thus the choice of labeling of P does not matter.

Theorem 2.17 *For each weakly rigid type there is a unique FG-extension of $\mathbb{C}(x)$ of this type (up to $\mathbb{C}(x)$-isomorphism).*

Proof. The existence follows from RET. For the uniqueness, assume $L_1/\mathbb{C}(x)$ and $L_2/\mathbb{C}(x)$ are both of the same weakly rigid type. We may assume that L_1 and L_2 lie in a common FG-extension $L/\mathbb{C}(x)$. Let $G = G(L/\mathbb{C}(x))$, $G_j = G(L_j/\mathbb{C}(x))$ for $j = 1, 2$. Let $\rho_j : G \to G_j$ be restriction.

For $p \in \mathbb{P}^1$ let C_p and $C_p^{(j)}$ denote the associated classes of G and G_j, respectively. Then $\rho_j(C_p) = C_p^{(j)}$ by Proposition 2.6(d). Label the branch points of L as $p_1, \ldots, p_r$. By RET there are generators $g_1, \ldots, g_r$ of G with $g_1 \cdot \ldots \cdot g_r = 1$ and $g_i \in C_{p_i}$ for $i = 1, \ldots, r$. Then $\rho_j(g_1), \ldots, \rho_j(g_r)$ are generators of G_j with the analogous properties.

Since L_1 and L_2 are of the same type, there exists an isomorphism $\epsilon : G_2 \to G_1$ mapping $C_p^{(2)}$ to $C_p^{(1)}$ for all p. Then the $\epsilon(\rho_2(g_1)), \ldots, \epsilon(\rho_2(g_r))$ satisfy the same properties as the $\rho_1(g_1), \ldots, \rho_1(g_r)$. By weak rigidity, there is an automorphism δ of G_1 mapping $\epsilon(\rho_2(g_i))$ to $\rho_1(g_i)$ for $i = 1, \ldots, r$. (Note that $\rho_j(g_i) = 1$ iff p_i is no branch point of L_j, so we can just drop those $\rho_j(g_i)$ since L_1 and L_2 have the same branch points.)

Then $\gamma := \delta\epsilon$ is an isomorphism $G_2 \to G_1$ mapping $\rho_2(g_i)$ to $\rho_1(g_i)$. Thus $\rho_1 = \gamma \circ \rho_2$. It follows that ρ_1 and ρ_2 have the same kernel. The fixed field in L of this kernel equals L_1 as well as L_2. Hence $L_1 = L_2$, and we are done. □

Trivially, the type of an abelian extension of $k(x)$ (i.e., with abelian Galois group) is weakly rigid. Thus the abelian extensions of $k(x)$ are uniquely determined by their type. (This is one reason that abelian extensions – as in number theory – are much easier to handle than non-abelian ones; e.g., the cyclic extensions of $\mathbb{C}(x)$ are constructed in Example 2.9.) The extensions of $\mathbb{C}(x)$ that are uniquely determined by their type form a natural class containing the abelian extensions. The surprising fact is that this class contains many extensions with (nonabelian) simple groups. See Chapter 3.

3

Descent of Base Field and the Rigidity Criterion

Given a Galois extension $L/k(x)$ and a subfield κ of k, we ask whether L comes from an extension of $\kappa(x)$ with the same Galois group. This is our basic descent problem. It is solved in Section 3.1 in the case that $k = \bar{\kappa}$ and L is rigid. Coupled with RET, this implies the basic rigidity criteria which we present in 3.2. Section 3.3 discusses applications of these criteria.

3.1 Descent

In this section k is an algebraically closed subfield of $\mathbb{C}$, and $\zeta_e = \exp(2\pi\sqrt{-1}/e)$. We let $L/k(x)$ be an FG-extension of degree n, with Galois group G. Further, κ is a subfield of k.

3.1.1 Fields of Definition

We say $L/k(x)$ (or just L, for short) is **defined over** κ if there exists a subfield L_κ of L, Galois over $\kappa(x)$ and regular over κ, with $[L_\kappa : \kappa(x)] = n$ $(=[L : k(x)])$.

Lemma 3.1 *Suppose L is defined over κ.*

(a) Let θ be a primitive element for $L_\kappa/\kappa(x)$, that is, $L_\kappa = \kappa(x)(\theta)$. Then $L = k(x)(\theta)$. Further, L is defined over each field κ' between κ and k, and we can take $L_{\kappa'} = \kappa'(x)(\theta)$.

(b) The group $G = G(L/k(x))$ is isomorphic to $G(L_\kappa/\kappa(x))$ via restriction to L_κ. Hence G occurs regularly over κ.

(c) The branch points $\neq \infty$ of L are algebraic over κ.

(d) Suppose G has trivial center or κ is algebraically closed. Then L_κ is unique, that is, there is only one $L_\kappa \subset L$ that is Galois over $\kappa(x)$ of degree n and regular over κ.

(e) If κ is algebraically closed, then the extensions $L/k(x)$ and $L_\kappa/\kappa(x)$ are (naturally) of the same type.

Proof. (a) The element θ satisfies an irreducible polynomial $F(y) \in \kappa(x)[y]$. This polynomial remains irreducible in $\kappa'(x)[y]$ since L_κ is regular over κ (Lemma 1.1). Thus the field $L_{\kappa'} = \kappa'(x)[\theta]$ has degree n over $k(x)$, and is regular over κ' (again by Lemma 1.1). Further, $L_{\kappa'}$ is Galois over $\kappa'(x)$ since it contains all roots of $F(y)$ (which already lie in L_κ). Thus L is defined over κ'. Finally, the field $k(x)(\theta)$ is a subfield of L of degree n over $k(x)$ (by the above, for $\kappa' = k$). Hence $L = k(x)(\theta)$. This proves (a).

(b) The group G permutes the roots of $F(y)$ in L. These roots generate L_κ over $\kappa(x)$, hence G leaves L_κ invariant. Thus restriction yields a homomorphism $G \to G(L_\kappa/\kappa(x))$. This homomorphism is injective by (a). Hence it is an isomorphism (because both groups have the same order n). This proves (b).

(c) We can assume the polynomial $F(y)$ is as in Proposition 2.6(c). Then its discriminant $D(x)$ is a nonzero element of $\kappa[x]$. Further, the branch points $\neq \infty$ of L are among the roots of $D(x)$. Hence (c).

(d) Suppose $\tilde{L}_\kappa$ is another subfield of L with the same properties. Let K be the subfield of L generated by L_κ and $\tilde{L}_\kappa$. Then K is Galois over $\kappa(x)$. Let κ' be the algebraic closure of κ in K. Then κ', hence also $\kappa'(x)$, is invariant under the group $G(K/\kappa(x))$. Hence $\kappa'(x)$ is Galois over $\kappa(x)$.

If θ' is a primitive element for $K/\kappa'(x)$ then the minimal polynomial for θ' over $\kappa'(x)$ remains irreducible over $k(x)$ (Lemma 1.1), hence has degree $\leq n = \deg(L/k(x))$. Thus $[K : \kappa'(x)] \leq n$. On the other hand, K contains the above element θ (since it contains L_κ). Hence $[K : \kappa'(x)] \geq [\kappa'(x)(\theta) : \kappa'(x)] = n$ (by (a)). It follows that $K = \kappa'(x)(\theta)$. In particular, if κ is algebraically closed then $K = \kappa(x)(\theta) = L_\kappa$, hence $L_\kappa = \tilde{L}_\kappa$.

Now assume G has trivial center. The field $K = \kappa'(x)(\theta)$ is generated by $\kappa'(x)$ and L_κ. Further, $\kappa'(x) \cap L_\kappa = \kappa(x)$ (by Lemma 1.1(i)), because L_κ is regular over κ. The same holds for $\tilde{L}_\kappa$ in place of L_κ. Hence by the Galois correspondence,

$$G(K/\kappa(x)) = G(K/L_\kappa) \times G' = G(K/\tilde{L}_\kappa) \times G'$$

(direct product) where $G' = G(K/\kappa'(x))$. We have $G' \cong G$ by (a) and (b), hence G' has trivial center (by assumption). It follows that $G(K/L_\kappa) = G(K/\tilde{L}_\kappa) =$ the full centralizer of G' in $G(K/\kappa(x))$. Hence $L_\kappa = \tilde{L}_\kappa$. This proves (d).

(e) Let θ be as in (a). Let $p \in \kappa \cup \{\infty\}$, and $\vartheta : L \to \Delta$ an embedding extending $\vartheta_p : k(x) \to k(t)$, where Δ is an FG-extension of $\Lambda = k((t))$ (as in Proposition 2.6). Set $\theta' = \vartheta(\theta)$. Then $\vartheta(\kappa(x)) = \kappa(t)$, hence $\vartheta(L_\kappa) = \kappa(t)(\theta')$. Thus the restriction of ϑ yields an embedding $\tilde{\vartheta}$ of L_κ into $\Delta_\kappa := \kappa((t))(\theta')$. Clearly, the restriction of the distinguished generator of $G(\Delta/\Lambda)$ to Δ_κ is the distinguished generator of $G(\Delta_\kappa/\kappa((t)))$. It follows that $g_{\tilde{\vartheta}} \in \tilde{G} = G(L_\kappa/\kappa(x))$ is the restriction of $g_\vartheta \in G$.

Thus the restriction isomorphism $G \to \tilde{G}$ maps the class C_p of G associated with p to the class $\tilde{C}_p$ of $\tilde{G}$ associated with p, for all $p \in \kappa \cup \{\infty\}$. Since the branch points of $L/k(x)$ (and of $L_\kappa/\kappa(x)$, of course) lie in $\kappa \cup \{\infty\}$ by (c), we see in particular that $L/k(x)$ has the same branch points as $L_\kappa/\kappa(x)$. Thus the restriction isomorphism $G \to \tilde{G}$ yields the claimed equality of types. □

Example 3.2 The quadratic extension $L = k(\sqrt{x})$ of $k(x)$ is defined over $\mathbb{Q}$, but there are infinitely many choices for $L_{\mathbb{Q}}$: For any square-free integer d we can take $L_{\mathbb{Q}} = \mathbb{Q}(\sqrt{dx})$. These fields are pairwise not $\mathbb{Q}(x)$-isomorphic. This shows that the hypothesis of trivial center in (d) is crucial.

Lemma 3.3

(i) L is defined over a finitely generated extension of κ.
(ii) If L_0 is an FG-extension of $\kappa(x)$, regular over κ, then there is an FG-extension $L/k(x)$, defined over κ, with L_κ being $\kappa(x)$-isomorphic to L_0.

Proof. (i) Let θ be a primitive element for $L/k(x)$, satisfying the irreducible polynomial $F(y) \in k(x)[y]$. Each root θ_i of $F(y)$ in L can be expressed as $\theta_i = f_i(\theta)$ with $f_i \in k(x)[y]$. The coefficients of F and of the f_i are elements of $k(x)$, hence lie in $\kappa'(x)$ for a finitely generated extension κ' of κ. Then $L_{\kappa'} := \kappa'(x)(\theta)$ is Galois over $\kappa'(x)$ because it contains all θ_i. Also the other required conditions hold (same arguments as for (a) above).

(ii) We can take L_0 in the form $L_0 = \kappa(x)[y]/(F)$ (for some irreducible $F(y) \in \kappa(x)[y]$). Then $L := k(x)[y]/(F)$ is an FG-extension of $k(x)$ of the same degree as $L_0/\kappa(x)$ (same arguments again). Thus L is defined over κ, with $L_\kappa = L_0$. □

Corollary 3.4 *Let k_0 be an algebraically closed subfield of $\mathbb{C}$. For each weakly rigid type $\mathcal{T}$ there is at most one FG-extension of $k_0(x)$ of this type (up to $k_0(x)$-isomorphism).*

Proof. Assume L_0 and L_0' are FG-extensions of $k_0(x)$ of type $\mathcal{T}$. By the previous Lemma there are FG-extensions L and L' of $\mathbb{C}(x)$ defined over k_0, such that L_{k_0} and L'_{k_0} are $k_0(x)$-isomorphic to L_0 and L_0', respectively. Then L and L' are also of type $\mathcal{T}$ (Lemma 3.1(e)). Since this type is weakly rigid, it follows that L and L' are $\mathbb{C}(x)$-isomorphic (Theorem 2.17). Each $\mathbb{C}(x)$-isomorphism

$L \to L'$ maps L_{k_0} onto L'_{k_0} by Lemma 3.1(d), inducing a $k_0(x)$-isomorphism $L_{k_0} \to L'_{k_0}$. Then L_0 and L'_0 are also $k_0(x)$-isomorphic. □

3.1.2 The Descent from $\bar{\kappa}$ to κ

Let α be an automorphism of k. View it as an automorphism of $k(x)$, fixing x. An **α-isomorphism** $\lambda : L \to L'$ from L to some FG-extension L' of $k(x)$ is a field isomorphism with $\lambda|_{k(x)} = \alpha$. (Thus a $k(x)$-isomorphism is an α-isomorphism for $\alpha = \text{id}$.) Such λ induces a group isomorphism $\lambda^* : G(L/k(x)) \to G(L'/k(x))$, $g \mapsto \lambda g \lambda^{-1}$.

Lemma 3.5 *Let $L/k(x)$ be an FG-extension. For each $\alpha \in Aut(k)$ there is an α-isomorphism λ from L to some FG-extension L' of $k(x)$. If L is defined over κ and $\alpha|_\kappa = id$, then we can take $L = L'$ and $\lambda^* = id$.*

Proof. Extend α to an automorphism of $k(x)[y]$ fixing x and y. Denote this automorphism by $f \mapsto f^\alpha$.

We can take L of the form $L = k(x)[y]/(F)$ (with $F(y) \in k(x)[y]$ irreducible). Let $L' = k(x)[y]/(F^\alpha)$. The map $f \mapsto f^\alpha$ induces a ring isomorphism $\lambda : L \to L'$. Hence L' is also an FG-extension of $k(x)$, and λ is an α-isomorphism.

If L is defined over κ we can take $F(y) \in \kappa(x)[y]$ such that $L_\kappa = \kappa(x)[y]/(F)$ (Lemma 3.1(a)). If further $\alpha|_\kappa = \text{id}$, then $F^\alpha = F$, hence $L = L'$. If θ is as in Lemma 3.1(a), and $g \in G$, then $\lambda^*(g)(\theta) = \lambda g \lambda^{-1}(\theta) = g(\theta)$ because λ fixes L_κ elementwise, and $g(\theta) \in L_\kappa$. Thus $\lambda^* = \text{id}$. □

Let $\bar{\kappa}$ denote the algebraic closure of κ in $\mathbb{C}$. Although we are not using infinite Galois theory in our main development, we will use the (usually infinite) Galois group $G(\bar{\kappa}/\kappa)$ (which is defined as in the finite case, as the group of all automorphisms of $\bar{\kappa}$ trivial on κ). This is only for notational convenience, so that we do not always need to say "Choose an FG-extension of κ large enough to contain ...." The group $G(\bar{\kappa}/\kappa)$ is naturally a profinite group, but we view it here only as a (an abstract) group. The only thing we need to know about it is that for every FG-extension κ' of κ (inside $\bar{\kappa}$) the restriction map $G(\bar{\kappa}/\kappa) \to G(\kappa'/\kappa)$ is surjective; that is, that every element of $G(\kappa'/\kappa)$ extends to an automorphism of $\bar{\kappa}$. This depends on Zorn's Lemma for general κ, but not for countable κ. Since our interest is on countable fields κ, we are not bothered by this technicality.

Our main descent criterion is:

Proposition 3.6 *Let $L/k(x)$ be an FG-extension whose Galois group G has trivial center. Suppose $k = \bar{\kappa}$. Then L is defined over κ if and only if for each $\alpha \in G(\bar{\kappa}/\kappa)$ there exists an α-automorphism λ of L with $\lambda^* = id$.*

Proof. The condition is necessary by the preceding Lemma. Let us prove it is sufficient. We know L is defined over an FG-extension κ_1 of κ (Lemma 3.3 and 3.1(a)). Set $L_1 = L_{\kappa_1}$.

Begin with some $\alpha_1 \in G(\kappa_1/\kappa)$, and extend it to $\alpha \in G(\bar{\kappa}/\kappa)$. By hypothesis, there is an α-automorphism λ of L with $\lambda^* = \text{id}$. We have $\lambda(L_1) = L_1$ by uniqueness of $L_1 = L_{\kappa_1}$ (Lemma 3.1(d)). Set $\lambda_1 = \lambda|_{L_1}$. We have shown that each $\alpha_1 \in G(\kappa_1/\kappa) \cong G(\kappa_1(x)/\kappa(x))$ (see Lemma 1.1) extends to $\lambda_1 \in \text{Aut}(L_1/\kappa(x))$. This means that L_1 is Galois over $\kappa(x)$ (since it is Galois over $\kappa_1(x)$).

Since $\lambda^* = \text{id}$ we have $\lambda g = g\lambda$ for every $g \in G$. Restricting to L_1 it follows that $\lambda_1 g_1 = g_1 \lambda_1$ for every $g_1 \in G_1 = G(L_1/\kappa_1(x))$ (Lemma 3.1(b)). Thus λ_1 lies in the centralizer C of G_1 in $H = G(L_1/\kappa(x))$. Since each $\alpha_1 \in G(\kappa_1/\kappa)$ extends to such $\lambda_1 \in C$, it follows that C maps surjectively to $G(\kappa_1(x)/\kappa(x)) \cong H/G_1$. This means that $H = G_1 C$. Further, $G_1 \cap C$ lies in the center of $G_1 \cong G$, hence $G_1 \cap C = 1$ (by hypothesis). It follows that H is the direct product of G_1 and C.

Let L_κ be the fixed field of C in L_1. Then L_κ is Galois over $\kappa(x)$ with group isomorphic to $H/C \cong G_1 \cong G$. Further, the intersection of L_κ (the fixed field of C) with $\kappa_1(x)$ (the fixed field of G_1) equals $\kappa(x)$ (the fixed field of $G_1 C = H$). Since $L_\kappa \subset L_1$, and L_1 is regular over κ_1, it follows that L_κ is regular over κ. This shows that L is defined over κ. □

Definition 3.7 *Let $\mathcal{T} = [H, P, (C_p)_{p \in P}]$ be a ramification type, and $n = |H|$. Then $\mathcal{T}$ is called κ-**rational** if $P \subset \bar{\kappa} \cup \{\infty\}$, and the following holds: For each $p \in P$ and $\alpha \in G(\bar{\kappa}/\kappa)$ we have $\alpha(p) \in P$ and*

$$C_{\alpha(p)} = C_p^m$$

where m is an integer such that $\alpha^{-1}(\zeta_n) = \zeta_n^m$.

A conjugacy class C in a group H is called **rational** if $C^m = C$ for each integer m prime to the order of H. (This is standard terminology in group theory.) As an important special case of the above Definition, we see $\mathcal{T}$ is κ-rational if $P \subset \kappa \cup \{\infty\}$ and each class C_p is rational.

It follows from Lemmas 2.8 and 3.5 that the condition of κ-rational type is necessary for L to be defined over κ. In the rigid case it is also sufficient:

Theorem 3.8 *Let $k = \bar{\kappa}$, and let $L/k(x)$ be an FG-extension. If the ramification type of L is rigid and κ-rational then L is defined over κ.*

Proof. We check the criterion from Proposition 3.6. By definition of a rigid type, $G = G(L/k(x))$ has trivial center.

Now fix some $\alpha \in G(\bar{\kappa}/\kappa)$. By Lemma 3.5 there is an α-isomorphism $\lambda : L \to L'$, for some FG-extension $L'/k(x)$. Let $G' = G(L'/k(x))$, and C_p, C'_p as in Lemma 2.8. By that Lemma and the hypothesis of κ-rational type, for $p, q \in \mathbb{P}^1_k$ with $q = \alpha(p)$ we have

$$C'_q = \lambda^*(C_p^m) = \lambda^*(C_q) \tag{3.1}$$

with m as in Definition 3.7. Thus L' is of the same type as L (via the isomorphism $\lambda^* : G \to G'$). Since this type is rigid, it follows that L and L' are $k(x)$-isomorphic (Corollary 3.4).

Let $\mu : L' \to L$ be a $k(x)$-isomorphism. Then $\chi := \mu\lambda$ is an α-automorphism of L. From (3.1) we get

$$\chi^*(C_q) = \mu^*(\lambda^*(C_q)) = \mu^*(C'_q) = C_q.$$

Thus χ^* is an automorphism of G fixing all classes C_q. By rigidity, it follows that χ^* is an *inner* automorphism (see the remarks after Definition 2.15). Thus $\chi^* = g^*$ for some $g \in G$. Hence $\psi = g^{-1}\chi$ is the desired α-automorphism of L with $\psi^* = \text{id}$. We have verified the criterion from Proposition 3.6. Hence L is defined over κ. □

Remark 3.9 We sketch some extensions of the results of this section. None of this will be used in our main development.

(a) Suppose we replace "rigid" by "weakly rigid" in the hypothesis of the theorem. Then it is still true that for each $\alpha \in G(\bar{\kappa}/\kappa)$ there exists an α-automorphism of L (as shown by the proof). The latter condition is clearly equivalent to L being Galois over $\kappa(x)$. Conjugation action of $G(L/\kappa(x))$ on its normal subgroup $G = G(L/k(x))$ yields a map $G(L/\kappa(x)) \to \text{Aut}(G)$. (This is just the map $\lambda \mapsto \lambda^*$.) We get an induced map $G(\bar{\kappa}/\kappa) \cong G(k(x)/\kappa(x)) \cong G(L/\kappa(x))/G \to \text{Aut}(G)/\text{Inn}(G)$. The image A of this map – a certain group of outer automorphisms of G – occurs as Galois group over κ. The group A can be determined in certain cases from the braid group action; see

Chapter 9. This leads to Galois realizations not obtainable from the basic rigidity criteria.

(b) We can further weaken the hypothesis of the Theorem. Define a *weakly κ-rational* type by replacing the condition $C_{\alpha(p)} = C_p^m$ in Definition 3.7 by the condition that $C_{\alpha(p)} = f(C_p^m)$ for some $f \in \mathrm{Aut}(H)$ (where f depends on α, but not on p). If L is of weakly rigid and weakly κ-rational type, we still get that L is Galois over $\kappa(x)$. Conversely, if L is Galois over $\kappa(x)$ then it is of weakly κ-rational type (by Lemma 2.8).

(c) The condition that L is Galois over $\kappa(x)$ is further equivalent to L being "weakly defined over κ," that is, to the existence of an extension $L_0/\kappa(x)$ of degree $n = [L : k(x)]$ (not necessarily Galois) that is regular over κ and contained in L.

Such L_0 is not canonical. It can be constructed as follows. Take $p \in \kappa$ a non-branch point of L, and let $\vartheta_p : k(x) \to k(t)$ as in 2.2.1. Then the tensor product $\tilde{L} = L \otimes_{k(x)} k((t))$ via ϑ_p is the direct sum of n fields $\cong k((t))$, and these are the only minimal ideals of $\tilde{L}$. The group $G(L/\kappa(x))$ acts naturally on $\tilde{L}$, permuting these minimal ideals, and its (normal) subgroup G permutes them sharply transitively. Hence the stabilizer Γ of any such minimal ideal is a complement to G in $G(L/\kappa(x))$. Finally, the fixed field of Γ in L is the desired L_0.

(d) The descent to κ without a condition on the automorphisms of $L/k(x)$ is not canonical (see (c); this is also implied by general principles of Galois cohomology). This suggests the following definition: A finite extension $M/k(x)$ (not necessarily Galois) is said to be defined over κ if it contains $M_0/\kappa(x)$, regular over κ and of the same degree as $M/k(x)$, and having (at least) the same number of automorphisms as $M/k(x)$. Most of the theory of this section can be generalized to this situation. We just look at one special case, namely that $M/k(x)$ has no nontrivial automorphism. This case often occurs in applications (cf. Remark 3.10).

Suppose M is embedded in the Galois extension $L/k(x)$. Then the condition $\mathrm{Aut}(M/k(x)) = 1$ translates into the condition that the group $U = G(L/M)$ is self-normalizing in $G = G(L/k(x))$. Suppose L is of weakly rigid and weakly κ-rational type. Then we know that $L/\kappa(x)$ is Galois. The group $\Omega = G(L/\kappa(x))$ acts via conjugation on its normal subgroup G. Suppose further that in this action, Ω preserves the conjugacy class of U in G. (This holds, e.g., if this class is invariant under all of $\mathrm{Aut}(G)$; a finer condition is obtained by using Lemma 2.8 again, as each element of Ω is an α-automorphism of L, for some α.) Then M is defined over κ.

Indeed, let Π be the normalizer of U in Ω. The group Ω acts on the set of G-conjugates of U (by hypothesis), and Π is the stabilizer of U in this action. Let M_0 be the fixed field of Π in L. Since G acts transitively on the above set,

$\Omega = G\Pi$, hence M_0 is regular over κ. Since U is self-normalizing in G, we have $G \cap \Pi = U$, hence $M_0 k = M$. Thus M is defined over κ.

Remark 3.10 The above has many applications besides those motivated by the Inverse Galois Problem (many of them appearing in the work of M. Fried). A glimpse of this follows.

Each rational function $f(y) \in \mathbb{C}(y)$ yields an extension $\mathbb{C}(y)/\mathbb{C}(x)$, where $x = f(y)$. This extension is usually not Galois, so we form its Galois closure $L_f/\mathbb{C}(x)$. The Galois group $G(L_f/\mathbb{C}(x))$ is called the *monodromy group* of f, denoted $\mathrm{Mon}(f)$, and viewed as a permutation group on the conjugates of y over $\mathbb{C}(x)$. Many questions about rational functions translate into group-theoretic questions about $\mathrm{Mon}(f)$. For example, $\mathrm{Mon}(f)$ is a primitive permutation group if and only if f is indecomposable (with respect to composition of functions). Several problems about rational functions, in particular, polynomials, have been solved by translating them into group theory via $\mathrm{Mon}(f)$ (see [Fr2], [Fr3], [Fr7], [FGS], [Mü]).

There is a purely group-theoretic condition on a finite permutation group G that determines whether G is the monodromy group of a rational function. This is the condition that G has so-called "genus 0" generators. Guralnick and Thompson conjectured that only finitely many nonabelian simple groups can occur as a composition factor of such a group G, apart from the alternating groups. This conjecture has nearly been proved, and in the process a lot of information on such "genus 0 groups" has been accumulated. For example, the monodromy groups of indecomposable polynomials f have been completely classified [Mü], using the classification of finite simple groups. Besides the symmetric, alternating, cyclic, and dihedral groups only a few exceptional groups occur.

If one restricts the coefficients of f to lie in a subfield κ of $\mathbb{C}$ then one is led to descent problems as studied in this chapter. For example, if $f \in \kappa(y)$ then the corresponding FG-extension $L_f/\mathbb{C}(x)$ is of weakly κ-rational type. Conversely, the criterion in (d) above can be used to show that there is a function f with coefficients in κ corresponding to certain group-theoretic constellations. This way the above classification result in [Mü] also yields the monodromy groups of indecomposable polynomials with *integer coefficients*. Besides the above infinite series only three groups occur, namely S_5 in its primitive action on 6 points, and for $q \in \{8, 9\}$ the group $P\Gamma L_2(q)$ acting on the $q + 1$ points of the projective line over the field with q elements. Also the corresponding polynomials $f \in \mathbb{Z}[y]$ of degree 6, 9 and 10, respectively, can be written down.

3.2 The Rigidity Criteria

Theorem 3.11 *Let $\mathcal{T} = [H, P, \mathbf{C}]$ be a rigid and κ-rational type. Then there exists a unique FG-extension $M/\mathbb{C}(x)$ of this type. It is defined over a purely transcendental extension $\kappa(t_1, \ldots, t_s)$ of κ. Thus the group H occurs regularly over κ.*

Proof. M exists and is unique by Theorem 2.17. It is defined over a finitely generated extension κ_1 of κ (Lemma 3.3). This κ_1 is finite over a purely transcendental extension $\kappa_0 = \kappa(t_1, \ldots, t_s)$ of κ (Remark 1.2). Set $k = \bar{\kappa}_0 (= \bar{\kappa}_1)$.

Then M is defined over k (Lemma 3.1(a)). Let $L = M_k$. Then $L/k(x)$ is also of type $\mathcal{T}$ (Lemma 3.1(e)). This $\mathcal{T}$ is κ-rational, hence κ_0-rational (since each element of $G(\bar{\kappa}_0/\kappa_0)$ restricts to an element of $G(\bar{\kappa}/\kappa)$). Thus L is of rigid and κ_0-rational type. It follows by Theorem 3.8 that L is defined over κ_0. Thus also M is defined over κ_0 (with $M_{\kappa_0} = L_{\kappa_0}$). Hence $H \cong G(M_{\kappa_0}/\kappa_0(x)) = G(M_{\kappa_0}/\kappa(t_1, \ldots, t_s, x))$ (Lemma 3.1(b)). Thus H occurs regularly over κ. Indeed, M_{κ_0} is regular over κ because it is regular over κ_0, and κ_0 is regular over κ. Further, $\kappa(t_1, \ldots, t_s, x)$ is purely transcendental over κ because x is transcendental over $\kappa_0 = \kappa(t_1, \ldots, t_s)$. □

Remark 3.12 Actually, M is defined over κ. To prove this requires an additional argument (showing that suitable specialization of the t_i preserves the type). Such an argument is used to prove Theorem 7.9 below. That theorem shows that M is defined over $\bar{\kappa}$. Then Theorem 3.8 yields that M is defined over κ.

Let us first consider the important special case of rational classes. Recall that a conjugacy class C in a group G is called rational if $C^m = C$ for each integer m prime to the order of G.

Corollary 3.13 (Rational Rigidity Criterion) *If a finite group H has rational conjugacy classes $C_1, \ldots, C_r$ that form a rigid tuple then H occurs regularly over $\mathbb{Q}$.*

Proof. Let $P \subset \mathbb{Q} \cup \{\infty\}$ be any set of r elements $p_1, \ldots, p_r$. Together with the given classes $C_1, \ldots, C_r$ this yields a type $\mathcal{T} = [H, P, \mathbf{C}]$ (as in Definition 2.16). This type is rigid by definition, and $\mathbb{Q}$-rational by the remarks after Definition 3.7. Now the claim follows from the theorem. □

See Corollary 3.16 below for the General Rigidity Criterion over $\mathbb{Q}$. (There we also recall some consequences of the conclusion.)

The name rigidity was coined by John Thompson. He used the Rational Rigidity Criterion in his famous paper [Th1] where he realized the Monster – largest sporadic simple group – as a Galois group over $\mathbb{Q}$. The Monster is an object of fascinating complexity, exemplifying the intrinsic difficulty in classifying the finite simple groups. However, the significance of the Monster goes far beyond group theory. Connections have been found to such distant areas as automorphic forms ("moonshine phenomenon") and mathematical physics (string theory). Thus it can rightfully be said that the Monster is the one single group that has attracted the most attention.

Taking all this together, one understands why Thompson's realization of the Monster M was a spectacular result. Imagine, there is actually a polynomial out there, with integer coefficients, that has Galois group M. Anybody who has taken a first course in Galois theory, and struggled to compute the Galois group of a polynomial of degree 4 or 5, say, can appreciate this result.

In the next section we explain the steps necessary to apply the rigidity criterion to M (and other simple groups).

Next to $\mathbb{Q}$, the field $\mathbb{Q}_{ab}$ (generated by all roots of unity) attracts the most attention in the work on the Inverse Galois Problem. One reason for this is Shafarevich's conjecture on the absolute Galois group of $\mathbb{Q}_{ab}$ (see Section 8.3.3). It would be a consequence of Shafarevich's conjecture that each finite group is a Galois group over $\mathbb{Q}_{ab}$. This is also not yet in reach, but most classes of simple groups have been realized over $\mathbb{Q}_{ab}$, using the rigidity criterion. It takes a particularly simple form over $\mathbb{Q}_{ab}$.

Corollary 3.14 (Rigidity Criterion over $\mathbb{Q}_{ab}$) *Suppose H has conjugacy classes $C_1, \ldots, C_r$ that form a rigid tuple. Then H occurs regularly over $\mathbb{Q}_{ab}$. Thus H is a Galois group over $\mathbb{Q}_{ab}$.*

Proof. Again the given classes yield a rigid and $\mathbb{Q}_{ab}$-rational type for any choice of $P \subset \mathbb{Q}_{ab} \cup \{\infty\}$. For the $\mathbb{Q}_{ab}$-rationality, note that $\alpha(\zeta_n) = \zeta_n$ for each $\alpha \in G(\bar{\mathbb{Q}}/\mathbb{Q}_{ab})$. Hence we can take $m = 1$ in Definition 3.7, and so $C_{\alpha(p)} = C_p = C_p^m$ for each $p \in P$. Thus H occurs regularly over $\mathbb{Q}_{ab}$ by Theorem 3.11. Then H is a Galois group over $\mathbb{Q}_{ab}$ since $\mathbb{Q}_{ab}$ is hilbertian (Theorem 1.13 and Corollary 1.28). □

Now we are heading for the General Rigidity Criterion that contains the two above as special cases.

Definition 3.15 *Let $(C_1, \ldots, C_r)$ be a tuple of conjugacy classes in a group H. We say it is κ-**rational** if $C_1^m, \ldots, C_r^m$ is a permutation of $C_1, \ldots, C_r$ for* *each integer m with the following property: There exists $\alpha \in G(\bar{\kappa}/\kappa)$ with $\alpha^{-1}(\zeta_n) = \zeta_n^m$, where $n = |H|$.*

If the tuple is κ-rational for $\kappa = \mathbb{Q}$ then we just call it rational. This means that for each integer m prime to the order of H the classes $C_1^m, \dots, C_r^m$ are a permutation of $C_1, \dots, C_r$.

Lemma 3.16 *If $(C_1, \dots, C_r)$ is κ-rational in H then there exists a finite set $P = \{p_1, \dots, p_r\} \subset \bar{\kappa}$ (of cardinality r) such that the type $\mathcal{T} = [H, P, (C_p)_{p \in P}]$ is κ-rational, where $C_{p_i} = C_i$, $i = 1, \dots, r$.*

Proof. Let $n = |H|$, and $S = G(\kappa(\zeta_n)/\kappa)$. Define an action of S of the set of conjugacy classes C of H as follows. For each $\sigma \in S$ there is an integer m with $\sigma^{-1}(\zeta_n) = \zeta_n^m$. Set $\sigma(C) = C^m$. This is well defined because if $\sigma^{-1}(\zeta_n) = \zeta_n^{m'}$ then $m' \equiv m \pmod{n}$, hence $C^{m'} = C^m$.

Now consider the given classes $C_1, \dots, C_r$. Let S_1 be the stabilizer of C_1 in S (for the action just defined). Let $\sigma_1, \dots, \sigma_\ell$ be a system of representatives for the cosets of S_1 in S. Then the classes $\sigma_1(C_1), \dots, \sigma_\ell(C_1)$ are all distinct, and are among $C_1, \dots, C_r$ (by κ-rationality). We may assume $C_i = \sigma_i(C_1)$ for $i = 1, \dots, \ell$.

Let κ_1 be the fixed field of S_1 in $\kappa(\zeta_n)$. Let p_1 be a primitive element for κ_1/κ, and set $p_i = \sigma_i(p_1)$ for $i = 1, \dots, \ell$. Then S permutes $p_1, \dots, p_\ell$, and $\mathrm{Stab}_S(p_1) = S_1 = \mathrm{Stab}_S(C_1)$. Hence the map $p_i \mapsto C_i$ ($i = 1, \dots, \ell$) is an isomorphism of S-sets. Defining $C_{p_i} = C_i$ for $i = 1, \dots, \ell$, we get

$$C_{\sigma(p)} = \sigma(C_p) = C_p^m$$

for $p \in \{p_1, \dots, p_\ell\}$, where m is associated with σ as above. By induction we may assume there exist $p_{\ell+1}, \dots, p_r$ such that the same holds for $p \in \{p_{\ell+1}, \dots, p_r\}$. Then $P = \{p_1, \dots, p_r\}$ is as desired. □

Combining the Lemma with Theorem 3.11 we obtain

Theorem 3.17 (General Rigidity Criterion) *Suppose κ is a subfield of $\mathbb{C}$, and H a finite group. If H has conjugacy classes $C_1, \dots, C_r$ that form a rigid and κ-rational tuple then H occurs regularly over κ.*

Corollary 3.18 (Rigidity Criterion over $\mathbb{Q}$) *Suppose H has conjugacy classes $C_1, \dots, C_r$ that form a rigid tuple. Assume further that for each integer m prime to the order of H the classes $C_1^m, \dots, C_r^m$ are a permutation of $C_1, \dots, C_r$. Then*

H occurs regularly over $\mathbb{Q}$ (hence over any field of characteristic 0). Thus H is a Galois group over $\mathbb{Q}$, more generally, over every finitely generated extension field of $\mathbb{Q}$, and even more generally, over every hilbertian field of characteristic 0.

Proof. The first assertion follows from the theorem. For the second assertion, recall that if H occurs regularly over $\mathbb{Q}$ then H occurs regularly over every extension field of $\mathbb{Q}$ (Corollary 1.15), that is, over every field of characteristic 0. Thus H is a Galois group over every hilbertian field of characteristic 0 (Corollary 1.15). Finally, recall that every finitely generated extension of $\mathbb{Q}$ is hilbertian (Corollary 1.11). □

3.3 Rigidity and the Simple Groups

If G is a finite group, we can choose a composition series in G and view G as being "built up" from its simple composition factors G_i. If G is a Galois group over a field κ, the composition series of G yields a tower of extensions of κ whose relative Galois groups are the simple groups G_i. Thus in order to realize G as Galois group over κ, one has to realize the simple G_i inductively over various extension fields of κ, not arbitrarily, however, but in such a way that these extensions fit together with the given composition series.

Some methods of piecing field extensions together like this are available; see Chapter 8. The technical term for this is "embedding problem." The first test cases are the solvable groups. All their composition factors are cyclic, hence easy to realize as a Galois group (see Section 8.2). The problem is to piece them together to get all solvable groups. Indeed, Shafarevich's realization of all solvable groups over $\mathbb{Q}$ is based on this idea. However, this does not generalize to $\mathbb{Q}(x)$: So far it is not known whether all solvable groups occur regularly over $\mathbb{Q}$ (not even over $\mathbb{Q}_{\mathrm{ab}}$).

Still, it seems natural to concentrate first on the simple groups. Besides the motivation from the general Inverse Galois Problem, it seems an intriguing task to investigate how the rich structure of simple groups would be mirrored in field theory via the Galois correspondence. Indeed, much work has been done trying to realize simple groups (and related groups) over $\mathbb{Q}$ and $\mathbb{Q}_{\mathrm{ab}}$. The picture is nearly complete over $\mathbb{Q}_{\mathrm{ab}}$, but over $\mathbb{Q}$ there is still a long way to go.

Let us begin with the alternating groups A_n (simple if $n \geq 5$). They were realized as Galois groups over $\mathbb{Q}$ by Hilbert's, first application of his irreducibility theorem. We follow Hilbert's proof. As usual in group theory, it also turns out in the Galois problem that it is not enough to consider just simple groups, but one must admit *almost simple groups*, groups between a (nonabelian) simple group and its automorphism group. In the case of A_n, we begin with a Galois realization of S_n.

3.3.1 The Alternating and Symmetric Groups

Suppose C_1, C_2, C_3 are conjugacy classes in a group H, and there exist generators g_1, g_2, g_3 of H with $g_i \in C_i$ and $g_1 g_2 g_3 = 1$. Clearly, the triple (C_1, C_2, C_3) is rigid in H if and only if H has trivial center and for each $g_2' \in C_2$ with $(g_1 g_2')^{-1} \in C_3$ and $\langle g_1, g_2' \rangle = H$ there is $h \in H$ with $hg_1h^{-1} = g_1$ and $hg_2'h^{-1} = g_2$.

Lemma 3.19 *Let $C^{(i)}$ be the class of i-cycles in S_n, $n \geq 3$. Then the classes $C^{(2)}, C^{(n-1)}$ and $C^{(n)}$ form a rigid triple in S_n.*

Proof. We know that S_n $(n \geq 3)$ has trivial center. If a subgroup of S_n contains an n-cycle and an $(n-1)$-cycle then it is doubly transitive (i.e., transitive on pairs of distinct letters from $1, \ldots, n$). Hence if this subgroup contains a transposition then it contains all transpositions, thus equals S_n.

Let $\tau = (n-1, n)$, $\sigma = (1, \ldots, n-1)$, and $\pi = (n-1, n, n-2, \ldots, 2, 1)$. Then $\sigma \cdot \tau \cdot \pi = 1$. These elements generate S_n by the previous paragraph. To complete the proof, it suffices to show that any transposition τ' such that $\sigma\tau'$ is an n-cycle is conjugate to τ under a power of σ. Clearly τ' must be of the form $\tau' = (j, n)$ for some $j = 1, \ldots, n-1$. Then σ^{n-1-j} maps j to $n-1$ (and fixes n). Thus σ^{n-1-j} conjugates τ' into τ. □

Since each conjugacy class of S_n is rational, it follows by the Rational Rigidity Criterion that S_n occurs regularly over $\mathbb{Q}$. Of course this is not too exciting, since S_n has already been realized in Example 1.17. However, the present realization of S_n yields one of A_n.

Lemma 3.20 *Suppose (C_1, C_2, C_3) is a rigid triple of rational classes of a finite group H. Then each subgroup of H of index 2 occurs regularly over $\mathbb{Q}$.*

Proof. Choose distinct $p_1, p_2, p_3 \in \mathbb{Q}$. Associating C_1, C_2, C_3 to these points yields a rigid and $\mathbb{Q}$-rational type $\mathcal{T}$ (as in the proof of 3.13). By Theorem 3.11 there is a purely transcendental field $\kappa_0 = \mathbb{Q}(t_1, \ldots, t_s) \subset \mathbb{C}$ and an FG-extension $M/\mathbb{C}(x)$ of type $\mathcal{T}$ defined over κ_0. Set $M_0 = M_{\kappa_0}$.

We have $G(M_0/\kappa_0(x)) \cong H$. Fix such an isomorphism. Then each subgroup U of H of index 2 corresponds to a quadratic extension N_0 of $\kappa_0(x)$ inside M_0. We can write $N_0 = \kappa_0(x)(\sqrt{f})$ where $f(x) \in \kappa_0[x]$ is a nonconstant square-free polynomial (i.e., $f(x)$ has no multiple irreducible factors in $\kappa_0[x]$, hence no multiple roots in $\mathbb{C}$). Here $f(x)$ is nonconstant because M_0 is regular over κ_0. Then $N := \mathbb{C}(x)(\sqrt{f})$ is a quadratic extension of $\mathbb{C}(x)$ inside M (Corollary 1.1(ii) or Example 2.9).

By Lemma 2.6(d), each branch point of $N/\mathbb{C}(x)$ is a branch point of $M/\mathbb{C}(x)$. Hence the branch points of N are among p_1, p_2, p_3. Since the number of branch points of N is even and ≥ 2 (Example 2.10), it follows that N has exactly two branch points, say p_1 and p_2. Thus f is of the form $f(x) = c(x - p_1)(x - p_2)$ with $c \in \kappa_0$ (since f has no multiple roots); see Example 2.10.

We have $N_0 = \kappa_0(x)(\sqrt{f}) = \kappa_0(x)(\sqrt{f}/(x - p_2)) = \kappa_0(x)(z)$, where $z^2 = c(x - p_1)/(x - p_2)$. Now z^2 is a linear fractional function of x, hence $x \in \kappa_0(z^2)$. Thus $N_0 = \kappa_0(x)(z) = \kappa_0(z)$ is a rational function field, and so $U \cong G(M_0/N_0) = G(M_0/\kappa_0(z))$ occurs regularly over κ_0. Then U occurs regularly over $\mathbb{Q}$ (as in the proof of Theorem 3.11). □

Corollary 3.21 *All symmetric and alternating groups occur regularly over* $\mathbb{Q}$.

Exercise Determine the type of the extension $M/\mathbb{C}(z)$ with group A_n obtained from Lemmas 3.19 and 3.20. Is this type rigid? (See [Ma1], p. 100.)

3.3.2 A Formula to Verify Rigidity

In general it is very difficult to check whether a given tuple of conjugacy classes is rigid. This problem falls into two parts. First, one needs to show that there are elements in those classes, with product 1, that generate the group. Second, one needs to show the uniqueness of these generators (up to conjugacy). For the latter, there is a very useful formula, involving the character values on the given conjugacy classes. Let us first recall the definition of the irreducible (complex) characters of a finite group H. They are complex-valued functions on H. Each irreducible representation $\lambda : H \to \mathrm{GL}_n(\mathbb{C})$ yields an irreducible character θ via $\theta(g) = \mathrm{tr}(\lambda(g))$ (where "tr" means trace). Clearly, equivalent representations yield the same character. Since there are only finitely many equivalence classes of irreducible representations of H, there are only finitely many irreducible characters $\theta_1, \ldots, \theta_s$ of H. Clearly, the character value $\theta_i(g)$ depends only on the conjugacy class C of g, so we also write $\theta_i(C)$.

Now we can state the announced formula. It is valid for any number r of conjugacy classes; most applications, however, have been in the case $r = 3$. Thus we state it only in this case, for simplicity.

Lemma 3.22 *Let $\theta_1, \ldots, \theta_s$ be the irreducible characters of H. Let C_1, C_2, C_3 be conjugacy classes of H. Set $c_i = |C_i|$. Then the number of triples*

$(g_1, g_2, g_3) \in C_1 \times C_2 \times C_3$ with $g_1 g_2 g_3 = 1$ equals

$$c_1 c_2 c_3 \sum_{i=1}^{s} \frac{\theta_i(C_1)\theta_i(C_2)\theta_i(C_3)}{\theta_i(1)^2}.$$

The proof requires only the very basic properties of characters. It can be found (in the case of an arbitrary number of classes) in [MM], Ch. I, for example.

Clearly, if H has trivial center, if condition (a) of Definition 2.15 holds and the above formula gives the value $|H|$ then the triple C_1, C_2, C_3 is rigid.

3.3.3 The Sporadic Groups

The next family of simple groups consists of the 26 sporadic groups. The rigidity criteria have been applied successfully to these groups, yielding Galois realizations over $\mathbb{Q}$ (resp., $\mathbb{Q}_{ab}$) for all sporadic groups except the Mathieu groups M_{23} and M_{24} (resp., for all sporadics). The group M_{24} was realized over $\mathbb{Q}$ by Matzat [Ma5], using the braid group action on generating systems of M_{24} of length $r = 4$. This leaves the group M_{23} as the only sporadic simple group that is not yet known to be a Galois group over $\mathbb{Q}$.

The point is that for all sporadic groups the character tables are known, and listed in the atlas of simple groups [Atlas]. These tables list the conjugacy classes C and irreducible characters θ_i, and the values $\theta_i(C)$. Thus the above formula can always be evaluated in the case of sporadic groups. Indeed, many rigid triples have been found.

Example 3.23 The Monster.

The Monster M has rational classes C_1, C_2, C_3 of elements of order 2, 3, and 29, respectively. The above formula actually yields that the number of triples $(g_1, g_2, g_3) \in C_1 \times C_2 \times C_3$ with $g_1 g_2 g_3 = 1$ equals $|M|$. It remains to prove that each such triple generates M. This is quite a difficult problem since the maximal subgroups of M have not yet been classified. Its solution uses the classification of simple groups, checking that no proper subgroup of M can contain such a triple (see [Th1]). With this done, M is known to occur regularly over $\mathbb{Q}$ (by the Rational Rigidity Criterion).

The intriguing story continues with the Mathieu groups $M_{11}, M_{12}, M_{22}, M_{23}$ and M_{24}, the smallest sporadics. (Excluding M_{22}, they are the only 4-transitive finite permutation groups other than S_n and A_n.) As mentioned above, the case of M_{23} is still open. Let us consider another example that is also due to Matzat ([Ma1], p. 120).

Example 3.24 The Mathieu group M_{12}.

M_{12} is the only 4-transitive permutation group of degree 12 other than A_{12} and S_{12}. View M_{12} as embedded in its automorphism group H. We have $[H : M_{12}] = 2$. Similarly as in the previous example, one finds that H admits a rigid triple (C_1, C_2, C_3) of rational classes C_i consisting of elements of order 2, 3, and 12, respectively. Thus H and M_{12} both occur regularly over $\mathbb{Q}$ (Rational Rigidity Criterion and Lemma 3.20).

3.3.4 The Lie Type Groups

The remaining simple groups are those of the Lie type. They fall into 16 families, 6 classical and 10 exceptional. The prototype is the family $\mathrm{PSL}_n(q)$ (= $\mathrm{SL}_n(q)$ modulo scalars), where $n \geq 2$ and q is a prime power. (Simple unless $(n, q) = (2, 2)$ or $= (2, 3)$.) Here $\mathrm{SL}_n(q)$ denotes the group of $n \times n$-matrices of determinant 1 over the field $\mathbb{F}_q$ of q elements.

In general, the Lie type groups comprise certain almost simple groups associated with the above families, and certain central extensions. With each Lie type group G there is associated a prime power q. We write $G = G(q)$ (and say $\mathbb{F}_q$ is the field of definition for the group).

In the work on Lie type groups, the limitations of the rigidity criteria became quickly apparent. If q is not a prime there are only very few cases known where the rigidity criteria realize $G(q)$ over $\mathbb{Q}$, and there are no such cases known if q is more than a third power of a prime. Also in the case that $q = p$ is a prime, only certain of the groups $G(p)$ have been realized over $\mathbb{Q}$, under various conditions on p, the Lie rank and the Lie family. See [MM], Chs. I and II, for the collection of known results. Recent work tries to find more general criteria by using the action of the braid group; see Chapter 9 (and [MM], Ch. III). For example, combining the braiding action with weak rigidity, the author has shown that the groups $\mathrm{PGL}_n(q)$ occur regularly over $\mathbb{Q}$, essentially for n even and $n \geq q$ (see Section 9.4).

Special interest concentrates on the groups $\mathrm{PSL}_2(q)$. One reason is the connection to modular curves (see [Se1], Ch. 5). Another reason is that the simple groups of "small" order – say the 10 smallest – are either alternating or of the form $\mathrm{PSL}_2(q)$. More precisely, the 10 smallest simple groups are A_5, A_6, A_7 and $\mathrm{PSL}_2(q)$ for q a prime power ≤ 19. (Note $A_5 \cong \mathrm{PSL}_2(4) \cong \mathrm{PSL}_2(5)$ and $A_6 \cong \mathrm{PSL}_2(9)$.) Among those only the group $\mathrm{PSL}_2(16)$ is not yet known to occur as a Galois group over $\mathbb{Q}$ (see [Se1], p. 88). We treat the example of the groups $\mathrm{PSL}_2(q)$ in some detail in Section 3.3.6.

The picture changes drastically if we look at Galois realizations over $\mathbb{Q}_{ab}$. Belyi has realized all classical simple Lie type groups over $\mathbb{Q}_{ab}$, by showing they have a rigid triple of conjugacy classes. The same has been done by Malle for most of the exceptional groups. Their methods are quite different (see [MM], Ch. II). Belyi uses the geometry of the natural module of the group, whereas Malle's approach relies on the formula in Lemma 3.21, requiring deep results from the character theory of Lie type groups, and also the classification of finite simple groups. Only very few cases remain open.

Finally, it should be remarked that nearly all applications of the rigidity criteria are in the case $r = 3$. This is not surprising: As the number of generators gets larger, and there is still only one relation, rigidity is less and less likely to hold. (Still, one can get examples of rigidity with r arbitrarily large, by taking direct products of groups with rigid generators.) There is one known case (due to Thompson [Th2]) of weakly rigid generators of $\mathrm{SL}_n(q)$ with r arbitrarily large. A generalized notion of rigidity, which involves other relations than just the single relation $g_1 \cdots g_r = 1$, has been introduced in [SV1]. So far, it has not yet led to new Galois realizations.

For our treatment of the groups $\mathrm{PSL}_2(q)$, it is convenient to use

3.3.5 A Criterion for Groups Modulo Center

There are many modifications of the rigidity criteria. Here we give a modification of the General Rigidity Criterion.

Let $(C_1, \ldots, C_r)$ be a tuple of conjugacy classes in a group H. We say it is **quasi-rigid** if it is weakly rigid and every automorphism of H fixing each C_i is inner (equivalently, it is weakly rigid and the automorphism γ in condition (b) of Definition 2.15 is always inner). Thus rigid is equivalent to quasi-rigid plus the condition of trivial center. A quasi-rigid type is defined analogously as in Definition 2.16.

Theorem 3.25 *If H has classes $C_1, \ldots, C_r$ that form a quasi-rigid and κ-rational tuple then $H/Z(H)$ occurs regularly over κ.*

Here $Z(H)$ denotes the center of H. The proof is the same as for the General Rigidity Criterion, replacing Theorem 3.8 by

Proposition 3.26 *Let $k = \bar{\kappa}$, and let $L/k(x)$ be an FG-extension with group G. Let L' be the fixed field of $Z(G)$. If the ramification type of L is quasi-rigid and κ-rational then L' is defined over κ.*

Proof. As in the proof of Theorem 3.8, for each $\alpha \in G(\bar{\kappa}/\kappa)$ we obtain an α-automorphism λ of L with $\lambda^* = \text{id}$. This means that each element of $G(\bar{\kappa}(x)/\kappa(x)) \cong G(\bar{\kappa}/\kappa)$ extends to an automorphism of L centralizing G. In particular, L is Galois over $\kappa(x)$.

Let θ be a generator of L over $k(x)$, and let $\theta_1, \ldots, \theta_t$ be the conjugates of θ over $\kappa(x)$. All these conjugates lie in L because L is Galois over $\kappa(x)$. Then $L_1 := \kappa(x)(\theta_1, \ldots, \theta_t)$ is an FG-extension of $\kappa(x)$. Let κ_1 be the algebraic closure of κ in L_1. Then L_1 is regular over κ_1, and it follows as in the proof of Lemma 3.1(d) that L is defined over κ_1 with $L_{\kappa_1} = L_1$. Thus $G_1 := G(L_1/\kappa_1(x))$ is isomorphic to G. Further, G_1 is normal in $H := G(L_1/\kappa(x))$ (cf. the proof of Lemma 3.1(d)).

We follow the proof of Proposition 3.6. Again we get $H = G_1C$, where C is the centralizer of G_1 in H. Let L_1^C be the fixed field of C in L_1. Then L_1^C is Galois over $\kappa(x)$ with group isomorphic to $H/C = (G_1C)/C \cong G_1/(G_1 \cap C) = G_1/Z(G_1)$. From $H = G_1C$ it follows that $\kappa(x) = \kappa_1(x) \cap L_1^C$, hence L_1^C is regular over κ. Further, $L_1^C \subset L_1^{Z(G_1)} \subset L^{Z(G)} = L'$ (using Lemma 3.1(b)). Thus L' is defined over κ, with $L'_\kappa = L_1^C$. □

3.3.6 An Example: The Groups $PSL_2(q)$

We consider only the case that q is odd. Compare with the treatment in [Ma1], III, §4.

Lemma 3.27

(a) *The nonidentity elements of $SL_2(q)$ of trace 2 fall into exactly two conjugacy classes C_1 and C_2. Let $u, v \neq 0$ in $\mathbb{F}_q$. Then the matrix $\begin{pmatrix}1 & u\\ 0 & 1\end{pmatrix}$ (resp., $\begin{pmatrix}1 & 0\\ v & 1\end{pmatrix}$) lies in C_1 if and only if u (resp., $-v$) is a square in $\mathbb{F}_q$.*

(b) *Let $U_1 = \begin{pmatrix}1 & 1\\ 0 & 1\end{pmatrix}$. Let $U \in SL_2(q)$ of trace 2, not upper triangular. Then there is $A \in SL_2(q)$ centralizing U_1 and conjugating U into a matrix of the form $\begin{pmatrix}1 & 0\\ c & 1\end{pmatrix}$.*

Proof. We first prove (b). Write $U = \begin{pmatrix}a & b\\ c & d\end{pmatrix}$. Then $c \neq 0$. Take A of the form $A = \begin{pmatrix}1 & m\\ 0 & 1\end{pmatrix}$. Then A centralizes U_1. We compute

$$AUA^{-1} = \begin{pmatrix}a + mc & *\\ c & *\end{pmatrix}.$$

Thus for $m = c^{-1}(1 - a)$ we get

$$AUA^{-1} = \begin{pmatrix}1 & 0\\ c & 1\end{pmatrix}.$$

(The right lower 1 comes from the condition of trace 2, and the right upper zero from the condition of determinant 1.) This proves (b).

Let $U \in \mathrm{SL}_2(q)$ of trace 2. If U is upper triangular then it is of the form $\left(\begin{smallmatrix}1 & u\\ 0 & 1\end{smallmatrix}\right)$ (since its characteristic polynomial is $(X-1)^2$). If U is not upper triangular then it is conjugate to some $\left(\begin{smallmatrix}1 & 0\\ c & 1\end{smallmatrix}\right)$ by (b). From

$$\begin{pmatrix}0 & 1\\ -1 & 0\end{pmatrix}\begin{pmatrix}1 & 0\\ v & 1\end{pmatrix}\begin{pmatrix}0 & 1\\ -1 & 0\end{pmatrix}^{-1} = \begin{pmatrix}1 & -v\\ 0 & 1\end{pmatrix} \tag{3.2}$$

it follows that in any case, U is conjugate to some $\left(\begin{smallmatrix}1 & u\\ 0 & 1\end{smallmatrix}\right)$. It remains to be seen when such a matrix is conjugate U_1. If $B \in \mathrm{SL}_2(q)$ conjugates U_1 into $\left(\begin{smallmatrix}1 & u\\ 0 & 1\end{smallmatrix}\right)$ then B must be upper triangular (since it fixes the only eigenspace of U_1). Now (a) follows from

$$\begin{pmatrix}w & *\\ 0 & w^{-1}\end{pmatrix} U_1 \begin{pmatrix}w & *\\ 0 & w^{-1}\end{pmatrix}^{-1} = \begin{pmatrix}1 & w^2\\ 0 & 1\end{pmatrix},$$

together with (3.2). □

Lemma 3.28 (Dickson) *Let q be a power of the odd prime p, and $\mathbb{F}_q = \mathbb{F}_p(c)$, $c \neq 0$. Then the matrices*

$$\begin{pmatrix}1 & 1\\ 0 & 1\end{pmatrix} \quad \textit{and} \quad \begin{pmatrix}1 & 0\\ c & 1\end{pmatrix}$$

generate the group $SL_2(q)$ unless $q = 9$.

Proof. If $q = p$ is a prime, this is a simple application of the Gauss algorithm (since left and right multiplication by powers of the two given matrices yield all elementary row and column operations). In the general case, this is a classical result of Dickson. The proof can be found in introductory books on group theory, for example, [Gor], Th. 8.4. □

Lemma 3.29 *Let q be a power of the odd prime p, and $q \neq 9$. Assume $\mathbb{F}_q = \mathbb{F}_p(\tau)$, where $\tau \neq 2$. Let $C(\tau)$ be the class of $SL_2(q)$ containing $\left(\begin{smallmatrix}\tau - 1 & 1\\ \tau - 2 & 1\end{smallmatrix}\right)^{-1}$. If $2-\tau$ is a nonsquare (resp., a square) in $\mathbb{F}_q$ then the triple $C_1, C_2, C(\tau)$ (resp., the triple $C_1, C_1, C(\tau)$) is quasi-rigid in $SL_2(q)$. Here C_1, C_2 are the two classes of unipotent elements from Lemma 3.27.*

Exercise Show $C(\tau)$ is the class consisting of all elements of trace τ, if $\tau \neq \pm 2$.

Proof. Let $H = \mathrm{SL}_2(q)$. The matrix $\begin{pmatrix} 1 & 0 \\ \tau-2 & 1 \end{pmatrix}$ lies in C_2 (resp., C_1) if $2-\tau$ is a nonsquare (resp., a square), by Lemma 3.25. Thus the equation

$$\begin{pmatrix} 1 & 1 \\ 0 & 1 \end{pmatrix} \begin{pmatrix} 1 & 0 \\ \tau-2 & 1 \end{pmatrix} = \begin{pmatrix} \tau-1 & 1 \\ \tau-2 & 1 \end{pmatrix}$$

shows there exists a triple (g_1, g_2, g_3) in $C_1 \times C_2 \times C(\tau)$ (resp., in $C_1 \times C_1 \times C(\tau)$) with $g_1 g_2 g_3 = 1$. This triple generates H by Dickson's theorem. Therefore, to prove quasi-rigidity it suffices to show that all $U \in C_2$ (resp., $U \in C_1$) with $U_1 U$ of trace τ and $\langle U_1, U \rangle = H$ are conjugate under the centralizer Γ of U_1 in H.

Since $\langle U_1, U \rangle = H$ the matrix U is not upper triangular. Hence U is conjugate under Γ to a matrix of the form $\begin{pmatrix} 1 & 0 \\ c & 1 \end{pmatrix}$ (Lemma 3.27(b)). Since $U_1 U$ has trace τ, $U_1 \begin{pmatrix} 1 & 0 \\ c & 1 \end{pmatrix}$ also has trace τ. Thus $c = \tau - 2$. This proves the claim. □

Thus for each odd $q \neq 9$, the group $\mathrm{SL}_2(q)$ admits a quasi-rigid tuple of conjugacy classes. Since each tuple of conjugacy classes is $\mathbb{Q}_{ab}$-rational, Theorem 3.25 applies with $\kappa = \mathbb{Q}_{ab}$. The center of $\mathrm{SL}_2(q)$ is $\{\pm 1\}$, hence we obtain the following Corollary for $q \neq 9$. For $q = 9$ we use the isomorphism $\mathrm{PSL}_2(9) \cong A_6$ and Corollary 3.21.

Corollary 3.30 *The group $PSL_2(q) = SL_2(q)/\{\pm 1\}$ occurs regularly over $\mathbb{Q}_{ab}$ for each odd prime power q.*

(As mentioned above, this also holds for even q, actually for all classical simple groups.) Galois realizations over $\mathbb{Q}$ are more difficult to obtain.

Corollary 3.31 *The group $PSL_2(p)$ occurs regularly over $\mathbb{Q}$ for each prime p with $p \not\equiv \pm 1 \pmod{24}$.*

Proof. The groups $\mathrm{PSL}_2(2) \cong S_3$ and $\mathrm{PSL}_2(3) \cong A_4$ occur regularly over $\mathbb{Q}$ by Corollary 3.19. Thus we may assume $p > 3$.

Let $\tau \in \mathbb{F}_p$ with $\tau \neq 2$, and $2-\tau$ a nonsquare. Then the classes $C_1, C_2, C(\tau)$ form a quasi-rigid triple in $H = \mathrm{SL}_2(p)$. This triple is $\mathbb{Q}$-rational if (and only if) the class $C(\tau)$ is rational. Indeed, if m is an integer $\not\equiv 0 \pmod p$ then (C_1^m, C_2^m) equals (C_1, C_2) or (C_2, C_1), corresponding to whether m is a square or a nonsquare mod p (Lemma 3.27). Thus if $C(\tau)$ is rational, then $\mathrm{PSL}_2(p) = H/Z(H)$ occurs regularly over $\mathbb{Q}$ (Theorem 3.25).

It remains to be seen for which p there is an element $\tau \in \mathbb{F}_p$ satisfying the above conditions. If 2 (resp., 3) is a nonsquare mod p then $\tau = 0$ (resp., $\tau = -1$) works: Then $C(\tau)$ consists of elements of order 4, resp., 3, and these elements are conjugate to their inverses, hence lie in a rational class.

The condition that 2 or 3 is a nonsquare mod p is equivalent to the condition $p \not\equiv \pm 1 \pmod{24}$ (by elementary number theory). Unfortunately, for other values of p there is no τ with the above properties. □

The above corollary was first proved by Shih using modular function fields. His approach also covers those p for which 7 is a nonsquare mod p. Malle has extended this further to include the case that 5 is a nonsquare (using his translation theory; see [MM], Ch. I). Together these results cover all primes <311. The only nonprime values of q for which $\mathrm{PSL}_2(q)$ is known to occur regularly over $\mathbb{Q}$ are $q = 4, 8, 9, 25$ and $q = p^2$ for p a prime with $p = 2$ mod 5 (see [Se1], §5.5, and [MM], Ch. II, Th. 7.3).

Exercise Show that for all primes $p > 3$ the group $\mathrm{PSL}_2(p)$ occurs regularly over $\mathbb{Q}(\sqrt{p(-1)^{(p-1)/2}})$ (= the unique quadratic number field contained in the field of pth roots of 1).

4
Covering Spaces and the Fundamental Group

The material in the first section is standard, and can be found in most introductory books on algebraic topology, for example, [Mas]. The goal is to provide a shortcut to the results needed here. We supply full proofs, except for the very first facts on homotopy.

The material in the second section is less standard, but certainly elementary. The first subsection (about coverings of a disc minus center) is preparatory. Then we turn to our main topic, the coverings of the punctured sphere. We study the behavior near a ramified point, in particular, we introduce the class of distinguished inertia group generators. This allows us to define the ramification type, in analogy to the definition in Chapter 2 for fields. The existence of coverings of prescribed ramification type (a topological version of Riemann's existence theorem) is the main result of this chapter. The proof is in the spirit of classical Riemann surface theory: It consists of a glueing process that uses the Galois group to index the sheets of the covering. As a by-product, this allows us to determine the fundamental group of a punctured sphere.

In this chapter, R and S denote topological spaces. From Section 4.1.2 on, S is a topological manifold.

4.1 The General Theory

4.1.1 Homotopy

For real t, s we let $[t, s]$, $[t, s[$ etc. denote the closed interval $\{t' : t \leq t' \leq s\}$, the half-open interval $\{t' : t \leq t' < s\}$, etc. The unit interval $[0, 1]$ is also denoted by I.

Definition 4.1 *A* **path** *in S is a continuous map $\gamma : I \to S$. We call $\gamma(0)$ and $\gamma(1)$ the* **initial point** *and* **endpoint** *of γ, respectively. The path γ is called* **closed** *if $\gamma(0) = \gamma(1)$.*

We denote by $C(S, p, q)$ the set of all paths γ in S with initial point p and endpoint q. Two paths γ_0 and γ_1 in $C(S, p, q)$ are called **homotopic** *if there exists a continuous map $\Gamma : I \times I \to S$ with*

$$\begin{aligned} \Gamma(0, t) &= \gamma_0(t), \Gamma(1, t) = \gamma_1(t) && \forall t \in I, \\ \Gamma(s, 0) &= p, \Gamma(s, 1) = q && \forall s \in I. \end{aligned}$$

Γ is called a **homotopy** *between γ_0 and γ_1.*

It is easy to see that homotopy of paths is an equivalence relation. We denote by $[\gamma]$ the equivalence class of γ under this relation, and call it the **homotopy class** of γ. Let $p \in S$. The set of homotopy classes of closed paths γ with initial and endpoint p is denoted by $\pi_1(S, p)$.

Now we define the natural product of two paths. To comply with usual functional notation, we write the product in reverse order (i.e., $\delta\gamma$ is the path that first traverses γ, then δ).

Definition 4.2 *Let γ and δ be paths with $\gamma(1) = \delta(0)$. Then we define the path product $\delta\gamma$ by*

$$\delta\gamma : I \to S, t \mapsto \begin{cases} \gamma(2t), & t \in \left[0, \frac{1}{2}\right] \\ \delta(2t - 1), & t \in \left]\frac{1}{2}, 1\right]. \end{cases}$$

One checks that if γ' (resp., δ') is homotopic to γ (resp., δ) then $\delta'\gamma'$ is homotopic to $\delta\gamma$. Thus the path product induces a product of homotopy classes via $[\delta][\gamma] = [\delta\gamma]$.

Define the inverse of a path γ to be the path γ^{inv} with $\gamma^{\text{inv}}(t) = \gamma(1 - t)$. Then the paths $\gamma\gamma^{\text{inv}}$ and $\gamma^{\text{inv}}\gamma$ are (defined and) homotopic to a constant path (easy to check!).

In particular, the above defines the product of any two classes in $\pi_1(S, p)$. One checks that this yields a group structure on $\pi_1(S, p)$. Its identity element is the class of the constant path p, and the inverse of $[\gamma]$ is the class of γ^{inv}.

Definition 4.3 *The group consisting of $\pi_1(S, p)$ together with the above product is called the* **fundamental group** *of S based at p.*

The groups $\pi_1(S, p)$ and $\pi_1(S, q)$ are isomorphic if $C(S, p, q) \neq \emptyset$. An isomorphism is given by the map $[\gamma] \mapsto [\omega\gamma\omega^{\text{inv}}]$ for any fixed $\omega \in C(S, p, q)$.

Example 4.4 A subset S of $\mathbb{R}^n$ is called **star-shaped** with respect to a point $p \in S$, if for any $q \in S$ the line segment $\{sp + (1-s)q \mid 0 \leq s \leq 1\}$ joining p and q is contained in S. If this holds then $\pi_1(S, p) = \{1\}$.

Indeed, for any $\gamma \in C(S, p, p)$ the map Γ defined by $\Gamma(s, t) = sp + (1-s)\gamma(t)$ is a homotopy between γ and the constant path p.

4.1.2 Coverings

Definition 4.5 *A* **topological manifold**, *or just* **manifold**, *is a Hausdorff space S such that each point of S has an open neighborhood that is homeomorphic to $\mathbb{R}^n$ for some $n \geq 1$.*

Remark 4.6

(a) We could also require that any point have an open neighborhood homeomorphic to an open ball in $\mathbb{R}^n$ (since such a ball is homeomorphic to $\mathbb{R}^n$).
(b) It is easy to check that a manifold is connected if and only if it is path-connected, that is, any two points can be joined by a path. Further, the connected components of a manifold are open and closed, and are themselves manifolds.

For the rest of this chapter, S is always a manifold.

Definition 4.7 *A surjective map $f : R \to S$ is called a* **covering** *if every $p \in S$ has a connected open neighborhood U such that the connected components V of $f^{-1}(U)$ are open in R, and f maps each such V homeomorphically onto U. Such U is called an* **admissible neighborhood** *(for f).*

Remark 4.8

(a) Clearly, each covering is continuous. If U is admissible then so is every connected open neighborhood $U' \subset U$. This proves the following: If $S' \subset S$ is open, then f restricts to a covering $f^{-1}(S') \to S'$.
(b) If $f : R \to S$ is a covering (of the manifold S) then R is also a manifold. Indeed, any point in S has an admissible neighborhood $U \cong \mathbb{R}^n$ (by (a)). Then all components of $f^{-1}(U)$ are $\cong \mathbb{R}^n$, hence R is locally homeomorphic to $\mathbb{R}^n$. For the Hausdorff property, note that any two distinct points of R with different (resp., same) f-image can be separated by neighborhoods of the form $f^{-1}(U)$, $f^{-1}(U')$ (resp., by different components of $f^{-1}(U)$).

In the literature, coverings are considered between more general spaces, but one needs to impose certain local properties of S to ensure that the basic properties of coverings hold.

Example 4.9 For $r > 0$ let $\mathbb{K}(r) := \{z \in \mathbb{C} \mid 0 < |z| < r\}$. Then $f : \mathbb{K}(r^{1/e}) \to \mathbb{K}(r)$, $z \mapsto z^e$ is a covering for each $e \in \mathbb{N}$.

Indeed, let ζ be a primitive eth root of unity. If V is a sufficiently small open disc around $b \in \mathbb{K}(r^{1/e})$ then the $V_j = \zeta^j V$ have disjoint closures ($j = 0, \dots, e-1$). Then $U = f(V)$ is a neighborhood of $p = f(b)$, and f maps each V_j homeomorphically onto U. Thus U is admissible for f.

Note that we have to exclude the point $z = 0$ because the map $z \mapsto z^e$ is not one-to-one in any neighborhood of 0.

4.1.3 The Homotopy Lifting Property

Lemma 4.10 *Let W be an open subset of $I \times I$ such that W contains $[0, t_0] \times I$ for some $t_0 \in I$, $t_0 < 1$. Then W contains $[0, t_1] \times I$ for some $t_1 > t_0$.*

Proof. Since $C := (I \times I) \setminus W$ is compact the same is true for $p_1(C)$, where p_1 is the projection on the first coordinate. But $p_1(C)$ is disjoint from $[0, t_0]$ by assumption, hence also from $[0, t_1]$ for some $t_1 > t_0$. □

Definition 4.11 *Let $f : R \to S$ be a covering and γ a path in S. A path $\tilde{\gamma}$ in R is called a* **lift** *of γ if $f \circ \tilde{\gamma} = \gamma$.*

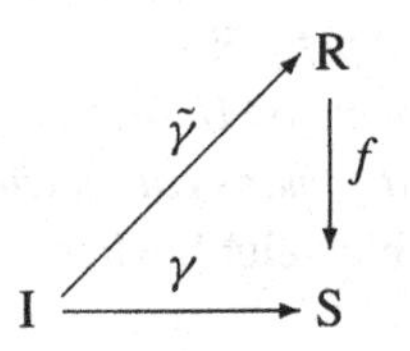

Recall we say a group G acts on a set M (from the left) if we are given a map $G \times M \to M$, $(g, m) \mapsto gm$, satisfying $1m = m$ and $(gh)m = g(hm)$ for all $m \in M$, $g, h \in G$ (where 1 is the identity element of G). If $m' = gm$ we say g sends m to m'. Further, G acts transitively on M if for all $m, m' \in M$ there exists $g \in G$ with $m' = gm$.

Theorem 4.12 *Let $f : R \to S$ be a covering and γ a path in S with initial point p.*

(a) For each $a \in f^{-1}(p)$ there exists a unique lift of γ to R with initial point a.
(b) Lifts of homotopic paths are homotopic if they have the same initial point.
(c) The group $\pi_1(S, p)$ acts on $f^{-1}(p)$ in the following way: $[\gamma]$ sends a_0 to a_1 where a_1 is the end point of the (unique) lift of γ with initial point a_0.

Suppose S is connected. Then this action of $\pi_1(S, p)$ on $f^{-1}(p)$ is transitive if and only if R is connected.

Proof. (a) Consider the set T of all $t \in I$ for which there exists a unique continuous map $\tilde{\gamma} : [0, t] \to R$ with $f\tilde{\gamma} = \gamma|_{[0,t]}$ and $\tilde{\gamma}(0) = a$. This set T is nonempty because it contains zero. Let t_0 be the supremum of T. Choose an admissible neighborhood U of $\gamma(t_0)$. There is $\epsilon > 0$ such that γ maps the interval $I' = I \cap [t_0 - \epsilon, t_0 + \epsilon]$ into U. Choose $t_1 \in T$ with $t_1 > t_0 - \epsilon$, and let $\tilde{\gamma} : [0, t_1] \to R$ be the map existing by definition of T. Set $I'' = I \cap [t_0 - \epsilon, t_1]$. Then $\tilde{\gamma}(I'')$ is (defined and) connected, and lies in $f^{-1}(U)$. Hence it lies in a (unique) component V of $f^{-1}(U)$. Thus $\tilde{\gamma}(t) = (f|_V)^{-1}(\gamma(t))$ for all $t \in I''$. Defining $\tilde{\gamma}(t)$ by the same formula for all $t \in I'$ extends $\tilde{\gamma}$ to a continuous map $I \cap [0, t_0 + \epsilon] \to R$ with $f\tilde{\gamma} = \gamma$ on $I \cap [0, t_0 + \epsilon]$. Clearly, this extension is unique. It follows that $\min(1, t_0 + \epsilon) \in T$. Since $t_0 = \sup(T)$ it follows that $t_0 = 1 \in T$. This proves (a).

(b) Let γ_0 and γ_1 be homotopic paths in $C(S, p, q)$, and let Γ be a homotopy between them. Furthermore let $a \in f^{-1}(p)$ and $\gamma_s(t) := \Gamma(s, t)$ for all $s, t \in I$. Each path γ_s has a unique lift $\tilde{\gamma}_s$ with initial point a. We claim that $\tilde{\Gamma} : I \times I \to R$, $(s, t) \mapsto \tilde{\gamma}_s(t)$ is a homotopy between $\tilde{\gamma}_0$ and $\tilde{\gamma}_1$.

Consider the set of all $t \in I$ such that $\tilde{\Gamma}$ is continuous on $I \times [0, t]$. This set is nonempty (contains zero). Let t_0 be its supremum. Let $s_0 \in I$ be arbitrary. Choose an admissible neighborhood U of $\Gamma(t_0, s_0)$. Since Γ is continuous there is $\epsilon > 0$ such that Γ maps the set

$$E = (I \times I) \cap]s_0 - \epsilon, s_0 + \epsilon[\times]t_0 - \epsilon, t_0 + \epsilon[$$

into U. Thus $\tilde{\Gamma}(E) \subset f^{-1}(U)$.

Since $\tilde{\Gamma}$ is continuous on $I \times [0, t_0[$ (and $\tilde{\Gamma}(I \times \{0\}) = \{a\}$), $\tilde{\Gamma}$ maps the connected set

$$E' = \begin{cases} (I \times I) \cap]s_0 - \epsilon, s_0 + \epsilon[\times]t_0 - \epsilon, t_0[& \text{for } t_0 > 0 \\ I \times \{0\} & \text{for } t_0 = 0 \end{cases}$$

into a (unique) component V of $f^{-1}(U)$. Since each $\tilde{\gamma}_s$ is continuous, it follows that $\tilde{\Gamma}(E) \subset V$.

Thus $\tilde{\Gamma}(\mathbf{e}) = (f|_V)^{-1}(\Gamma(\mathbf{e}))$ for all $\mathbf{e} \in E$. It follows that $\tilde{\Gamma}$ is continuous on E, that is, in a neighborhood of the point (t_0, s_0). Let W be the set of all points of $I \times I$ where $\tilde{\Gamma}$ is continuous. It follows that W contains an open neighborhood of $I \times [0, t_0]$. By the lemma, if $t_0 < 1$ then W contains $[0, t_1] \times I$ for some $t_1 > t_0$. This contradicts the definition of t_0. Hence $t_0 = 1$, and $\tilde{\Gamma}$ is continuous on all of $I \times I$.

In particular, $\tilde{\Gamma}(I \times \{1\})$ is connected. But it is also discrete since contained in $f^{-1}(q)$, where $q = \gamma_0(1) = \gamma_1(1)$. Hence it consists of exactly one point. This shows that $\tilde{\Gamma}$ is a homotopy between $\tilde{\gamma}_0$ and $\tilde{\gamma}_1$.

(c) Each path γ in S has a unique lift to S with initial point a_0 by (a). The endpoint a_1 of this lift does not change if we replace γ by a homotopic path, by (b). Hence the effect of $[\gamma]$ on a_0 is well defined.

The identity element of the group $\pi_1(S, p)$ is represented by the constant path $\gamma = p$. Clearly, the lift of a constant path is constant, hence the identity of $\pi_1(S, p)$ acts trivially. Finally, the relation $[\delta\gamma]a_0 = [\delta]([\gamma]a_0)$ follows from the obvious fact that the lift of $\delta\gamma$ with initial point a_0 is $\tilde{\delta}\tilde{\gamma}$, where $\tilde{\gamma}$ is the lift of γ with initial point a_0 and $\tilde{\delta}$ is the lift of δ with initial point $a_1 = \tilde{\gamma}(1)$.

If R is connected then for every two points $a_0, a_1 \in f^{-1}(p)$ there is $\omega \in C(R, a_0, a_1)$. Then $[f \circ \omega] \in \pi_1(S, p)$ and $[f \circ \omega]a_0 = a_1$. Thus $\pi_1(S, p)$ acts transitively on $f^{-1}(p)$. Conversely, assume this action is transitive and let $a', b' \in R$ be arbitrarily chosen. We have to show that $C(R, a', b')$ is not empty. Since S is path-connected we can find $\gamma_1 \in C(S, f(a'), p)$ and $\gamma_2 \in C(S, f(b'), p)$. The endpoint a (resp., b) of the lift $\tilde{\gamma}_1$ (resp., $\tilde{\gamma}_2$) of γ_1 (resp., γ_2) with initial point a' (resp., b') is a point of $f^{-1}(p)$. By assumption there is $\tilde{\gamma} \in C(R, a, b)$. Then the path $\tilde{\gamma}_2^{\text{inv}}\tilde{\gamma}\tilde{\gamma}_1$ connects a' and b'. □

Corollary 4.13 *Let $f : R \to S$ be a covering and suppose S is connected. Let R_1 be a connected component of R. Then f restricts to a covering $R_1 \to S$. If $f^{-1}(p) \subset R_1$ for some $p \in S$, then $R = R_1$ is connected.*

Proof. We first show that $f_1 := f|_{R_1}$ is surjective. Let $a \in R_1$ and $p = f(a)$. Then for any $q \in S$ there is $\gamma \in C(S, p, q)$. The lift $\tilde{\gamma}$ of γ via f with initial point a is a path in R_1 (since R_1 is a component of R). The endpoint b of $\tilde{\gamma}$ satisfies $f(b) = q$. This shows that f_1 is surjective.

Let U be an admissible neighborhood for f. Each component V of $f^{-1}(U)$ satisfies either $V \subset R_1$ or $V \cap R_1 = \emptyset$. In the former case, V is also a component of $f_1^{-1}(U)$, and all components of $f_1^{-1}(U)$ are of this form. Hence U is admissible for f_1, and f_1 is a covering.

By (c) above, $\pi_1(S, p)$ acts transitively on $f_1^{-1}(p)$. If $f^{-1}(p) = f_1^{-1}(p)$ then $\pi_1(S, p)$ acts transitively on $f^{-1}(p)$, hence R is connected (again by (c)). □

The next corollary shows the importance of the action of the fundamental group $\pi_1(S, p)$ on $f^{-1}(p)$: This action determines the covering up to equivalence.

Corollary 4.14 *Let $f_i : R_i \to S$ be a covering for $i = 1, 2$, with R_i connected. Let $b_i \in R_i$ with $f_1(b_1) = f_2(b_2) =: p$. Suppose for each $[\gamma] \in \pi_1(S, p)$ we* *have* $[\gamma]b_1 = b_1$ *if and only if* $[\gamma]b_2 = b_2$. *Then there exists a homeomorphism* $\alpha : R_1 \to R_2$ *with* $f_2 \circ \alpha = f_1$ *and* $\alpha(b_1) = b_2$.

Proof. For $b \in R_1$ choose $\delta_1 \in C(R_1, b_1, b)$. Let $\delta = f_1 \circ \delta_1$, and let δ_2 be the lift of δ via f_2 with initial point b_2. Define $\alpha(b)$ to be the endpoint of δ_2.

To see that this is well defined, consider some $\delta_1' \in C(R_1, b_1, b)$. Define δ' and δ_2' analogously. Then $\gamma_1 := \delta_1^{\text{inv}}\delta_1'$ and $\gamma := \delta^{\text{inv}}\delta'$ are closed paths, based at b_1 and p, respectively. Further, γ_1 is the lift of γ with initial point b_1. Thus $[\gamma]b_1 = b_1$. The hypothesis then yields $[\gamma]b_2 = b_2$, which means that γ lifts via f_2 to a closed path γ_2 based at b_2. Then $\delta_2\gamma_2$ is the lift of $\delta\gamma$ with initial point b_2, hence is homotopic to δ_2'. Thus $\delta_2'(1) = \delta_2\gamma_2(1) = \delta_2(1)$. This shows that α is well defined. By symmetry of the construction it is clear that α has an inverse (same construction with R_1 and R_2 interchanged). Thus α is bijective.

It is also clear by construction that $f_2 \circ \alpha = f_1$. Let $q \in S$. We can choose a path-connected neighborhood U of q which is admissible for f_1 and f_2. Then α maps $f_1^{-1}(U)$ onto $f_2^{-1}(U)$. Let $b \in f_1^{-1}(q)$, and let V be the component of $f_1^{-1}(U)$ containing b. Each point of V can be connected to b_1 by a path that follows first a fixed path δ_1 from b_1 to b, then continues on a path inside V beginning at b. Using these paths to construct the α-images of the points of V, one sees immediately that $\alpha(V)$ lies in a component V' of $f_2^{-1}(U)$. Using the inverse construction for α^{-1}, it follows that $\alpha(V) = V'$. Then $\alpha|_V = (f_2|_{V'})^{-1} \circ f_1|_V$, hence $\alpha|_V : V \to V'$ is homeomorphic. Thus α is locally homeomorphic and bijective, hence homeomorphic. □

Remark The homeomorphism α is unique by Lemma 4.17 (iii) below.

Corollary 4.15 *If* $\pi_1(S, p) = \{1\}$ *then each covering* $f : R \to S$ *with connected* R *is a homeomorphism.*

Proof. Apply the previous corollary with $f_1 = f$, $f_2 = \text{id}: S \to S$. Its hypothesis is trivially satisfied if $\pi_1(S, p) = \{1\}$, hence it follows that $f = \alpha$ is a homeomorphism. □

4.1.4 Galois Coverings and the Group of Deck Transformations

Definition 4.16

(a) *Two coverings* $f : R \to S$ *and* $\tilde{f} : \tilde{R} \to S$ *of the same space* S *are called* **equivalent** *if there exists a homeomorphism* $\alpha : R \to \tilde{R}$ *with* $\tilde{f}\alpha = f$.

(b) A **deck transformation** *of the covering $f : R \to S$ is a homeomorphism $\alpha : R \to R$ with $f\alpha = f$. The deck transformations form a group under composition, denoted by Deck(f) or Deck(R/S) (if f is understood).*

Clearly, the group Deck(f) acts on each fiber $f^{-1}(p)$, $p \in S$; that is, if $b \in f^{-1}(p)$ then also $\alpha(b) \in f^{-1}(p)$ for each $\alpha \in$ Deck(f). Recall that $\pi_1(S, p)$ also acts naturally on $f^{-1}(p)$ (Theorem 4.12).

Lemma 4.17 *Let $f : R \to S$ be a covering.*

(i) Let $p, q \in S$ and $\gamma \in C(S, p, q)$. For each $b \in f^{-1}(p)$ let ${}^{\gamma}b$ be the endpoint of the lift of γ with initial point b. Then the map $b \mapsto {}^{\gamma}b$ is a bijection between $f^{-1}(p)$ and $f^{-1}(q)$. This bijection commutes with the action of Deck(f), that is, $\alpha({}^{\gamma}b) = {}^{\gamma}(\alpha(b))$ for all $\alpha \in$ Deck(f).
(ii) The action of Deck(f) on $f^{-1}(p)$ commutes with that of $\pi_1(S, p)$.
(iii) Suppose R is connected. If $\alpha \in$ Deck(f) fixes a point b of R then $\alpha = id$.
(iv) Suppose R is connected, and Deck(f) has a subgroup G that acts transitively on some fiber $f^{-1}(p)$. Then $G =$ Deck(f).

Proof.

(i) The map $f^{-1}(q) \to f^{-1}(p)$, $a \mapsto {}^{\gamma^{-1}}a$, is inverse to the map $b \mapsto {}^{\gamma}b$. Hence both maps are bijective. The other claim in (i) follows from the following fact: If $\gamma \in C(S, p, q)$ and $\tilde{\gamma}$ is the lift of γ with initial point b, then $\alpha \circ \tilde{\gamma}$ is the lift of γ with initial point $\alpha(b)$.
(ii) This is the special case $p = q$ of (i).
(iii) Let $b' \in R$ be arbitrary, and $\tilde{\gamma} \in C(R, b, b')$. Let $\gamma = f \circ \tilde{\gamma} \in C(S, f(b), f(b'))$. Then $\tilde{\gamma}$ is the lift of γ with initial point b, hence $b' = {}^{\gamma}b$ in the notation of (i). If $\alpha(b) = b$ we get $\alpha(b') = \alpha({}^{\gamma}b) = {}^{\gamma}(\alpha(b)) = {}^{\gamma}b = b'$. Thus $\alpha = \text{id}$.
(iv) Let $b \in f^{-1}(p)$. For each $\alpha \in$ Deck(f) there is $\beta \in G$ with $\beta(\alpha(b)) = b$. Then $\beta\alpha = \text{id}$ by (iii), hence $\alpha = \beta^{-1} \in G$. □

Assume S is connected. Then it follows from (i) that all fibers $f^{-1}(p)$, $p \in S$, have the same cardinality. This cardinality is called the **degree** of f. Moreover, if Deck(f) acts transitively on one fiber $f^{-1}(p)$, then it acts transitively on each fiber $f^{-1}(p)$, $p \in S$.

Definition 4.18 *A covering $f : R \to S$ is called a* **Galois covering** *if R (hence also S) is connected and Deck(f) acts transitively on some (hence each) fiber $f^{-1}(p)$, $p \in S$.*

Proposition 4.19 *Let $f : R \to S$ be a Galois covering, and set $H = Deck(f)$.*

(a) The degree n of f equals the order of H. (Here n is a possibly infinite cardinal. We are mostly concerned with the case that n is finite.) For each admissible neighborhood U in S (resp., for each $p \in S$) the set $f^{-1}(U)$ (resp., $f^{-1}(p)$) has n components (resp., n elements), and these are permuted transitively by H.

(b) Let $b \in R$ and $p = f(b)$. Then there is a (unique) surjective homomorphism

$$\Phi_b : \pi_1(S, p) \to H = Deck(f)$$

such that $\Phi_b([\gamma])$ maps $[\gamma]b$ to b, for each $[\gamma] \in \pi_1(S, p)$. (Recall that $[\gamma]b$ is the endpoint of the lift of γ with initial point b.)

Proof. Let $b \in f^{-1}(p)$. Consider the map $\theta_b : H \to f^{-1}(p), \alpha \mapsto \alpha(b)$. This map is injective by (iii) above. (Indeed, if $\alpha(b) = \alpha'(b)$ then $\alpha^{-1}\alpha'$ fixes b, hence is the identity.) Since f is Galois, θ_b is surjective. Hence θ_b is bijective. This implies the first claim in (a).

We have $|f^{-1}(p)| = n$ by definition of the degree. Since each component of $f^{-1}(U)$ contains exactly one element of $f^{-1}(p)$ for $p \in U$ (by definition of admissible neighborhood), the rest of (a) follows.

By Theorem 4.12 (c), the group $\pi_1(S, p)$ acts transitively on $f^{-1}(p)$. Thus the map $\Psi_b : \pi_1(S, p) \to f^{-1}(p), [\gamma] \mapsto [\gamma]b$, is surjective. Hence $\Phi'_b := \theta_b^{-1} \circ \Psi_b$ is a surjection $\pi_1(S, p) \to H$. This Φ'_b is not a group homomorphism in general (it is an "anti"-homomorphism), but the map $\Phi_b : [\gamma] \mapsto (\Phi'_b([\gamma]))^{-1}$ is a homomorphism (as seen by a routine check using (ii) above). Clearly, Φ_b has the property claimed in (b), and is unique with this property by (iii) above. □

If the n in part (a) is finite we say f is a **finite Galois covering**.

4.2 Coverings of the Punctured Sphere

4.2.1 The Coverings of the Disc Minus Center

We are going to determine the fundamental group and the connected coverings of the punctured disc $\mathbb{K}(r) := \{z \in \mathbb{C} : 0 < |z| < r\}$, for any $r > 0$. Each $\mathbb{K}(r)$ is homeomorphic to $\mathbb{K} := \mathbb{K}(1)$, thus it would suffice to study $\mathbb{K}$.

We fix a square root of -1 in $\mathbb{C}$, and denote it by $\sqrt{-1}$. We let exp denote the natural exponential function.

Proposition 4.20 *Let $\mathbb{H} = \{z \in \mathbb{C} : Re(z) < 0\}$ be the left half plane. The map $f_\infty : \mathbb{H} \to \mathbb{K}, z \mapsto \exp(z)$ is a Galois covering. Its deck transformation* *group* Λ *is infinite cyclic, and consists of all maps* $\lambda_m : z \mapsto z + m2\pi\sqrt{-1}$, $m \in \mathbb{Z}$. *Let* $b \in \mathbb{H}$, *and* $p = f_\infty(b) \in \mathbb{K}$. *The associated map* $\Phi_b : \pi_1(\mathbb{K}, p) \to \Lambda$ *(see Proposition 4.19) is an isomorphism. It maps the class* $[\gamma]$ *of the path* $\gamma(t) = p\exp(2\pi\sqrt{-1}t)$, $t \in I$, *to the map* $z \mapsto z - 2\pi\sqrt{-1}$. *Hence the fundamental group* $\pi_1(\mathbb{K}, p)$ *is infinite cyclic, generated by* $[\gamma]$.

Proof. We know that $\exp(z) = \exp(z')$ if and only if $z' = \lambda_m(z)$ for some $m \in \mathbb{Z}$. To check that f_∞ is a covering, let $z \in \mathbb{H}$ and $q = f_\infty(z)$. Let V be an open disc of radius ≤ 1 around z, contained in $\mathbb{H}$. Then the discs $\lambda_m(V)$, $m \in \mathbb{Z}$, are pairwise disjoint. Further, $U := f_\infty(V)$ is open in $\mathbb{K}$, and f_∞ restricts to a homeomorphism $\lambda_m(V) \to U$ for each m. (The closure $\bar{V}$ of V is compact, and f_∞ restricts to a bijective continuous, hence homeomorphic map $\lambda_m(\bar{V}) \to \bar{U}$.) Since $f_\infty^{-1}(U) = \bigcup_{m\in\mathbb{Z}} \lambda_m(V)$, it follows that U is admissible for f_∞. This proves that f_∞ is a covering.

Clearly, each λ_m is a deck transformation of f_∞. Since the group formed by the λ_m is transitive on the fiber $f_\infty^{-1}(p)$, it follows that f_∞ is Galois, and $\mathrm{Deck}(f_\infty) = \{\lambda_m : m \in \mathbb{Z}\}$ (Lemma 4.17(iv)).

If $[\delta] \in \ker(\Phi_b)$ then δ lifts to a closed path $\tilde{\delta}$ with initial point b. By Example 4.4 we have $\pi_1(\mathbb{H}, b) = \{1\}$, hence $\tilde{\delta}$ is homotopic in $\mathbb{H}$ to the constant path b. Applying f_∞ to such a homotopy we see that δ is homotopic in $\mathbb{K}$ to the constant path p. Thus $[\delta] = 1$. This shows that Φ_b is injective, hence isomorphic.

Let $\tilde{\gamma}$ be the path in $\mathbb{H}$ with $\tilde{\gamma}(t) = b + 2\pi\sqrt{-1}t$, $t \in I$. Then $\gamma = f_\infty \circ \tilde{\gamma}$ is the path $\gamma(t) = p\exp(2\pi\sqrt{-1}t)$ from the proposition. Thus $[\gamma]b = \tilde{\gamma}(1) = b + 2\pi\sqrt{-1}$. By definition of Φ_b it follows that the deck transformation $\Phi_b([\gamma])$ maps $b + 2\pi\sqrt{-1}$ to b. Thus $\Phi_b([\gamma])$ is the map $z \mapsto z - 2\pi\sqrt{-1}$. Hence the proposition. □

Let μ_e be the group of eth roots of 1 in $\mathbb{C}$.

Lemma 4.21 *The map* $f_e : \mathbb{K}(r^{1/e}) \to \mathbb{K}(r)$, $z \mapsto z^e$ *is a Galois covering of degree* e, *for each* $e \in \mathbb{N}$. *Its deck transformation group is cyclic of order* e, *and consists of all maps* $z \mapsto \zeta z$, $\zeta \in \mu_e$.

Proof. In Example 4.9 we showed that f_e is a covering. Clearly, the maps $z \mapsto \zeta z$ form a subgroup of $\mathrm{Deck}(f_e)$ that acts transitively on each fiber. Now the claim follows from Lemma 4.17(iv). □

Corollary 4.22 *Let* $f : E \to \mathbb{K}(r)$ *be a covering of finite degree* e, *with* E *connected.*

(a) *Then f is equivalent to the covering $f_e : \mathbb{K}(r^{1/e}) \to \mathbb{K}(r), z \mapsto z^e$. Thus there is a homeomorphism $\varphi : E \to \mathbb{K}(r^{1/e})$ with $\varphi(u)^e = f(u)$ for all $u \in E$. We abbreviate the latter by $\varphi^e = f$. This φ is unique up to multiplication by an eth root ζ of 1, that is, any φ' with the same property is of the form $\varphi' = \zeta\varphi$.*

(b) *The group Deck(f) is cyclic of order e. It has a unique element σ with the following property: For each homeomorphism $\varphi : E \to \mathbb{K}(r^{1/e})$ with $\varphi^e = f$ we have $\varphi \circ \sigma^{-1} = \zeta_e\varphi$, where $\zeta_e = \exp(2\pi\sqrt{-1}/e)$. This σ generates Deck(f); it is called the* **distinguished generator**.

(c) *Let $u \in E$ and $p = f(u)$. Let σ as in (b). Then $\gamma(t) = p\exp(2\pi\sqrt{-1}t)$, $t \in I$, is a closed path in $\mathbb{K}(r)$ based at p, and the lift of γ via f with initial point $\sigma(u)$ has endpoint u.*

(d) *Let $0 < \hat{r} < r$, $\hat{E} := f^{-1}(\mathbb{K}(\hat{r}))$ and $\hat{f} := f|_{\hat{E}}$. Then $\hat{E}$ is connected and $\hat{f} : \hat{E} \to \mathbb{K}(\hat{r})$ is a covering of degree e. The distinguished generator of Deck(f) restricts to that of Deck($\hat{f}$).*

Proof. (a) Let $u \in E$ and $p = f(u)$. Recall that the group $\Gamma := \pi_1(\mathbb{K}(r), p)$ acts transitively on $f^{-1}(p)$ (Theorem 4.12 (c)). Since $f^{-1}(p)$ has cardinality e, the stabilizer $\Gamma_u = \{\theta \in \Gamma : \theta u = u\}$ is a subgroup of Γ of index e. Applying this to the covering f_e, we also get that the stabilizer $\Gamma_{u'}$ of any $u' \in f_e^{-1}(p)$ has index e in Γ. But Γ is cyclic (by the proposition, and since $\mathbb{K}(r)$ is homeomorphic $\mathbb{K}(1)$). It follows that $\Gamma_u = \Gamma_{u'}$. Hence the desired φ exists by Corollary 4.14. This means that f is equivalent to f_e (Definition 4.16).

The claim about φ' follows from the fact that $\alpha := \varphi'\varphi^{-1}$ is a deck transformation of f_e, hence is of the form $z \mapsto \zeta z$ with $\zeta \in \mu_e$ (Lemma 4.21).

(b) Clearly, the map $g \mapsto \varphi \circ g \circ \varphi^{-1}$ is an isomorphism from Deck(f) to Deck(f_e). It is independent of the choice of φ. Indeed, for φ' and α as above we have $\varphi' \circ g \circ (\varphi')^{-1} = \alpha \circ (\varphi \circ g \circ \varphi^{-1}) \circ \alpha^{-1} = \varphi \circ g \circ \varphi^{-1}$ because Deck(f_e) is abelian.

Multiplication by ζ_e^{-1} is a generator of Deck(f_e) by the lemma. Let σ be the inverse image of this generator in Deck(f) (under the above isomorphism). It is clear that it has the property claimed in (b), and is unique with this property.

(c) Let $u' = \varphi(\sigma(u)) \in f_e^{-1}(p)$. Define the path $\tilde{\gamma}$ in $\mathbb{K}(r^{1/e})$ by $\tilde{\gamma}(t) = u'\exp(2\pi\sqrt{-1}t/e)$, $t \in I$. Then $f_e \circ \tilde{\gamma}$ is the path γ from (c). Thus $\tilde{\gamma}$ is the lift of γ via f_e with initial point u'. Applying φ^{-1} it follows that $\varphi^{-1} \circ \tilde{\gamma}$ is the lift of γ via f with initial point $\sigma(u)$. Its endpoint is $\varphi^{-1} \circ \tilde{\gamma}(1) = \varphi^{-1}(u'\exp(2\pi\sqrt{-1}/e)) = \varphi^{-1}(\zeta_e\varphi(\sigma(u))) = u$ by (b).

(d) Follows from (a). □

4.2.2 Coverings of the Punctured Sphere – Behavior Near a Ramified Point

Let $\mathbb{P}^1 = \mathbb{C} \cup \{\infty\}$ as in Chapter 2. Now we make $\mathbb{P}^1$ into a topological space. For $p \in \mathbb{P}^1$, define the open disc $D(p, r)$ around p of radius $r > 0$ as follows:

$$D(p,r) = \begin{cases} \{z \in \mathbb{C} : |z-p| < r\} & \text{if } p \in \mathbb{C} \\ \{z \in \mathbb{C} : |z| > r^{-1}\} \cup \{\infty\} & \text{if } p = \infty. \end{cases}$$

Define a subset of $\mathbb{P}^1$ to be open if it contains a disc around each of its points. This makes $\mathbb{P}^1$ into a topological space, homeomorphic to the 2-sphere via stereographic projection: Recall the familiar picture where a sphere resting on the origin of the complex plane is projected onto the plane from its North pole. (The North pole is sent to ∞, and the discs on the sphere around the North pole are mapped to the discs $D(\infty, r)$ defined above.)

In the rest of this chapter we study finite Galois coverings of the punctured sphere $\mathbb{P}^1 \setminus P$, where P is a finite set. In the next chapter it is shown that these coverings correspond to the finite Galois extensions of $\mathbb{C}(x)$. For the rest of this chapter, P denotes a finite subset of $\mathbb{P}^1$.

Proposition 4.23 *Let $f : R \to \mathbb{P}^1 \setminus P$ be a finite Galois covering. Fix some $p \in P$.*

1. *Let $D = D(p, r)$ be a disc around p not containing other elements of P. Thus $D^* := D \setminus \{p\}$ is contained in $\mathbb{P}^1 \setminus P$. Let $\kappa_p : D^* \to \mathbb{K}(r)$ be the homeomorphism sending z to $z - p$ (resp., $1/z$) if $p \neq \infty$ (resp., $p = \infty$). Then for each connected component E of $f^{-1}(D^*)$ the map $f_E = \kappa_p \circ f|_E$ is a covering $f_E : E \to \mathbb{K}(r)$ (of finite degree). We call E a* **circular component** *of level r over p.*
2. *Let $0 < \hat{r} < r$. Then there is a 1-1 correspondence between the circular components E of level r and the circular components $\hat{E}$ of level $\hat{r}$ over p, given by inclusion (i.e., for each E there is exactly one $\hat{E}$ with $\hat{E} \subset E$, and vice versa). If $\hat{E} \subset E$ then $f_{\hat{E}}$ is the restriction of f_E to $\hat{E}$, and $\hat{E} = f_E^{-1}(\mathbb{K}(\hat{r}))$.*
3. *The group $H :=$ Deck(f) permutes the components E of $f^{-1}(D^*)$ transitively. Let H_E be the stabilizer of E in H. Restricting the action of H_E to E yields an isomorphism $H_E \to$ Deck(f_E). Thus H_E is cyclic. Let $h_E \in H_E$ be the element corresponding to the distinguished generator of Deck(f_E) (see Corollary 4.22 (b)). We call h_E the* **distinguished generator** *of H_E.*
4. *Let $h \in H$ and $E' = h(E)$. Then $h h_E h^{-1} = h_{E'}$. Hence the h_E form a conjugacy class C_p of H. The class C_p depends only on p (and f), but not*

on the choice of the disc D. Let e be the common order of the elements of C_p. Then e equals the degree of the covering $f_E : E \to \mathbb{K}(r)$, for any component E of $f^{-1}(D^*)$. In particular, $C_p = \{1\}$ if and only if f_E is a homeomorphism.

5. *Let $p^* \in D^*$ and $\bar{p} = \kappa_p(p^*)$. Let $\lambda(t) = \kappa_p^{-1}(\bar{p}\exp(2\pi\sqrt{-1}t))$, a closed path in D^* based at p^* (winding once around p in counterclockwise direction). Let b be any point of R, and $q_0 = f(b)$. Further, let δ be a path in $\mathbb{P}^1 \setminus P$ joining q_0 to p^*. Then $\gamma := \delta^{\text{inv}}\lambda\delta$ is a closed path in $\mathbb{P}^1 \setminus P$ based at q_0, and the map $\Phi_b : \pi_1(\mathbb{P}^1 \setminus P, q_0) \to H$ (from Proposition 4.19) sends $[\gamma]$ to an element of the class C_p.*

Proof. (1) The covering f restricts to a covering $f^{-1}(D^*) \to D^*$ (Remark 4.8(a)). Composing this covering with the homeomorphism κ_p we obtain a covering $f^{-1}(D^*) \to \mathbb{K}(r)$. It restricts to a covering $E \to \mathbb{K}(r)$ (Corollary 4.13).

(2) Let $\hat{X} := f^{-1}(D(p, \hat{r}) \setminus \{p\})$. The space $\tilde{E} := f_E^{-1}(\mathbb{K}(\hat{r})) = E \cap \hat{X}$ is connected by Corollary 4.22(d); it is open and closed in $\hat{X}$, hence is a component of $\hat{X}$. This means $\tilde{E}$ is a circular component of level $\hat{r}$ (contained in E).

Conversely, assume $\hat{E} \subset E$ is a circular component of level $\hat{r}$ (over p). Then clearly $f_{\hat{E}} = f_E|_{\hat{E}}$, hence $\hat{E} \subset \tilde{E}$. Thus $\hat{E} = \tilde{E}$ (since both are components of $\hat{X}$). We have proved that E contains exactly one circular component of level $\hat{r}$ (over p). Conversely, each such component lies in $f^{-1}(D^*)$, hence in some component E of $f^{-1}(D^*)$. This proves (2).

(3) The group H acts on $f^{-1}(D^*)$, hence permutes the components E of $f^{-1}(D^*)$. Let H_E be the stabilizer of E. If $h \in H$ maps a point of E into some other component E' then $h(E) = E'$ (since the components of $f^{-1}(D^*)$ are pairwise disjoint). In particular, if h maps a point of E into E then $h \in H_E$.

For $p^* \in D^*$ the set $F_E = f^{-1}(p^*) \cap E$ is a fiber of the covering f_E. Since H acts transitively on $f^{-1}(p^*)$ (by the definition of a Galois covering), any two points of F_E can be mapped into each other by an element $h \in H$. Then $h \in H_E$ by the above, that is, H_E acts transitively on F_E. Clearly, restriction yields a homomorphism $H_E \to \text{Deck}(f_E)$. This homomorphism is injective by Lemma 4.17 (iii). Its image is a subgroup of $\text{Deck}(f_E)$ that acts transitively on the fiber F_E. Each such subgroup is all of $\text{Deck}(f_E)$ by Lemma 4.17(iv). Hence restriction yields an isomorphism $H_E \to \text{Deck}(f_E)$.

Since H acts transitively on $f^{-1}(p^*)$ (and $f^{-1}(p^*)$ meets each E), it also follows that H permutes the components E transitively.

(5) The lift $\tilde{\delta}$ of δ via f with initial point b has its endpoint b^* in some component E of $f^{-1}(D^*)$. Let $\tilde{\lambda}$ be the lift of λ via f with initial point $h_E(b^*)$. Then $\tilde{\lambda}$ equals the lift of the path $\bar{p}\exp(2\pi\sqrt{-1}t)$ via f_E, with initial point $h_E(b^*)$. Hence the endpoint of $\tilde{\lambda}$ is b^* (Corollary 4.22(c)).

The path $(h_E \circ \tilde{\delta})$ is the lift of δ with initial point $h_E(b)$. Hence $\tilde{\delta}^{\text{inv}}\tilde{\lambda}(h_E \circ \tilde{\delta})$ is the lift of γ with initial point $h_E(b)$. Clearly, it has endpoint b. Hence $\Phi_b([\gamma]) = h_E \in C_p$.

(4) Continue the notation from the proof of (5). Let $h \in H$. Then $h \circ \tilde{\lambda}$ is the lift of λ via f with initial point $h(h_E(b^*)) \in h(E) = E'$. The path $h \circ \tilde{\lambda}$ has endpoint $h(b^*)$. As in (5) we get that $h_{E'}$ maps the endpoint of $h \circ \tilde{\lambda}$ to the initial point. Thus $h_{E'}(h(b^*)) = h(h_E(b^*))$. Thus $h_E^{-1}h^{-1}h_{E'}h$ fixes b^*, hence is the identity (Lemma 4.17 (iii)). This proves $hh_Eh^{-1} = h_{E'}$.

In the situation of (2), with $\hat{E} \subset E$, we have $h_{\hat{E}} = h_E$ by Corollary 4.22(d). Hence the class C_p does not depend on the radius r of the disc D.

The equality $e = \deg(f_E)$ follows from Corollary 4.22(b). The rest follows from Corollary 4.22(a). □

Definition 4.24 *Let $f : R \to \mathbb{P}^1 \setminus P$ be a finite Galois covering. We say that $r > 0$ is sufficiently small if $D(p, r) \cap P = \{p\}$ for all $p \in P$. Now fix $p \in P$. We define a relation on the set of circular components over p (of sufficiently small level): $E \equiv \hat{E}$ if $E \subset \hat{E}$ or $\hat{E} \subset E$. By* (2) *this is an equivalence relation. The equivalence classes are called the* **ideal points** *of R over p.*

Fix a sufficiently small r. Then by (2), each ideal point over p is represented by exactly one circular component of level r (over p). Thus the number of ideal points over p equals the number of components E of $f^{-1}(D^*)$ (in the notation of the proposition). Since $H = \text{Deck}(f)$ permutes these components transitively (by (3)), it follows that the number of ideal points over p is $\leq |H| = \deg(f)$.

Intuitively, we think of the ideal points as the "missing centers" of the components E (which are homeomorphic to a disc minus center, by Corollary 4.22). Now we put these centers in.

Proposition 4.25 *Let $f : R \to \mathbb{P}^1 \setminus P$ be a finite Galois covering. Let $\bar{R}$ be the (disjoint) union of R and all ideal points over all $p \in P$. Define a subset V of $\bar{R}$ to be open if $V \cap R$ is open in R, and the following holds: For each ideal point $\pi \in V$ there is $E \in \pi$ with $E \subset V$. This makes $\bar{R}$ into a connected compact Hausdorff space. The covering f extends to a continuous surjective map $\bar{f} : \bar{R} \to \mathbb{P}^1$ with $\bar{f}(\pi) = p$ for each ideal point π over p. Further, each $\alpha \in$ Deck(f) extends uniquely to a homeomorphism $\bar{\alpha} : \bar{R} \to \bar{R}$ with $\bar{f} \circ \bar{\alpha} = \bar{f}$.*

Proof. It is clear that $\bar{R}$ is a topological space, and the induced topology on R coincides with the original topology on R. Also it is clear that $\bar{f}$ is surjective. For later use, note that if π is an ideal point then for each $E \in \pi$ the set $E \cup \{\pi\}$ is an open neighborhood of π, and each neighborhood of π contains such a set.

Let $p \in P$, and let $\pi_1, \dots, \pi_m$ be the ideal points over p. Then for each $j = 1, \dots, m$ and sufficiently small r there is exactly one $E_j \in \pi_j$ of level r (see the remarks before the proposition), and

$$\bar{f}^{-1}(D(p,r)) = \bigcup_{j=1}^{m} E_j \cup \{\pi_j\}. \tag{4.1}$$

Now we show that $\bar{f}$ is continuous. Since f is continuous, it suffices to show that for each $p \in P$ and for each sufficiently small r the set $\bar{f}^{-1}(D(p,r))$ is open in $\bar{R}$. This follows from (4.1).

Consider $\alpha \in \mathrm{Deck}(f)$. For each $p \in P$, α permutes the components E_j appearing in (4.1). Define $\bar{\alpha}(\pi_j) = \pi_k$ if $\alpha(E_j) = E_k$. This extends α to a bijection $\bar{\alpha} : \bar{R} \to \bar{R}$ with $\bar{f} \circ \bar{\alpha} = \bar{f}$. This $\bar{\alpha}$ is a homeomorphism since it permutes the basic neighborhoods $E_j \cup \{\pi_j\}$ (and since α is homeomorphic). It is unique since R is dense in $\bar{R}$.

Now we check that $\bar{R}$ is Hausdorff: As in Remark 4.8, any two points of $\bar{R}$ with distinct $\bar{f}$-images can be separated by neighborhoods of the form $\bar{f}^{-1}(U)$. Since R is Hausdorff, it only remains to check that any two distinct ideal points π, π' over the same $p \in P$ can be separated by disjoint neighborhoods. Indeed, we can take the neighborhoods $E \cup \{\pi\}$, $E' \cup \{\pi'\}$, where $E \in \pi$ and $E' \in \pi'$.

$\bar{R}$ is connected because R is connected and dense in $\bar{R}$. It remains to show that $\bar{R}$ is compact. Note first that the topology on $\bar{R}$ has a countable basis: Take all components of all $f^{-1}(U)$, where U is an admissible neighborhood for f of the form $U = D(a,r) \subset \mathbb{P}^1 \setminus P$, where $a \in \mathbb{Q}$ or $a = \infty$, and $r \in \mathbb{Q}$. Further, take all sets $E \cup \{\pi\}$, where π is an ideal point and $E \in \pi$ is of sufficiently small level $r \in \mathbb{Q}$. Each open subset of $\bar{R}$ is a union of sets of the above form.

Since $\bar{R}$ has a countable basis, we can use the compactness criterion requiring that every sequence have a limit point. Let $(a_n)_{n \in \mathbb{N}}$ be a sequence in $\bar{R}$. Then the $\bar{f}(a_n)$ form a sequence in the compact space $\mathbb{P}^1$. Hence this sequence has a limit point $p \in \mathbb{P}^1$. We consider the case $p \in P$. (The case $p \notin P$ is similar, using admissible neighborhoods of p.)

Let $\pi_1, \dots, \pi_m$ be the ideal points over p. We claim that one of the π_j is a limit point of (a_n). Assume wrong. Then each π_j has a neighborhood of the form $E_j \cup \{\pi_j\}$ ($E_j \in \pi_j$) that contains no a_n. We may assume all E_j are of the same level r. (Take the minimum level.) Then (4.1) holds, hence it follows that $D(p,r)$ contains no $\bar{f}(a_n)$. This contradiction completes the proof. □

4.2.3 Coverings of Prescribed Ramification Type

First a simple lemma.

Lemma 4.26 *Let the X_j $(j \in J)$ be open subsets of a topological space X such that X is the union of the X_j. Then each path δ in X is homotopic to the product of (finitely many) paths δ_ν such that each δ_ν is a path in some X_j.*

Proof. Each inverse image $\delta^{-1}(X_j)$ is open in I, hence is a countable union of (disjoint) intervals. The union of these intervals over all j equals I. Since I is compact, already finitely many of these intervals cover I, say $I_1, \ldots, I_s$. Assume s is minimal; then no I_ν is contained in the union of the others.

We can order the I_ν such that $\inf(I_\nu) < \inf(I_{\nu+1})$ for $\nu = 1, \ldots, s-1$. Then $I_\nu \cap I_{\nu+1} \neq \emptyset$ (by minimality). Choose t_ν in this set. Let $t_0 = 0$ and $t_s = 1$. Then the interval $[t_\nu, t_{\nu+1}]$ lies in $I_{\nu+1}$ for $\nu = 0, \ldots, s-1$. Hence δ maps $[t_\nu, t_{\nu+1}]$ entirely into some X_j. Restricting δ to $[t_\nu, t_{\nu+1}]$ and re-parametrizing suitably, we obtain the desired δ_ν. □

The choice of one square root of -1 equips the complex plane $\mathbb{C}$ with an orientation, so that we can speak of traversing a circle in clockwise or counterclockwise direction (where the path $\exp(2\pi\sqrt{-1}t)$ traverses counterclockwise, and $\exp(-2\pi\sqrt{-1}t)$ clockwise).

Theorem 4.27 *Let $p_1, \ldots, p_n \in \mathbb{C}$ be $n \geq 1$ (distinct) points, and set $S = \mathbb{C} \setminus \{p_1, \ldots, p_n\}$. Choose a point $q_0 \in S$ such that the line q_0p_i contains no p_j for all $i \neq j$. Write $p_i = q_0 + \rho_i \exp(\sqrt{-1}\vartheta_i)$ with $\rho_i \in \mathbb{R}_+$ and $0 \leq \vartheta_i < 2\pi$ (for $i = 1, \ldots, n$). Label the p_i such that $\vartheta_1 > \vartheta_2 > \cdots$. Choose rays $M_1, \ldots, M_n$ in the complex plane $\mathbb{C}$ emanating from q_0, such that each connected component of $\mathbb{C} \setminus (M_1 \cup \cdots \cup M_n)$ contains exactly one p_i. Let S_i be the component containing p_i. Let D_i be a disc around p_i whose closure is contained in S_i. Let γ_i be a path in $S_i \cup \{q_0\}$ going on a straight line from q_0 towards p_i until it reaches the boundary of D_i, then traveling once around this boundary in counterclockwise direction, and returning to q_0 again on the line q_0p_i (Figure 4.1). Then we have:*

(a) The paths $\gamma_1, \ldots, \gamma_n$ are closed paths in S based at q_0, and their classes generate the fundamental group $\pi_1(S, q_0)$.
(b) Let G be a group with generators $g_1, \ldots, g_n$. Then there exists a Galois covering $f : R \to S$, an isomorphism θ : Deck$(f) \to G$ and a point $b \in f^{-1}(q_0)$ such that the composition of θ with the surjection $\Phi_b : \pi_1(S, q_0) \to$ Deck(f) (see Proposition 4.19) maps $[\gamma_i]$ to g_i (for $i = 1, \ldots, n$). Now identify G with Deck(f) via the isomorphism θ. If G is finite then

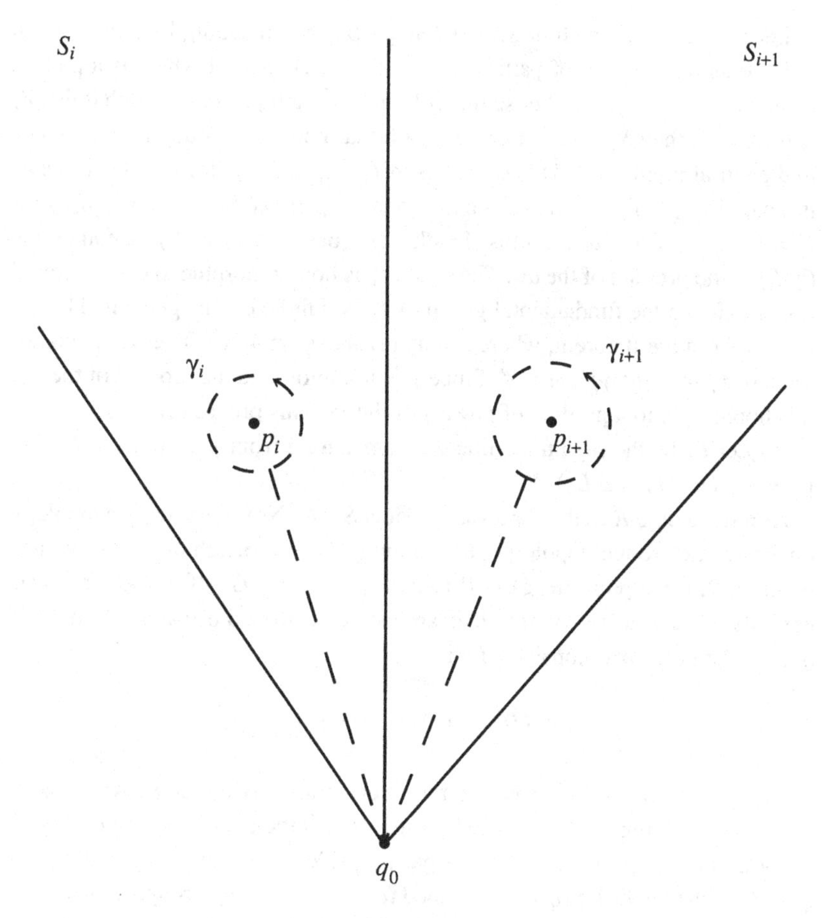

Fig. 4.1.

$f : R \to \mathbb{P}^1 \setminus \{p_1, \ldots, p_n, \infty\}$ is a finite Galois covering, and for the associated classes $C_i := C_{p_i}$ and C_∞ of G (see Proposition 4.23(4)) we have

$$g_i \in C_i \qquad \text{and} \qquad (g_1 \cdots g_n)^{-1} \in C_\infty$$

for $i = 1, \ldots, n$.

Proof. (a) Enlarge each S_i slightly to $S_i' = \{a \in \mathbb{C} : \text{dist}(a, S_i) < \epsilon\}$, such that $S_i' \cap D_j = \emptyset$ for $i \neq j$. The S_i' form an open cover of $\mathbb{C}$, hence the $S_i' \setminus \{p_i\}$ form an open cover of S.

Let γ be a closed path in S, based at q_0. By the preceding Lemma, we can write γ as the product of paths δ_ν ($\nu = 1, \ldots, s$) each of which is a path in some $T_\nu := S'_{i_\nu} \setminus \{p_{i_\nu}\}$. Choose the order in this path product such that first δ_1 is traversed, then δ_2, etc. Let κ_ν be a path that goes on a straight line from q_0 to the initial point of δ_ν. This point lies in T_ν and in $T_{\nu-1}$ (for $\nu > 1$), hence κ_ν is a path in $T_\nu \cap T_{\nu-1}$. Let κ_{s+1} be the constant path q_0. Set $\omega_\nu = \kappa_{\nu+1}^{\text{inv}} \delta_\nu \kappa_\nu$ for $\nu = 1, \ldots, s$. Then ω_ν is a closed path in T_ν, based at q_0, and γ is homotopic (in S) to the product of the ω_ν. The space T_ν is homeomorphic to the punctured disc $\mathbb{K}$. Hence the fundamental group of T_ν is infinite cyclic, generated by the path γ_i from the theorem, where $i = i_\nu$ (Proposition 4.20). Thus ω_ν is homotopic in T_ν to some γ_i^m, $m \in \mathbb{Z}$. Since γ is homotopic to the product of the ω_ν, it is homotopic to a product of powers of the γ_i. This proves (a).

(b) Let L_i be the ray on the line $q_0 p_i$ emanating from p_i with $q_0 \notin L_i$. Let $Q = S \setminus (L_1 \cup \ldots \cup L_n)$.

As a set, define R as the Cartesian product $S \times G$. Now we specify a topology on R (not the product topology), by defining a basis of neighborhoods for each point of R. Fix a point $(q, g) \in R$ (where $q \in S$, $g \in G$). If $q \in Q$, the basis consists of all $B \times \{g\}$, where B is an open disc around q contained in Q. If $q \in L_i$ then the basis consists of all

$$\hat{D}_g = (D^- \times \{g\}) \cup (D^+ \times \{gg_i^{-1}\})$$

where D is an open disc around q not intersecting any line $q_0 p_j$ with $j \neq i$, and not containing p_i. Further, D^+ (resp., D^-) is the open (resp., half-closed) half-disc on the "positive" (resp., "negative") side of L_i, consisting of all points $q' \in D$ such that the line $q_0 q'$ is obtained from the line $q_0 p_i$ through a rotation in counterclockwise (resp., clockwise) direction by an angle ϑ with $0 < \vartheta < \pi/2$ (resp., $0 \leq \vartheta < \pi/2$) (Figure 4.2).

Clearly, for any two such neighborhoods of (q, g) we have that one of them is contained in the other. Hence we get a topology on R by defining a subset of R to be open if it contains one of these neighborhoods for each of its points. We view R as a topological space in this way.

Claim 1 The map $f : R \to S$, $(q, g) \mapsto q$, is a covering.

Proof. We have $f^{-1}(B) = \bigcup_{g \in G} B \times \{g\}$ and $f^{-1}(D) = \bigcup_{g \in G} \hat{D}_g$ (in the above notation). Clearly, each $B \times \{g\}$ (resp., $\hat{D}_g$) maps homeomorphically to B (resp., D). Further, the $B \times \{g\}$ (resp., $\hat{D}_g$) are the components of $f^{-1}(B)$ (resp., $f^{-1}(D)$). Hence B (resp., D) is admissible for f.

Claim 2 For each $h \in G$, the map $\alpha_h : R \to R$, $(q, g) \mapsto (q, hg)$, is a deck transformation of f. We have $\alpha_{hh'} = \alpha_h \circ \alpha_{h'}$.

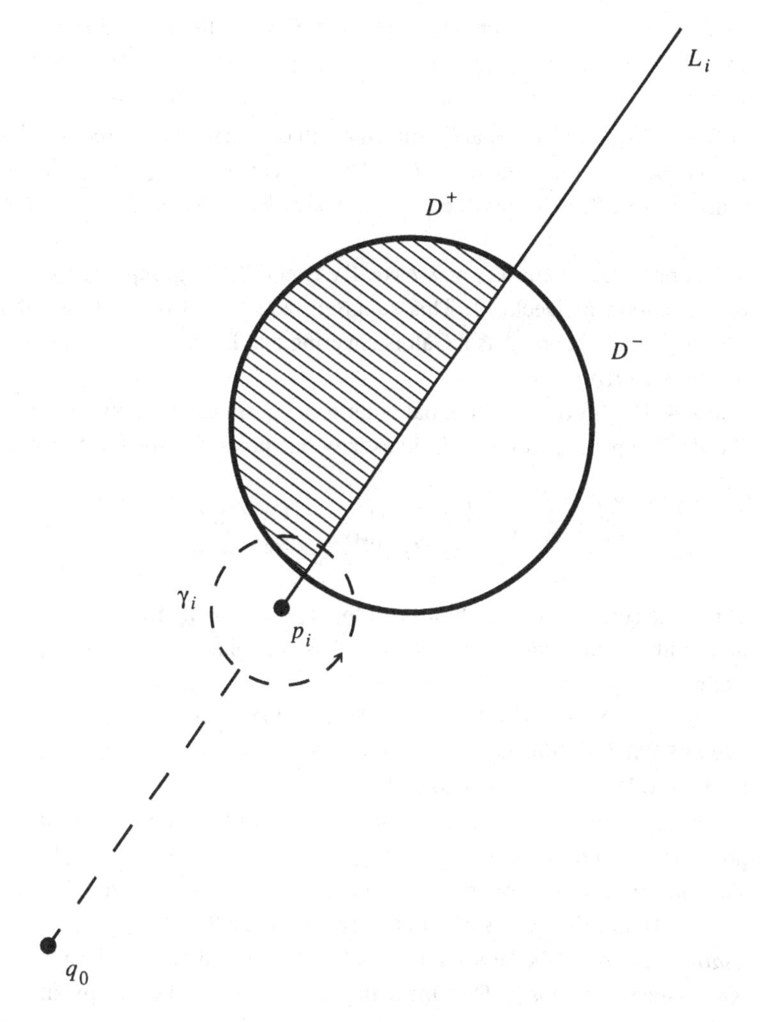

Fig. 4.2.

Proof. The second assertion is clear. It implies that α_h is bijective, with inverse $\alpha_{h^{-1}}$. Further, α_h permutes the neighborhoods defining the topology on R: $\alpha_h(B \times \{g\}) = B \times \{hg\}$ and $\alpha_h(\hat{D}_g) = \hat{D}_{hg}$. It follows that α_h is a homeomorphism. We also have $f \circ \alpha_h = f$, hence α_h is a deck transformation.

Claim 3 The map $f : R \to S$ is a Galois covering. Its deck transformation group is isomorphic to G, via the map $g \mapsto \alpha_g$.

Proof. The main point is to show that R is connected. Each of its subsets $Q \times \{g\}$ is homeomorphic to Q, hence connected. Let C be the connected component of R containing $Q \times \{1\}$. Let D be a disc around $q \in L_i$ as above. Then there is a path in $\hat{D}_1$ that joins $Q \times \{1\}$ to $Q \times \{g_i^{-1}\}$. Thus C is also the component of R containing $Q \times \{g_i^{-1}\}$. Since $\alpha_{g_i^{-1}}$ maps $Q \times \{1\}$ to $Q \times \{g_i^{-1}\}$, it follows that C is fixed by all $\alpha_{g_i^{-1}}$. Thus C is fixed by all α_g, $g \in G$ (because the g_i generate G). Hence C contains all $Q \times \{g\}$. Thus C is dense in R. But connected components are closed, hence $C = R$ and R is connected.

The proof of Claim 3 is now easy. By Claim 2, the group $\{\alpha_g : g \in G\}$ is a subgroup of Deck(f). This subgroup acts transitively on each fiber $f^{-1}(q)$, $q \in S$, hence f is a Galois covering and Deck$(f) = \{\alpha_g : g \in G\}$ (Lemma 4.17(iv)).

Claim 4 The lift of γ_i with initial point $b = (q_0, 1)$ has endpoint (q_0, g_i^{-1}).

Proof. The path $\gamma_i(t)$ meets L_i in exactly one point, say for $t = t_i$. Define

$$\tilde{\gamma}_i(t) = \begin{cases} (\gamma_i(t), 1) & \text{for } t \leq t_i \\ \left(\gamma_i(t), g_i^{-1}\right) & \text{for } t > t_i. \end{cases}$$

Then $\tilde{\gamma}_i$ is continuous by definition of the topology on R. Hence $\tilde{\gamma}_i$ is the lift of γ_i with initial point $(q_0, 1)$. We see it has endpoint (q_0, g_i^{-1}). This proves Claim 4.

From now on we identify G with Deck(f) via the isomorphism $g \mapsto \alpha_g$.

Claim 5 We have $\Phi_b([\gamma_i]) = g_i$ for $i = 1, \ldots, n$. Thus g_i lies in the class C_i of $G =$ Deck(f) associated with p_i.

Proof. By definition of Φ_b, the deck transformation $\Phi_b([\gamma_i])$ maps the endpoint (q_0, g_i^{-1}) of $\tilde{\gamma}_i$ to the initial point $b = (q_0, 1)$. Hence $\Phi_b([\gamma_i]) = g_i$. The path γ_i has the properties of the path γ from Proposition 4.23(5), for $p = p_i$. Hence $\Phi_b([\gamma_i]) \in C_{p_i} = C_i$. This proves Claim 5.

Claim 6 Let $\rho > 0$ be large enough such that q_0 and all p_i lie inside the circle K_0 around 0 of radius ρ. Then the path $\gamma_\infty = \gamma_1 \cdots \gamma_n$ is homotopic in S to a path γ' going from q_0 on a straight line until it reaches K_0, then traveling once around K_0 in counterclockwise direction, and returning to q_0 on the same line.

Proof. Let D be an open disc around q_0 containing all p_i. Let $S_i^* := S_i \cap D$ (a segment of D). Clearly, γ_i is homotopic in S to a path γ_i^* that travels once around the boundary of S_i^* in counterclockwise direction, beginning and ending at q_0 (see Fig. 4.1). Thus γ_∞ is homotopic to $\gamma_1^* \cdots \gamma_n^*$. Recall our convention that in this path product, first γ_n^* is traversed, then γ_{n-1}^* etc. Thus $\gamma_1^* \cdots \gamma_n^*$ is homotopic in S to a path γ^* that goes on a straight line from q_0 to the boundary of D, then travels once around this boundary in counterclockwise direction, and returns to q_0 on the same line. Finally, γ^* is homotopic in S to a path γ' as in Claim 6.

Claim 7 $(g_1 \cdots g_n)^{-1} \in C_\infty$.

Proof. Let γ' be as in Claim 6. Then the path $\gamma := (\gamma')^{\text{inv}}$ has the properties from Proposition 4.23(5), for $p = \infty$. Hence $\Phi_b([(\gamma')^{\text{inv}}]) \in C_\infty$. On the other hand, $\Phi_b([\gamma']) = \Phi_b([\gamma_\infty]) = g_1 \cdots g_n$ by Claim 5 and 6. This proves Claim 7, and thereby the Theorem. □

Remark 4.28 Heuristically, R is obtained by glueing the "sheets" $Q \times \{g\}$, indexed by the elements g of G, along the rays $L_i \times \{g\}$; when moving in counterclockwise direction around p_i, one passes at L_i from the sheet with index g (which contains $D^- \times \{g\}$) into the sheet with index gg_i^{-1}. The group G permutes the sheets naturally, by left multiplication of the indices.

Recall that the topology on the Riemann sphere $\mathbb{P}^1 = \mathbb{C} \cup \{\infty\}$ makes it homeomorphic to the unit sphere S^2 in Euclidean 3-space.

Corollary 4.29 *In the notation of the theorem, the fundamental group $\pi_1(S, q_0)$ of $S = \mathbb{C} \setminus \{p_1, \ldots, p_n\}$ is freely generated by the classes of $\gamma_1, \ldots, \gamma_n$. It follows that for any $(n+1)$ distinct points $p_1, \ldots, p_{n+1}$ on the sphere S^2 (where $n \geq 1$), the fundamental group of $S^2 \setminus \{p_1, \ldots, p_{n+1}\}$ is free of rank n.*

Proof. The second assertion follows immediately from the first, because $S^2 \setminus \{p_{n+1}\}$ is homeomorphic to $\mathbb{C}$. To prove the first assertion, let G be the free group on generators $g_1, \ldots, g_n$. By part (b) of the theorem there exists a homomorphism $\pi_1(S, q_0) \to G$ sending $[\gamma_i]$ to g_i. Since G is free, there also exists a homomorphism $G \to \pi_1(S, q_0)$ sending g_i to $[\gamma_i]$. The composition of the two homomorphisms in either way is the identity, because it fixes the generators g_i and $[\gamma_i]$, respectively. Hence both homomorphisms are isomorphic. □

Remark 4.30 The spaces S^2 and $S^2 \setminus \{p\}$ (for any $p \in S^2$) have trivial fundamental group.

For $S^2 \setminus \{p\} \cong \mathbb{R}^2$ the claim follows from Example 4.4. If γ is a closed path in S^2 that avoids some point $p \in S^2$ then γ is actually a path in $S^2 \setminus \{p\} \cong \mathbb{R}^2$, hence is null-homotopic. Finally, one can show that each closed path in S^2 is homotopic to a path whose image is not all of S^2. (Use Lemma 4.26.)

We can now give the topological version of Riemann's existence theorem, the main result of this chapter. First we need the definition of topological ramification type.

Definition 4.31 *Let $f : R \to \mathbb{P}^1 \setminus P$ be a finite Galois covering. Let $H =$ Deck(f), and for $p \in P$ let C_p be the associated conjugacy class of H. Let $P' = \{p \in P : C_p \neq \{1\}\}$. We define the* **ramification type** *of f to be the class of the triple $(H, P', (C_p)_{p \in P'})$. (Recall our notion of classes of triples from Definition 2.12.)*

Let us scc how the ramification type behaves under a change of coordinates. Consider a homeomorphism $g : \mathbb{P}^1 \to \mathbb{P}^1$ of the form $z \mapsto z - p_0$, $\infty \mapsto \infty$ (for $p_0 \in \mathbb{C}$), or of the form $z \mapsto 1/z$, $0 \mapsto \infty$, $\infty \mapsto 0$. Then in the notation of the above Definition, $\tilde{f} := g \circ f : R \to \mathbb{P}^1 \setminus g(P)$ is another finite Galois covering, with deck transformation group equal to $H = \text{Deck}(f)$.

Further, $\tilde{f}$ is of ramification type $[H, g(P'), (C_{g^{-1}(q)})_{q \in g(P')}]$. Indeed, let $p \in P$, and let γ be a path as in Proposition 4.23(5) (winding once around p in counterclockwise direction). Then $\tilde{\gamma} := g \circ \gamma$ is a path winding once around $g(p)$ in counterclockwise direction, since g is orientation-preserving. Let $b \in f^{-1}(q_0)$, where q_0 is the initial point of γ. The lift of $\tilde{\gamma}$ via $\tilde{f}$ equals the lift of γ via f, both with initial point b. The (unique) element of H mapping the endpoint of this lift to its initial point b lies in the class of $H = \text{Deck}(f)$ associated with p, as well as in the class of $H = \text{Deck}(\tilde{f})$ associated with $g(p)$ (by Proposition 4.23(5) or Corollary 4.22(c)). This proves the claim on the ramification type of $\tilde{f}$.

Theorem 4.32 (RET – The Topological Version) *Let $\mathcal{T} = [G, P, (K_p)_{p \in P}]$ be a ramification type. Let $r = |P|$, and label the elements of P as $p_1, \ldots, p_r$. Then there exists a finite Galois covering of $\mathbb{P}^1 \setminus P$ of ramification type $\mathcal{T}$ if and only if there exist generators $g_1, \ldots, g_r$ of G with $g_1 \cdots g_r = 1$ and $g_i \in K_{p_i}$ for $i = 1, \ldots, r$.*

Proof. By the remarks before the theorem, we can use a change of coordinates to assume that $\infty \in P$, say $p_r = \infty$.

First, assume G has generators $g_1, \ldots, g_r$ as in the theorem. By Theorem 4.27(b), applied with $n = r-1$, there is a finite Galois covering $f : R \to \mathbb{P}^1 \setminus P$ and an identification of Deck(f) with G such that g_i lies in the class C_{p_i} of Deck(f) associated with p_i ($i = 1, \ldots, n$). Further, $g_r = (g_1 \cdots g_n)^{-1} \in C_\infty = C_{p_r}$. Thus $C_p = K_p$ for all $p \in P$. Hence f has ramification type $\mathcal{T}$.

Now suppose conversely that $f : R \to \mathbb{P}^1 \setminus P$ is a finite Galois covering of type $\mathcal{T}$. Then we may assume $G = \text{Deck}(f)$ and $K_p = C_p$, the class of G associated with $p \in P$. Let $S = \mathbb{P}^1 \setminus P$. Take again $n = r - 1$, and choose $q_0 \in S$ and paths $\gamma_1, \ldots, \gamma_n$ as in Theorem 4.27. Set $\gamma_r = (\gamma_1 \cdots \gamma_{r-1})^{\text{inv}}$. Fix some

$b \in f^{-1}(q_0)$, and consider the surjective homomorphism $\Phi_b : \pi_1(S, q_0) \to G$ from Proposition 4.19. As in the proof of Theorem 4.27 we get that $g_i := \Phi_b([\gamma_i])$ lies in the class C_{p_i} for $i = 1, \ldots, r$. These elements $g_1, \ldots, g_r$ generate G by Theorem 4.27(a). Clearly, $g_1 \ldots g_r = 1$. This completes the proof. □

Remark 4.33 The definition of ramification type requires that all occurring conjugacy classes be nontrivial. Thus the above theorem only considers the case (for simplicity) that $C_p \neq 1$ for all $p \in P$ (which means that there is "true ramification" at each point of P). However, the proof shows more generally that if $f : R \to \mathbb{P}^1 \setminus P$ is any finite Galois covering and $P' = \{p \in P : C_p \neq \{1\}\}$, labeled as $P' = \{p_1, \ldots, p_r\}$, then there exist generators $g_1, \ldots, g_r$ of Deck(f) with $g_1 \cdots g_r = 1$ and $g_i \in C_{p_i}$ for $i = 1, \ldots, r$.

5

Riemann Surfaces and Their Function Fields

In this chapter we develop the interplay between finite field extensions of $\mathbb{C}(x)$ and branched covers of the Riemann sphere. This allows us to reduce the algebraic version of Riemann's existence theorem to a theorem about the existence of meromorphic functions on Riemann surfaces (the analytic version of RET). The proof of the analytic version is given in the next chapter.

In more detail, let $L/\mathbb{C}(x)$ be an extension of finite degree n, and let y be a primitive element for this extension. Then y satisfies an equation $F(x, y) = \sum_{i=1}^{n} p_i(x)y^i = 0$ for polynomials $p_i(x)$. Viewing the $p_i(x)$ as functions on $\mathbb{C}$, we see that after excluding finitely many values $x_0 \in \mathbb{C}$ (where the polynomial $F(x_0, y)$ has multiple roots), the roots y of $F(x, y) = 0$ exist locally on $\mathbb{C}$ as analytic functions. It is usually not possible, however, to continue these roots to global functions on $\mathbb{C}$ (or on $\mathbb{C}$ minus finitely many points). By classical Riemann surface theory, y exists as a globally defined function on a Riemann surface Y covering the Riemann sphere $\mathbb{P}^1$. Then L gets identified with the field of meromorphic functions on Y. Removing finitely many points from Y we obtain an (unramified) topological covering of $\mathbb{P}^1 \setminus \{p_1, \ldots, p_r\}$.

We follow the reverse procedure, beginning with a topological covering $R \to \mathbb{P}^1 \setminus \{p_1, \ldots, p_r\}$ of given ramification type $\mathcal{T}$ (as constructed in the last chapter). We add finitely many points to obtain the compact Riemann surface $\bar{R}$. The analytic version of RET guarantees that $\bar{R}$ carries "enough" meromorphic functions. It follows that the induced function field extension $\mathcal{M}(\bar{R})/\mathcal{M}(\mathbb{P}^1) = \mathcal{M}(\bar{R})/\mathbb{C}(x)$ also has type $\mathcal{T}$. This proves the existence of Galois extensions of $\mathbb{C}(x)$ of given ramification type – the (main part of the) algebraic version of RET.

5.1 Riemann Surfaces

Let Y be a topological space. A **coordinate map** on Y is a pair (V, φ), where $V \subset Y$ is open and $\varphi : V \to U$ is a homeomorphism from V onto an open subset $U \subset \mathbb{C}$. Two coordinate maps (V_j, φ_j) $(j = 1, 2)$ are called **compatible**, if $\varphi_1\varphi_2^{-1} : \varphi_2(V_1 \cap V_2) \to \varphi_1(V_1 \cap V_2)$ is biholomorphic. An **atlas** on Y is a collection of compatible coordinate maps (V_i, φ_i) such that Y is the union of the V_i. Two atlases are called **equivalent** if their union is again an atlas. A **complex structure** on Y is an equivalence class of atlases on Y. A **Riemann surface** is a connected Hausdorff-space Y together with a complex structure on Y. When we speak of a coordinate map on a Riemann surface, we mean a coordinate map that is compatible with the complex structure.

Example 5.1

(i) Each connected open subset $D \subset \mathbb{C}$ carries a natural structure as Riemann surface (with an atlas consisting of just the single coordinate map (D, id)).
(ii) In the last chapter we made $\mathbb{P}^1 = \mathbb{C} \cup \{\infty\}$ into a topological space, homeomorphic to the sphere. Consider the open subsets $V_1 := \mathbb{P}^1 \setminus \{\infty\} = \mathbb{C}$, $V_2 := \mathbb{P}^1 \setminus \{0\} = \mathbb{C}^* \cup \{\infty\}$. We define coordinate maps $\varphi_i : V_i \to \mathbb{C}$ by

$$\varphi_1 = \mathrm{id}_{\mathbb{C}}, \qquad \varphi_2 : z \mapsto \begin{cases} \frac{1}{z} & \text{if } z \in \mathbb{C}^* \\ 0 & \text{if } z = \infty. \end{cases}$$

Clearly, they are compatible, hence form an atlas on $\mathbb{P}^1$. The resulting Riemann surface is called the **Riemann sphere**.

Definition 5.2

(i) A continuous map $f : Y \to Z$ between two Riemann surfaces is called **analytic**, *if for all coordinate maps (V, φ) on Y and (V', φ') on Z with $f(V) \subset V'$ the mapping $\varphi' f \varphi^{-1} : \varphi(V) \to \varphi'(V')$ is holomorphic. (Clearly, it suffices to check this for the coordinate maps from a single atlas on Y and Z.)*
(ii) Let Y be a Riemann surface. A **meromorphic function** *on Y is an analytic map $Y \to \mathbb{P}^1$, different from the constant map ∞. Let $\mathcal{M}(Y)$ denote the set of all meromorphic functions on Y.*

Basic complex analysis shows that for connected open $D \subset \mathbb{C}$, a meromorphic function on the Riemann surface D is the same thing as a meromorphic function on the plane region D (in the classical sense), that is, a function that has a Laurent series expansion $\sum_{i=N}^{\infty} a_i(z-p)^i$ around each $p \in D$ (where $N \in \mathbb{Z}$). It follows that if (V, z) is a coordinate map on a Riemann surface Y and $v_0 \in V$, then each meromorphic function g on Y has an expansion of the form $g(u) = \sum_{i=N}^{\infty} a_i(z(u) - z(v_0))^i$ around v_0. We abbreviate this as $g = \sum_{i=N}^{\infty} a_i(z - z(v_0))^i$ (on a neighborhood of v_0).

Let $f, f' \in \mathcal{M}(Y)$. The points of Y where f takes the value ∞ are called the **poles** of f. The poles of f form a discrete subset of Y. (Indeed, any coordinate neighborhood of a pole on which the Laurent series for f converges contains no other pole.) In particular, if Y is compact then f has only finitely many poles.

We first define the functions $f \pm f'$ and $f \cdot f'$ (pointwise operations) at those points of Y which are not a pole of f or f'. By adding, subtracting, or multiplying the Laurent series for f and f' in local coordinates around each pole we see that $f \pm f'$ and $f \cdot f'$ extend (uniquely) to meromorphic functions on Y. This makes $\mathcal{M}(Y)$ into a ring.

If $f \in \mathcal{M}(Y)$ vanishes identically on some coordinate neighborhood of Y then $f = 0$ globally. (This is because Y is connected: The set of all $y \in Y$ such that f vanishes on a neighborhood of y is open and closed, by the identity theorem for power series.) Hence each $f \neq 0$ in $\mathcal{M}(Y)$ can be written locally as $(z - z(a))^n g(z)$, where g has no zero or pole. It is now clear how to define $1/f$, and see it is a meromorphic function on Y. Thus $\mathcal{M}(Y)$ is a field, called the **field of meromorphic functions on** Y.

Example 5.3 (The Field of Meromorphic Functions on $\mathbb{P}^1$***)*** Clearly, the identity function z on $\mathbb{P}^1$ is meromorphic: $z \in \mathcal{M}(\mathbb{P}^1)$. Since $\mathcal{M}(\mathbb{P}^1)$ is a field, it contains the field $\mathbb{C}(z)$ of rational functions in z. Actually,

$$\mathcal{M}(\mathbb{P}^1) = \mathbb{C}(z).$$

Indeed, since $\mathbb{P}^1$ is compact, each $f \in \mathcal{M}(\mathbb{P}^1)$ has only finitely many poles $p_1, \ldots, p_s$. We may assume that ∞ is not among them. (Replace f by $1/f$ if necessary.) For $i = 1, \ldots, s$ there is $f_i \in \mathbb{C}(z)$ such that $f - f_i$ has no pole at p_i, and f_i has p_i as its only pole. (Take for f_i the sum of all terms involving negative powers of $z - p_i$ in the Laurent series for f around p_i). Then the function $g = f - \sum_{i=1}^s f_i$ is meromorphic on $\mathbb{P}^1$, and has no pole. Hence g restricts to a bounded holomorphic function on $\mathbb{C}$, which is constant by Liouville's theorem. Thus $f = g + \sum_{i=1}^s f_i \in \mathbb{C}(z)$.

Here is the analytic version of Riemann's existence theorem. We give a proof in the next chapter.

Theorem 5.4 (RET – The Analytic Version) *Let Y be a compact Riemann surface. Then for any distinct points $p_1, \ldots, p_s$ in Y and complex numbers $c_1, \ldots, c_s$ there is $g \in \mathcal{M}(Y)$ with $g(p_i) = c_i$ for $i = 1, \ldots, s$.*

We conclude this general section with a simple lemma about meromorphic functions in the plane.

Lemma 5.5 *Let $D \subset \mathbb{C}$ open and connected, $p_0 \in D$ and $D_0 = D \setminus \{p_0\}$. View $\mathcal{M}(D)$ as a subfield of $\mathcal{M}(D_0)$ (via restriction). Then $\mathcal{M}(D)$ is algebraically closed in $\mathcal{M}(D_0)$.*

Proof. Let $g \in \mathcal{M}(D_0)$ be algebraic over $\mathcal{M}(D)$. Then g satisfies an equation $a_n g^n + a_{n-1} g^{n-1} + \cdots + a_0 = 0$ with $a_i \in \mathcal{M}(D)$ not all zero (and $n \geq 1$). Multiplying this equation by a sufficiently large power of $z - p_0$ we may assume that none of the a_i has a pole at p_0. Further, we may assume $a_n = 1$ (Lemma 1.3). Then

$$g^n = -a_{n-1} g^{n-1} - \cdots - a_0.$$

Since all a_i are bounded near p_0, it follows that g is also bounded near p_0. Hence $g \in \mathcal{M}(D)$ by Riemann's extension theorem. □

5.2 The Compact Riemann Surface Arising from a Covering of the Punctured Sphere

5.2.1 Construction of an Atlas

In this section, P is a finite subset of $\mathbb{P}^1$, and $f : R \to \mathbb{P}^1 \setminus P$ a finite Galois covering. Set $H = \text{Deck}(f)$.

Lemma 5.6 *Consider open subsets $U \subset \mathbb{P}^1 \setminus P$ that are admissible for f and satisfy $\{0, \infty\} \not\subset U$. For each component V of $f^{-1}(U)$, the map f restricts to a homeomorphism $V \to U$ (by definition). Set $\varphi = f|_V$ if $\infty \notin U$, and $\varphi = 1/(f|_V)$ if $\infty \in U$. Then the (V, φ) are coordinate maps that form an atlas on R. This makes R into a Riemann surface such that the map $f : R \to \mathbb{P}^1$ is analytic. Further, each $\alpha \in H$ becomes an analytic map $\alpha : R \to R$.*

Proof. If (V_1, φ_1) and (V_2, φ_2) are coordinate maps of the above type, then $\varphi_1 \varphi_2^{-1} : \varphi_2(V_1 \cap V_2) \to \varphi_1(V_1 \cap V_2)$ is either the identity or the map $z \mapsto 1/z$, hence biholomorphic. Thus all these coordinate maps are compatible, hence form an atlas on R. (Clearly they cover R.) In these local coordinates, each $\alpha \in H$ becomes the identity and f becomes the identity or the map $z \mapsto 1/z$. Hence f and α are analytic. □

Let $\bar{f} : \bar{R} \to \mathbb{P}^1$ be the extension of f constructed in Proposition 4.25. From now on we just write $f = \bar{f}$ for simplicity. Consider an ideal point π of R, and let $p = f(\pi)$. For each $E \in \pi$, of level r, consider the covering $f_E = \kappa_p \circ f|_E$: $\mathbb{E} \to \mathbb{K}(r)$ of finite degree e (from Proposition 4.23 (1), (4)). By Corollary 4.22(a) there is a homeomorphism $\varphi : E \to \mathbb{K}(r^{1/e})$ with $\varphi^e = f_E$. It extends to a homeomorphism φ_π from $V_\pi := E \cup \{\pi\}$ to $D(0, r^{1/e}) = \mathbb{K}(r^{1/e}) \cup \{0\}$, sending π to 0. (φ_π is a homeomorphism because the basic neighborhoods $\hat{E} \cup \{\pi\}$ of π correspond to the discs $D(0, \hat{r}^{1/e})$ under φ; see Proposition 4.23(2).)

We have constructed coordinate maps (V_π, φ_π) around each ideal point π. Together with the coordinate maps (V, φ) from Lemma 5.6, they cover $\bar{R}$.

Lemma 5.7 *The above coordinate maps form an atlas on $\bar{R}$. This makes $\bar{R}$ into a compact Riemann surface such that the map $f : \bar{R} \to \mathbb{P}^1$ is analytic. Further, each $\alpha \in H$ extends uniquely to an analytic homeomorphism $\bar{\alpha} : \bar{R} \to \bar{R}$ with $f \circ \bar{\alpha} = f$.*

Proof. First we show that any (V_π, φ_π) is compatible with any coordinate map (V, φ) of the other type. In the above notations we have $\varphi_\pi^e = f_E = \kappa_p \circ f$ on $V \cap V_\pi$, where $p = f(\pi)$, and κ_p is as in Proposition 4.23(1). Hence $\varphi\varphi_\pi^{-1} : \varphi_\pi(V \cap V_\pi) \to \varphi(V \cap V_\pi)$ is the map $z \mapsto \kappa_p^{-1}(z^e)$ (resp., $z \mapsto 1/\kappa_p^{-1}(z^e)$) if $\infty \notin f(V)$ (resp., if $\infty \in f(V)$). This map is holomorphic with nonvanishing derivative (since $0 \notin \varphi_\pi(V \cap V_\pi)$) and homeomorphic (since φ_π and φ are homeomorphic); hence it is biholomorphic. This proves compatibility.

It remains to show that any two coordinate maps of the form (V_π, φ_π) and $(V'_{\pi'}, \varphi'_{\pi'})$ are compatible. We can assume $\pi = \pi'$ (otherwise V_π and $V'_{\pi'}$ are compatible because their intersection is covered by coordinate maps of the other type). Then we can further assume $V_\pi \subset V'_\pi$ (Proposition 4.23(2)). Let $V_\pi = E \cup \{\pi\}$ with $E \in \pi$, of level r. Then $\varphi'_\pi\varphi_\pi^{-1} : D(0, r^{1/e}) \to D(0, r^{1/e})$ is multiplication by some root ζ of unity (Corollary 4.22(a)). Since the map $z \mapsto \zeta z$ is biholomorphic, this proves compatibility. Now we have shown that we obtain an atlas on $\bar{R}$ as claimed in the Lemma, making $\bar{R}$ into a compact Riemann surface.

Let us show that $f : \bar{R} \to \mathbb{P}^1$ is analytic. For each ideal point π not over ∞ (resp., lying over ∞), we have to show that the map $f\varphi_\pi^{-1}$ (resp., $f\varphi_\pi^{-1}$ composed with $z \mapsto 1/z$) is analytic on $D(0, r^{1/e})$. But this map turns out as $z \mapsto \kappa_p^{-1}(z^e)$ (resp., as $z \mapsto z^e$). Hence f is analytic around ideal points. At the other points, it is analytic by the previous lemma.

We know that each $\alpha \in H$ extends uniquely to a homeomorphism $\bar{\alpha} : \bar{R} \to \bar{R}$ with $f\bar{\alpha} = f$ (Proposition 4.25). We also know from the previous lemma that α is analytic on R. Consider one of the above coordinate maps (V_π, φ_π) around an ideal point π. Then $(\bar{\alpha}(V_\pi), \varphi_\pi \circ \bar{\alpha}^{-1})$ is again such a coordinate map. In these local coordinates, $\bar{\alpha}$ becomes the map $z \mapsto \zeta z$ for some root ζ of unity (as above). Thus $\bar{\alpha}$ is analytic. □

Clearly, the $\bar{\alpha}$ ($\alpha \in H$) form a group (under composition), isomorphic to H under the map $\alpha \mapsto \bar{\alpha}$. From now on we also drop the bar from $\bar{\alpha}$, and write $\alpha = \bar{\alpha}$.

This identifies H with a group of analytic isomorphisms of $\bar{R}$. (Here analytic isomorphism means an analytic homeomorphism whose inverse is also analytic.) This group acts transitively on each fiber $f^{-1}(p)$, $p \in \mathbb{P}^1$. For $p \notin P$ this is in the definition of a Galois covering. For $p \in P$ it follows from Proposition 4.23(3) (since the ideal points over p correspond to the components E of $f^{-1}(D^*)$, with D^* as in Proposition 4.23).

Lemma 5.8 *Suppose $g \in \mathcal{M}(\bar{R})$ satisfies $g \circ \alpha = g$ for all $\alpha \in H$. Then $g = g' \circ f$ for some $g' \in \mathcal{M}(\mathbb{P}^1)$.*

Proof. Since H acts transitively on each fiber $f^{-1}(z)$, $z \in \mathbb{P}^1$, it follows that g takes the same value on all points in $f^{-1}(z)$. Define this value to be $g'(z)$. If $U \subset \mathbb{P}^1 \setminus (P \cup \{\infty\})$ is admissible for f, and V a component of $f^{-1}(U)$, then on U we have $g' = g \circ (f|_V)^{-1}$. Since $f|_V : V \to U$ is a coordinate map on R, it follows that g' is meromorphic on U. Varying U it follows that g' is meromorphic on $\mathbb{P}^1 \setminus (P \cup \{\infty\})$.

Clearly g' is continuous at each $p \in P \cup \{\infty\}$ (as a map $\mathbb{P}^1 \to \mathbb{P}^1$). Hence g' is meromorphic at p by Riemann's extension theorem. Thus $g' \in \mathcal{M}(\mathbb{P}^1)$. □

5.2.2 The Identification between Topological and Algebraic Ramification Type

Now we get to the main result of this section. Its proof uses the analytic version of RET.

Theorem 5.9 *Let P be a finite subset of $\mathbb{P}^1$, and $f : R \to \mathbb{P}^1 \setminus P$ a finite Galois covering. Let $f : \bar{R} \to \mathbb{P}^1$ be its extension (constructed above) to an analytic map on the compact Riemann surface $\bar{R}$. View $H = Deck(f)$ as a group of analytic isomorphisms of $\bar{R}$ (as above). Let $\mathcal{M} = \mathcal{M}(\bar{R})$ be the field of meromorphic functions on $\bar{R}$, and let $\mathbb{C}(f)$ be the subfield generated by f and the constant ($\mathbb{C}$-valued) functions. Then for each $\alpha \in H$, the map*

$$\iota_\alpha : \mathcal{M} \to \mathcal{M}, \qquad g \mapsto g \circ \alpha^{-1}$$

is an automorphism of $\mathcal{M}$. The field $\mathcal{M}$ is Galois over $\mathbb{C}(f)$, and the map

$$\iota : H = Deck(f) \to G(\mathcal{M}/\mathbb{C}(f)), \qquad \alpha \mapsto \iota_\alpha$$

is an isomorphism.

Proof. For $g \in \mathcal{M}$ we have $g \circ \alpha^{-1} \in \mathcal{M}$ since α is an analytic isomorphism of $\bar{R}$. Thus the map ι_α is well defined. Clearly, it is an automorphism of the field $\mathcal{M}$ that is the identity on the subfield $\mathbb{C}(f)$.

It is also clear that $\iota_{\alpha\beta} = \iota_\alpha \iota_\beta$. Thus the map $\alpha \mapsto \iota_\alpha$ is a homomorphism from H onto a subgroup G of $\mathrm{Aut}(\mathcal{M}/\mathbb{C}(f))$. The fixed field of G consists of those $g \in \mathcal{M}$ with $g \circ \alpha = g$ for all $\alpha \in H$. Such g is of the form $g = g' \circ f$ with $g' \in \mathcal{M}(\mathbb{P}^1)$ (Lemma 5.8). Since $\mathcal{M}(\mathbb{P}^1) = \mathbb{C}(z)$ (Example 5.3) it follows that $g \in \mathbb{C}(f)$.

We have proved that $\mathbb{C}(f)$ is the fixed field of G in $\mathcal{M}$. Hence $\mathcal{M}$ is Galois over $\mathbb{C}(f)$ with Galois group G (by Artin's theorem). It remains to show that the map $\iota : H \to G$ is injective. This depends on the analytic version of RET. (We have to know that $\mathcal{M}$ is large enough; without RET, it is not even clear how to exclude the case $\mathcal{M} = \mathbb{C}(f)$, in which case G would be trivial.)

Let $\alpha \in H$ with $\iota_\alpha = \mathrm{id}$. Let $a \in \bar{R}$, and $p = f(a)$. By the analytic version of RET, there is $g \in \mathcal{M}$ taking pairwise distinct values on $f^{-1}(p)$. Since $b := \alpha(a) \in f^{-1}(p)$ and $g(b) = \iota_\alpha(g)(b) = g(\alpha^{-1}(b)) = g(a)$ it follows that $a = b$. Thus $\alpha = \mathrm{id}$. This proves that ι is injective. Hence the theorem. □

Addendum *For each $p \in P$ let C_p^{top} be the conjugacy class of $H = \mathrm{Deck}(f)$ associated with p in Proposition 4.23. For each $p \in \mathbb{P}^1$ let C_p^{alg} be the conjugacy class of $G = G(\mathcal{M}/\mathbb{C}(f))$ associated with p in Proposition 2.6, for $x = f$. Then $C_p^{\mathrm{alg}} = \{1\}$ if $p \notin P$. For each $p \in P$ the isomorphism $\iota : H \to G$ maps C_p^{top} onto C_p^{alg}. This identifies the topological ramification type of the finite Galois covering $f : R \to \mathbb{P}^1 \setminus P$ with the algebraic ramification type of the finite Galois extension $\mathcal{M}/\mathbb{C}(f)$.*

Proof. Let $\Lambda = \mathbb{C}((t))$ be the field of formal Laurent series over $\mathbb{C}$ (see Section 2.1.1).

Case 1 $p \in \mathbb{P}^1 \setminus P$.

Let $v \in f^{-1}(p)$. There is a coordinate map (V, φ) as in Lemma 5.6 with $v \in V$. If $p \neq \infty$ we can assume $\infty \notin \varphi(V)$ and $\varphi = f|_V$. If $p = \infty$ we have $\varphi = 1/(f|_V)$. Each $g \in \mathcal{M}$ has a Laurent series expansion around v of the form

$$g = \sum_{i=N}^{\infty} a_i (\varphi - \varphi(v))^i$$

(see Section 5.1). Mapping g to $\sum_{i=N}^{\infty} a_i t^i$ yields a ring homomorphism

$$\vartheta : \mathcal{M} \to \Lambda = \mathbb{C}((t)).$$

This homomorphism is clearly nonzero, hence an embedding of fields. We need to compute $\vartheta(f)$. If $p \neq \infty$ then $f = \varphi = p + (\varphi - \varphi(v))$ on V, hence $\vartheta(f) = p + t$. If $p = \infty$ then $f = 1/\varphi = 1/(\varphi - \varphi(v))$ on V, hence $\vartheta(f) = t^{-1}$.

Thus $\vartheta : \mathcal{M} \to \Lambda$ extends the map $\vartheta_p : \mathbb{C}(f) \to \mathbb{C}(t)$ from Section 2.2.1 (for $x = f$). Hence the ramification index e of the extension $\mathcal{M}/\mathbb{C}(f)$ at p is 1, and thus $C_p^{\text{alg}} = \{1\}$ (Proposition 2.6(b)).

Case 2 $p \in P$.

Let $\pi \in f^{-1}(p)$ (an ideal point). Let (V_π, φ_π) be a coordinate map around π as in Section 5.2.1. In particular, $V_\pi = E \cup \{\pi\}$ and $\varphi_\pi^e = \kappa_p \circ f$ on V_π. Each $g \in \mathcal{M}$ has a Laurent series expansion around π of the form

$$g = \sum_{i=N}^{\infty} b_i \varphi_\pi^i.$$

Mapping g to $\sum_{i=N}^{\infty} b_i \tau^i$ yields an embedding

$$\vartheta : \mathcal{M} \to \Lambda_e = \mathbb{C}((\tau))$$

where $\Lambda_e = \mathbb{C}((\tau))$ is as in Section 2.1.3. Recall that we view $\Lambda = \mathbb{C}((t))$ as a subfield of $\Lambda_e = \mathbb{C}((\tau))$, where $t = \tau^e$.

Again ϑ extends the map $\vartheta_p : \mathbb{C}(f) \to \Lambda$. Indeed, from $\varphi_\pi^e = \kappa_p \circ f$ we get $f = \kappa_p^{-1} \circ \varphi_\pi^e = p + \varphi_\pi^e$ if $p \neq \infty$, and $f = 1/\varphi_\pi^e$ if $p = \infty$. Thus $\vartheta(f) = p + \tau^e = p + t$ in the former case, and $\vartheta(f) = \tau^{-e} = t^{-1}$ in the latter case. This shows that ϑ extends ϑ_p.

Let ω be the distinguished generator of $G(\Lambda_e/\Lambda)$ (in the sense of Section 2.2.1). By Lemma 2.3, ω maps $\sum_{i=N}^{\infty} b_i \tau^i$ to $\sum_{i=N}^{\infty} (b_i \zeta_e^i) \tau^i$. By definition of the class C_p^{alg} (Proposition 2.6), it contains the element

$$\vartheta^{-1} \circ \omega \circ \vartheta.$$

By definition of the class C_p^{top} (Proposition 4.23), it contains the distinguished generator h_E of the stabilizer of E in H. Hence the claim follows from

$$\iota(h_E) = \vartheta^{-1} \circ \omega \circ \vartheta. \tag{5.1}$$

It remains to prove (5.1). By definition, h_E fixes E (hence V_π), and induces the distinguished generator of the covering $f_E = \kappa_p \circ f : E \to \mathbb{K}(r)$. Since $\varphi_\pi^e = f_E$ on E we get $\varphi_\pi \circ h_E^{-1} = \zeta_e \varphi_\pi$ (on E, hence also on V_π) by Corollary 4.22(b).

For $g \in \mathcal{M}$ as above we have $\iota(h_E)(g) = g \circ h_E^{-1} = \sum_{i=N}^{\infty} b_i (\varphi_\pi \circ h_E^{-1})^i = \sum_{i=N}^{\infty} b_i (\zeta_e \varphi_\pi)^i = \sum_{i=N}^{\infty} (b_i \zeta_e^i) \varphi_\pi^i$ (in a neighborhood of π). From this and the above description of ω we get (5.1). Hence the claim. □

Remark 5.10 The stabilizer of E in H equals the stabilizer of the point $\pi \in f^{-1}(p)$ represented by E. This stabilizer is usually called the *inertia group* of π. Recall that as π ranges over $f^{-1}(p)$, the distinguished generators of these inertia groups range over the conjugacy class C_p^{top} of H. Thus C_p^{top} is *the class of distinguished inertia group generators over p*.

Now "one half" of the algebraic version of RET follows (actually the more difficult half, since it requires the analytic version).

Corollary 5.11 *Let $P = \{p_1, \ldots, p_r\}$ be a finite subset of $\mathbb{P}^1$ (of cardinality r). Let G be a finite group with generators $g_1, \ldots, g_r$ satisfying $g_1 \cdots g_r = 1$ and $g_i \neq 1$ for all i. Let K_{p_i} be the conjugacy class of g_i in G. Then there exists a finite Galois extension of $\mathbb{C}(x)$ of ramification type $\mathcal{T} = [G, P, (K_p)_{p \in P}]$.*

Proof. By Theorem 4.32 there exists a finite Galois covering $f : R \to \mathbb{P}^1 \setminus P$ of ramification type $\mathcal{T}$. By Theorem 5.9 and its addendum, the associated finite Galois extension $\mathcal{M}/\mathbb{C}(f)$ has the same ramification type. □

5.3 Constructing Generators of $G(L/\mathbb{C}(x))$

Theorem 5.12 *Let $L/\mathbb{C}(x))$ be a finite Galois extension. Then there exists a finite subset P of $\mathbb{P}^1$ and a finite Galois covering $f : R \to \mathbb{P}^1 \setminus P$ such that there is a $\mathbb{C}$-linear isomorphism $L \to \mathcal{M}(\bar{R})$ sending x to f. Here f is identified with its extension to the compact Riemann surface $\bar{R}$ (as above).*

Proof. Choose some $F(x, y) \in \mathbb{C}[x, y]$, irreducible as a polynomial in y over $\mathbb{C}(x)$, such that L is a splitting field of this polynomial over $\mathbb{C}(x)$. Let n be the y-degree of F. By Lemma 1.6 there are only finitely many $p \in \mathbb{C}$ for which the polynomial $F(p, y) \in \mathbb{C}[y]$ has less than n (distinct) roots. Let P be the set of those p, together with ∞.

Let R' be the set of all $(u, v_1, \ldots, v_n) \in \mathbb{C}^{n+1}$ with $u \in \mathbb{P}^1 \setminus P$ and $F(u, v_1) = \cdots = F(u, v_n) = 0$ and $v_1, \ldots, v_n$ all distinct. (Then $v_1, \ldots, v_n$ are exactly the roots of $F(u, y)$.) View R' as a topological space, with the induced topology from $\mathbb{C}^{n+1}$.

Claim 1 The map $f' : R' \to \mathbb{P}^1 \setminus P$, $(u, v_1, \ldots, v_n) \mapsto u$, is a covering. The symmetric group S_n acts naturally as a group of deck transformations of f' (permuting $v_1, \ldots, v_n$).

Proof. For each $u_0 \in \mathbb{P}^1 \setminus P$ there are holomorphic functions $\psi_1, \ldots, \psi_n$, defined in a neighborhood U of u_0, such that $\psi_1(u), \ldots, \psi_n(u)$ are exactly the roots of $F(u, y) \in \mathbb{C}[y]$, for each $u \in U$ (Theorem 1.18). For each $\sigma \in S_n$ set

$$V_\sigma = \{(u, \psi_{\sigma(1)}(u), \ldots, \psi_{\sigma(n)}(u)) : u \in U\}.$$

Then $(f')^{-1}(U)$ is the disjoint union of the V_σ. The map $U \to V_\sigma$, $u \mapsto (u, \psi_{\sigma(1)}(u), \ldots, \psi_{\sigma(n)}(u))$, is inverse to the map $f'|_{V_\sigma} : V_\sigma \to U$. Hence both maps are homeomorphisms. Taking U compact and connected, we get the same for V_σ. It follows that the V_σ are open in $(f')^{-1}(U)$ (since they are disjoint). Hence they are the connected components of $(f')^{-1}(U)$. Thus if D is an open disc around u_0 contained in U then D is admissible for f'. Hence f' is a covering. The other assertion in Claim 1 is clear.

Claim 2 Let R be a connected component of R'. Then the restriction of f' yields a finite Galois covering $f : R \to \mathbb{P}^1 \setminus P$.

Proof. $f = f'|_R$ is a covering by Corollary 4.13. Let $u \in \mathbb{P}^1 \setminus P$ and $v, v' \in f^{-1}(u)$. By Claim 1 there is $\alpha \in \mathrm{Deck}(f')$ mapping v to v'. Then α maps R into itself (since R is the component of R' containing v and v'). The same holds for α^{-1}, hence α restricts to a deck transformation of f. Thus $\mathrm{Deck}(f)$ acts transitively on $f^{-1}(u)$. Hence f is a Galois covering. Clearly, f has finite degree ($\leq \deg(f') = n!$).

Claim 3 For $i = 1, \ldots, n$, the function $g_i : R \to \mathbb{C}$, $(u, v_1, \ldots, v_n) \mapsto v_i$, extends to a meromorphic function g_i on $\bar{R}$, which satisfies the equation $F(f, g_i) = 0$.

Proof. Each point of R has a neighborhood of the form V_σ, and the map $f|_{V_\sigma}$ is a coordinate map on R (in the complex structure defined on R in Lemma 5.6). In this local coordinate, the function g_i is represented by the above holomorphic function $\psi_{\sigma(i)}$. This shows that g_i is meromorphic (even holomorphic) on R.

View $\mathcal{M}(\bar{R})$ as a subfield of $\mathcal{M}(R)$ (via restriction). Then $\mathcal{M}(\bar{R})$ is algebraically closed in $\mathcal{M}(R)$: Indeed, if $h \in \mathcal{M}(R)$ is algebraic over $\mathcal{M}(\bar{R})$ then h is meromorphic at each of the (finitely many) points of $\bar{R} \setminus R$ by Lemma 5.5. Hence $h \in \mathcal{M}(\bar{R})$.

For $v = (u, v_1, \ldots, v_n) \in R$ we have $F(u, v_i) = 0$. Hence $F(f(v), g_i(v)) = 0$ for all $v \in R$. Thus f and g_i, viewed as elements of $\mathcal{M}(R)$, satisfy the equation $F(f, g_i) = 0$. Since $f \in \mathcal{M}(\bar{R})$ and $\mathcal{M}(\bar{R})$ is algebraically closed in $\mathcal{M}(R)$, it follows that $g_i \in \mathcal{M}(\bar{R})$.

Claim 4 The functions $g_1, \ldots, g_n$ generate $\mathcal{M}(\bar{R})$ over $\mathbb{C}(f)$.

Proof. By Theorem 5.9, every element of $G(\mathcal{M}(\bar{R})/\mathbb{C}(f))$ is of the form $\iota_\alpha, \alpha \in H$. Thus it suffices to prove the following: If ι_α fixes $g_1, \ldots, g_n$ then $\alpha = \mathrm{id}$.

Indeed, if ι_α fixes $g_1, \ldots, g_n$ then for all $v \in R$ we have $g_i(v) = g_i(\alpha^{-1}(v))$, $i = 1, \ldots, n$. Since $f(v) = f(\alpha^{-1}(v))$ also, it follows that $v = \alpha^{-1}(v)$ (by definition of f and the g_i). Hence $\alpha = \mathrm{id}$.

Conclusion. Since the g_i satisfy the irreducible polynomial $F(f, y)$ over $\mathbb{C}(f)$, it follows from Claim 4 that $\mathcal{M}(\bar{R})$ is a splitting field of this polynomial over $\mathbb{C}(f)$. (Note that we know $\mathcal{M}(\bar{R})$ is Galois over $\mathbb{C}(f)$.) Hence the Theorem. □

Corollary 5.13 *Let $L/\mathbb{C}(x)$ be a finite Galois extension, with branch points $p_1, \ldots, p_r$. Then there exist generators $g_1, \ldots, g_r$ of $G = G(L/\mathbb{C}(x))$ with $g_1 \cdots g_r = 1$ and $g_i \in C_{p_i}$ for $i = 1, \ldots, r$. Here C_p denotes the class of G associated with $p \in \mathbb{P}^1$.*

Proof. Given $L/\mathbb{C}(x)$, let $f : R \to \mathbb{P}^1 \setminus P$ be as in the theorem. By Theorem 5.9 addendum, the branch points $p_1, \ldots, p_r$ of L are exactly those elements $p \in P$ with $C_p^{\text{top}} \neq 1$. Hence by Remark 4.33, the group Deck(f) has generators $h_1, \ldots, h_r$ with $h_1 \cdots h_r = 1$ and $h_i \in C_{p_i}^{\text{top}}$ for all i. Then the images of $h_1, \ldots, h_r$ under the isomorphism ι from Theorem 5.9 addendum yield generators of $G(\mathcal{M}(\bar{R})/\mathbb{C}(f))$ with the analogous properties. This proves the claim, since $L \cong \mathcal{M}(\bar{R})$. □

Corollary 5.11 and 5.13 together imply the algebraic version of RET (Theorem 2.13). Thus we have derived the algebraic version from the analytic version. This was the main goal of Chapter 5. We have proved (or nearly proved) further important results along the way, which we summarize below. They will not be used before Chapter 10.

5.4 Digression: The Equivalence between Coverings and Field Extensions

Theorem 5.14 *Let G be a finite group. Let $P \subset \mathbb{P}^1$ finite, and $q \in \mathbb{P}^1 \setminus P$. There is a natural* 1–1 *correspondence between the following objects.*

1. *The $\mathbb{C}(x)$-isomorphism classes of Galois extensions $L/\mathbb{C}(x)$ with Galois group isomorphic to G and with branch points contained in P.*
2. *The equivalence classes of Galois coverings $f : R \to \mathbb{P}^1 \setminus P$ with deck transformation group isomorphic to G.*
3. *The normal subgroups of the fundamental group $\pi_1(\mathbb{P}^1 \setminus P, q)$ with quotient isomorphic to G.*

The correspondence between (1) and (2) is given by associating with f those Galois extensions $L/\mathbb{C}(x)$ for which there is a $\mathbb{C}$-linear isomorphism $L \to \mathcal{M}(\bar{R})$ sending x to f (where $\bar{R}$ is as constructed in Lemma 5.7). The correspondence between (2) and (3) is given by associating with f the kernel of the surjection $\Phi_b : \pi_1(\mathbb{P}^1 \setminus P, q) \to$ Deck(f) (see Proposition 4.19), for any $b \in f^{-1}(q)$.

Proof. The correspondence between (2) and (3) is a special case of a general fact in covering space theory. First note that the kernel of Φ_b does not depend on the choice of b. (Actually, for another $b' \in f^{-1}(q)$ the map $\Phi_{b'}$ is the composition of Φ_b with an inner automorphism of $\mathrm{Deck}(f)$.) It is clear that equivalent coverings yield the same normal subgroup. Conversely, two coverings corresponding to the same normal subgroup are equivalent by Corollary 4.14. It remains to show that each normal subgroup $\mathcal{N}$ of $\Gamma = \pi_1(\mathbb{P}^1 \setminus P, q)$ with $\Gamma/\mathcal{N} \cong G$ is of the form $\ker(\Phi_b)$, for some f as in (2). If $\infty \in P$ this follows from Theorem 4.27, taking $g_1, \ldots, g_n$ to be the images of $\gamma_1, \ldots, \gamma_n$ in G. (The choice of q is of no relevance here, using the isomorphism between $\pi_1(S, q)$ and $\pi_1(S, p)$ from 4.1.1.) The general case follows by a change of coordinates.

We have now established the 1–1 correspondence between the objects in (2) and (3). In particular, the number of objects in (2) equals that in (3). This number λ can be computed as follows. Let $n = |P| - 1$. Then by Corollary 4.29 the group Γ is free of rank n, say on generators $\gamma_1, \ldots, \gamma_n$. Thus there is a 1–1 correspondence between the surjective homomorphisms $\varphi : \Gamma \to G$ and the generating systems $g_1, \ldots, g_n$ of G of length n. (This correspondence is given by setting $g_i = \varphi(\gamma_i)$.) Two such homomorphisms φ, φ' have the same kernel iff $\varphi' = \alpha\varphi$ for some $\alpha \in \mathrm{Aut}(G)$. For the corresponding generating systems $g_1, \ldots, g_n$ and $g'_1, \ldots, g'_n$ of G this means that $(g'_1, \ldots, g'_n) = (g_1, \ldots, g_n)^\alpha$, where α acts componentwise on tuples of group elements. Thus the number λ of objects in (3) equals the number of $\mathrm{Aut}(G)$-orbits on generating systems of G of length n.

Now we turn to the correspondence between (1) and (2). In Theorem 5.9, with each f as in (2) there has been associated some $L/\mathbb{C}(x)$ as in (1). It is clear that equivalent coverings yield the same $L/\mathbb{C}(x)$, up to $\mathbb{C}(x)$-isomorphism. Thus we have a well-defined map from the objects in (2) to those in (1). This map is surjective by Theorem 5.12. If we can show that the number of objects in (2) is the same as that in (1), and is finite, then this surjection yields the claimed 1–1 correspondence. The number of objects in (2) is the above λ. In Remark 7.4 it is derived (purely algebraically) from the algebraic version of RET that λ also equals the number of objects in (1). Hence the claim. □

Remark 5.15 Corresponding objects in (1) and (2) are of the same ramification type (Theorem 5.9 addendum).

6

The Analytic Version of Riemann's Existence Theorem

In this chapter we prove

Theorem 6.1 (RET – The Analytic Version) *Let Y be a compact Riemann surface. For any distinct points $a_1, \ldots, a_n$ of Y and complex numbers $c_1, \ldots, c_n$ there exists a meromorphic function f on Y with $f(a_i) = c_i$ for $i = 1, \ldots, n$.*

This is a deep result since the definition of Riemann surface yields the existence of functions only locally, on coordinate patches. We follow the proof given in Forster's book [Fo]. It constructs globally defined functions using a limiting process, phrased in the language of Hilbert spaces, and a patching process, phrased in the language of cohomology.

By focusing on this theorem, we can avoid much of the more general framework of Forster's book. In particular, we do not need to develop any cohomology theory beyond the definition of a cocycle and a coboundary. Yet, the usefulness of those ideas should become apparent in the present application.

We also provide proofs for the required results from the theory of Hilbert spaces. We need the existence of orthogonal complements and Banach's theorem.

All vector spaces occurring in this chapter are complex vector spaces. For a complex number α we let $\bar{\alpha}$ denote its complex conjugate, and $|\alpha| = \sqrt{\alpha\bar{\alpha}}$ its absolute value.

6.1 Abstract Hilbert Spaces

6.1.1 Continuous Linear Maps and Orthogonal Complements

Let H be a (complex) vector space. A *hermitian form* on H is a bi-additive map $\langle , \rangle : H \times H \to \mathbb{C}$ satisfying $\langle \alpha h, h' \rangle = \alpha \langle h, h' \rangle$ and $\langle h', h \rangle = \overline{\langle h, h' \rangle}$ for all $h, h' \in H, \alpha \in \mathbb{C}$. Then $\langle h, h \rangle$ is a real number. The hermitian form is called *positive definite* if $\langle h, h\rangle$ is strictly positive for each $h \neq 0$ in H. We then write $\|h\| = \sqrt{\langle h, h\rangle}$ for each $h \in H$.

Let $\langle , \rangle$ be a positive definite hermitian form on H. Then H becomes a metric space with the distance function $d(h, h') = \|h - h'\|$. (Same proof as in the finite-dimensional case, using the Cauchy–Schwarz inequality.) We say H is *complete* relative $\langle , \rangle$ if this metric space is complete, that is, every Cauchy sequence converges. When speaking of topological notions like closed subspaces, continuous linear maps, etc., we always refer to the topology on H associated with this metric.

Definition 6.2 *A* **Hilbert space** *is a vector space H equipped with a positive definite hermitian form, and complete with respect to this form.*

Remark 6.3

(i) A closed (linear) subspace of a Hilbert space H is again a Hilbert space (with hermitian form restricted from H).
(ii) Translation $h \mapsto h + h_0$ ($h_0 \in H$) and dilatation $h \mapsto rh$ ($r \neq 0$ in $\mathbb{C}$) are homeomorphisms $H \to H$.

Lemma 6.4 *A linear map $\phi : H \to H'$ between Hilbert spaces is continuous if and only if there is a constant $C > 0$ such that $\|\phi(h)\| \leq C\|h\|$ for all $h \in H$.*

Proof. Assume first that such C exists. If $h = \lim h_n$ in H then from

$$\|\phi(h) - \phi(h_n)\| = \|\phi(h - h_n)\| \leq C\|h - h_n\|$$

it follows that $\phi(h) = \lim \phi(h_n)$. Hence ϕ is continuous.

For the converse, now assume ϕ is continuous. Then the inverse image under ϕ of the open ball $\{h' \in H' : \|h'\| < 1\}$ is open in H. Hence there is $r > 0$ such that each $h_0 \in H$ with $\|h_0\| < r$ satisfies $\|\phi(h_0)\| < 1$. For any $h \neq 0$ in H we have $\|\frac{r}{2\|h\|}h\| = \frac{r}{2} < r$, hence $\|\phi(\frac{r}{2\|h\|}h)\| < 1$. The latter means $\|\phi(h)\| < \frac{2}{r}\|h\|$. Thus $C = \frac{2}{r}$ works. □

Remark 6.5 The map $\|\cdot\| : H \to \mathbb{C}$ is continuous (but not linear). This follows from the inequality $\|h\| - \|h'\| \leq \|h - h'\|$, which in turn follows from the triangle inequality of the associated metric space. (Neither for this, nor for the preceding lemma, do we need the completeness of H. However, completeness is crucial for the following result.)

Proposition 6.6 (Existence of Orthogonal Complements) *Let E be a closed subspace of a Hilbert space H. Let F be the set of all $f \in H$ with $\langle f, e\rangle = 0$ for each $e \in E$. Then F is a closed subspace of H, and each $h \in H$ decomposes uniquely as $h = e + f$ with $e \in E$, $f \in F$. We have $\|h\|^2 = \|e\|^2 + \|f\|^2$.*

Proof. Clearly, F is a linear subspace of H, and $E \cap F = 0$. This implies uniqueness of the decomposition $h = e + f$. Indeed, if $h = e' + f'$ is another such decomposition then $e - e' = f' - f$ lies in $E \cap F = 0$. Hence $e = e'$, $f = f'$.

Fix $h \in H$, and let $\delta = \inf\{\|h - e\| : e \in E\}$. Then there is a sequence (e_n) in E with $\delta = \lim \|h - e_n\|$. We claim (e_n) is a Cauchy sequence. Consider the identity

$$\frac{1}{2}\|e_n - e_m\|^2 = \|h - e_n\|^2 + \|h - e_m\|^2 - 2\left\|h - \frac{1}{2}(e_n + e_m)\right\|^2.$$

Since $\frac{1}{2}(e_n + e_m)$ lies in E we have $\|h - \frac{1}{2}(e_n + e_m)\|^2 \geq \delta^2$, hence

$$\frac{1}{2}\|e_n - e_m\|^2 \leq (\|h - e_n\|^2 - \delta^2) + (\|h - e_m\|^2 - \delta^2).$$

It follows that (e_n) is a Cauchy sequence, hence converges to some $e \in E$. Then for $f := h - e$ we have $\|f\| = \delta$ (using Remark 6.5).

For each $e' \in E$ and $\alpha \neq 0$ in $\mathbb{R}$ we have

$$\delta^2 \leq \|h - (e - \alpha e')\|^2 = \|f + \alpha e'\|^2 = \delta^2 + 2\alpha \operatorname{Re}\langle f, e'\rangle + \alpha^2\|e'\|^2;$$

hence

$$0 \leq \frac{2}{\alpha} \operatorname{Re}\langle f, e'\rangle + \|e'\|^2.$$

For each $e' \neq 0$ there is some α violating this, unless $\operatorname{Re}\langle f, e'\rangle = 0$. Thus we get $\operatorname{Re}\langle f, e'\rangle = 0$, and – replacing e' by $\sqrt{-1}e'$ – also $\operatorname{Im}\langle f, e'\rangle = 0$. Hence $\langle f, e'\rangle = 0$ for all $e' \in E$, that is, $f \in F$. This proves the existence of the decomposition $h = e + f$. Since $\langle e, f\rangle = 0$ we get $\|h\|^2 = \|e\|^2 + \|f\|^2$.

From the existence and uniqueness of the decomposition $h = e + f$ it follows that $H = E \oplus F$ (vector space direct sum). Thus projection $\pi : H \to E$ (sending h to e) is linear, and $\|\pi(h)\| = \|e\| \leq \sqrt{\|e\|^2 + \|f\|^2} = \|h\|$. Hence π is continuous by Lemma 6.4. Thus $F = \pi^{-1}(0)$ is a closed subspace of H. □

So far things behave just as in the case of finite-dimensional Hilbert spaces (familiar from Linear Algebra). This extends to the following.

Definition 6.7 (Direct Sum of Hilbert Spaces) *The direct sum of Hilbert spaces $H_1, \dots, H_s$ is the vector space direct sum $H = H_1 \oplus \cdots \oplus H_s$, equipped with the hermitian form*

$$\langle (h_1, \dots, h_s), (h'_1, \dots, h'_s) \rangle = \sum_{i=1}^{s} \langle h_i, h'_i \rangle.$$

One easily checks that H is again a Hilbert space.

6.1.2 Banach's Theorem

In a metric space, the open (resp., closed) ball with center c and radius r is the set of all elements of distance $<r$ (resp., $\leq r$) from c.

The following lemma is often phrased using the notion of Baire category.

Lemma 6.8 *Let H be a (nonempty) complete metric space, and let U_n ($n \in \mathbb{N}$) be dense open subsets of H. Then $\bigcap_{n=1}^{\infty} U_n$ is nonempty.*

Proof. We construct a sequence of closed balls B_n of radius $r_n > 0$ with $B_n \subset U_n \cap B_{n-1}$, such that $\lim r_n = 0$. Then the centers of the B_n form a Cauchy sequence. The limit of this Cauchy sequence lies in each B_n, thus in each U_n. Hence the claim.

It remains to construct the B_n. Set $U_0 = H$, and let B_0 be any closed ball of positive radius. Now assume $B_0, \dots, B_{n-1}$ ($n \in \mathbb{N}$) have already been constructed, satisfying the desired properties. Let B^0_{n-1} be the open ball with same radius and center as B_{n-1}. Then $U_n \cap B^0_{n-1}$ is open and nonempty, hence it contains an open ball of positive radius $r \leq r_{n-1}$. Let B_n be the closed ball of radius $r_n = r/2$ and same center. Then the B_n are as desired. □

The following is a special case of Banach's theorem.

Theorem 6.9 *Suppose $\phi : H \to H'$ is a continuous bijective linear map between Hilbert spaces H, H'. Then its inverse $\phi^{-1} : H' \to H$ is continuous.*

Proof. Let $B(r) = \{h \in H : \|h\| < r\}$ and $B'(r) = \{h' \in H' : \|h'\| < r\}$ denote the open ball around 0 in H (resp., H') of radius r. Then $c + B(r)$ is the ball around $c \in H$ of radius r. Further, for subsets S of H or H' we let $\overline{S}$ denote the (topological) closure.

From $H = \bigcup_{n=1}^{\infty} B(n)$ we get $H' = \bigcup_{n=1}^{\infty} \phi(B(n))$. Let $C_n = \overline{\phi(B(n))}$, and $U_n = H' \setminus C_n$. Then the U_n are open in H', and their intersection is empty.

Hence the lemma implies that not all U_n are dense in H'. Thus C_n contains an open ball $c' + B'(r')$ for some n and some $r' > 0$. Let $c = \phi^{-1}(c')$. Then

$$B'(r') \subset C_n - c' = \overline{\phi(B(n))} - c' = \overline{\phi(B(n) - c)} \subset \overline{\phi(B(m))}$$

for some $m > 0$ (using Remark 6.3 (ii)). Summarizing, $B'(r') \subset \overline{\phi(B(m))}$. For real $t > 0$ we have $tB(m) = B(tm)$. Together with Remark 6.3 it follows that $B'(tr') \subset \overline{\phi(B(tm))}$. Hence for $r := r'/m$ we get

Claim 1 There is $r > 0$ with $B'(r) \subset \overline{\phi(B(1))}$. More generally, $B'(tr) \subset \overline{\phi(B(t))}$ for each $t > 0$.

We are going to show

Claim 2 $\overline{\phi(B(1))} \subset \phi(B(3))$.

From Claims 1 and 2 we get $\phi^{-1}(B'(r)) \subset B(3)$. As in the proof of Lemma 6.4, this implies that ϕ^{-1} is continuous. Hence the theorem.

It remains to prove Claim 2. Fix some $h' \in \overline{\phi(B(1))}$. We construct a sequence $h_0, h_1, h_2, \ldots$ in H with the following properties:

$$(a)\ \ \|h_i - h_{i+1}\| \le 2^{-i} \qquad \text{and} \qquad (b)\ \ h' \in \overline{\phi(h_i + B(2^{-i}))}$$

for $i = 0, 1, \ldots$. Set $h_0 = 0$. Now assume $h_0, \ldots, h_i$ have already been constructed, satisfying (a) and (b). Let r be as in Claim 1. It follows from (b) that the open ball $h' + B'(r2^{-(i+1)})$ around h' has nonempty intersection with $\phi(h_i + B(2^{-i}))$. Choose an element g' in this intersection, and set $h_{i+1} = \phi^{-1}(g')$.

Since $g' \in \phi(h_i + B(2^{-i}))$ we get $h_{i+1} \in h_i + B(2^{-i})$. Hence (a) holds. Further, $g' \in h' + B'(r2^{-(i+1)}) \subset h' + \overline{\phi(B(2^{-(i+1)}))}$ by Claim 1. Hence $h' \in g' - \overline{\phi(B(2^{-(i+1)}))} = \phi(h_{i+1}) + \overline{\phi(B(2^{-(i+1)}))} = \overline{\phi(h_{i+1} + B(2^{-(i+1)}))}$ (using Remark 6.3). This completes the construction of the sequence (h_i).

For $\nu \ge \mu$ in $\mathbb{N}$ we get $\|h_\mu - h_\nu\| \le \sum_{i=\mu}^{\nu-1} \|h_i - h_{i+1}\| \le \sum_{i=\mu}^{\infty} 2^{-i} = 2^{-\mu+1}$ from (a). Thus (h_i) is a Cauchy sequence, hence converges to some $h \in H$. Since $\|h_\nu\| = \|h_0 - h_\nu\| \le 2$ (by setting $\mu = 0$), we have $\|h\| \le 2$, hence $h \in B(3)$.

Since ϕ is continuous, there is $C > 0$ as in Lemma 6.4. Then $\phi(B(2^{-i})) \subset B'(C2^{-i})$ for all i. Thus (b) yields $h' - \phi(h_i) \in \overline{\phi(B(2^{-i}))} \subset \overline{B'(C2^{-i})}$. Hence $\|h' - \phi(h_i)\| \le C2^{-i}$. Thus $h' = \lim \phi(h_i) = \phi(\lim h_i) = \phi(h)$. Since $h \in B(3)$, this proves Claim 2, and thereby the theorem. □

6.2 The Hilbert Spaces $L^2(D)$

For the rest of this chapter, "function" always means $\mathbb{C}$-valued function.

6.2.1 Square Integrable Functions

Let $D \subset \mathbb{C}$ be open. For a continuous (complex) function f on D, the function $|f|^2 = f\bar{f}$ is continuous, real-valued, and nonnegative on D. We define $\|f\|_D \in \mathbb{R}_+ \cup \{\infty\}$ by

$$\|f\|_D^2 = \int_D |f|^2$$

where the integral is the usual Lebesgue integral in the plane (see, e.g., [Ru]). Here we identify $\mathbb{C}$ with the real plane $\mathbb{R}^2$ in the usual way.

If $\|f\|_D < \infty$ then we say f is *square integrable* (over D). From the basic properties of the integral it follows that the square integrable continuous functions on D form a vector space (with pointwise addition and scalar multiplication). We are mainly interested in the subspace of holomorphic functions.

Definition 6.10 *$L^2(D)$ denotes the vector space of square integrable holomorphic functions on D.*

Clearly, if D has finite area $\int_D 1$ then every bounded holomorphic function on D lies in $L^2(D)$.

Consider again a continuous function h on D, and write it as $h = u + \sqrt{-1}v$ (for real-valued functions u and v). If $\int_D |h| < \infty$ then also $\int_D |u|$ and $\int_D |v|$ are finite (since $|u| \leq |h|$ and $|v| \leq |h|$); then the (Lebesgue) integrals of u and v over D exist (and are real numbers), and we set

$$\int_D h = \int_D u + \sqrt{-1} \int_D v.$$

From the inequality

$$2|a\bar{b}| \leq |a|^2 + |b|^2, \qquad a, b \in \mathbb{C}$$

it follows that for $f, g \in L^2(D)$ the function $h = f\bar{g}$ satisfies $\int_D |h| < \infty$. Hence the integral of $h = f\bar{g}$ over D is defined (and is a complex number). We set

$$\langle f, g \rangle_D = \int_D f\bar{g}.$$

From the basic properties of the integral it follows that $\langle , \rangle_D$ is a positive-definite hermitian form on $L^2(D)$. The associated norm is the above $\|\cdot\|_D$. We call the associated metric on $L^2(D)$ the $\|\cdot\|_D$-metric. We are going to show that it makes $L^2(D)$ into a Hilbert space.

6.2.2 Functions on a Disc

Let $D(a,r) = \{z \in \mathbb{C} : |z-a| < r\}$ denote the open disc in $\mathbb{C}$ with radius r and center a. When we write this, it is understood that $r > 0$ and $a \in \mathbb{C}$.

Lemma 6.11 *Let $D = D(a,r)$. The functions*

$$g_n(z) = (z-a)^n, \qquad n = 0, 1, \ldots$$

lie in $L^2(D)$ and satisfy

$$\langle g_n, g_m \rangle_D = \begin{cases} \pi r^{2n+2}(n+1)^{-1} & \text{if } n = m \\ 0 & \text{if } n \neq m. \end{cases}$$

Proof. The g_n are bounded on D, hence lie in $L^2(D)$. For $n \geq m$ we compute

$$\begin{aligned} \langle g_n, g_m \rangle_D &= \int_D |z-a|^{2m}(z-a)^{n-m} \\ &= \int_0^r \int_0^{2\pi} \rho^{2m+1}(\rho\cos\theta + \sqrt{-1}\rho\sin\theta)^{n-m}\, d\theta\, d\rho \end{aligned}$$

using polar coordinates. If $n > m$ then

$$\begin{aligned} &\int_0^{2\pi} (\cos\theta + \sqrt{-1}\sin\theta)^{n-m}\, d\theta \\ &\quad = \int_0^{2\pi} (\cos(n-m)\theta + \sqrt{-1}\sin(n-m)\theta)\, d\theta \\ &\quad = 0 \end{aligned}$$

because we integrate over full periods. The case $n = m$ is clear. □

Lemma 6.12 *Let $D = D(a,r)$. Let $f \in L^2(D)$, with Taylor expansion $f(z) = \sum_{n=0}^{\infty} c_n(z-a)^n$ on D. If f extends to a holomorphic function on some $D(a,r')$ with $r' > r$ then*

$$\|f\|_D^2 = \sum_{n=0}^{\infty} |c_n|^2 \pi r^{2n+2}(n+1)^{-1}.$$

Proof. Since f is holomorphic on $D' = D(a, r')$, we have $f(z) = \sum_{n=0}^{\infty} c_n (z-a)^n$ for all $z \in D'$, and this series converges uniformly on D. Thus for each $\epsilon > 0$ we have $|f - f_N| \leq \epsilon$ on D for large N, where $f_N(z) = \sum_{n=0}^{N} c_n(z-a)^n$ is the Nth partial sum. Hence $\|f - f_N\|_D^2 = \int_D |f - f_N|^2 \leq \int_D \epsilon^2 = \epsilon^2 r^2 \pi$ for large N.

It follows that (f_N) converges to f in the $\|\cdot\|_D$-metric. Since $\|\cdot\|_D$ is continuous in this topology (Remark 6.5) we get $\|f\|_D^2 = \lim_{N\to\infty} \|f_N\|_D^2 = \lim_{N\to\infty} \|\sum_{n=0}^{N} c_n g_n\|_D^2 = \lim_{N\to\infty} \sum_{n=0}^{N} |c_n|^2 \pi r^{2n+2}(n+1)^{-1}$ by the preceding lemma. Hence the claim. □

The hypothesis that f extends to D' was made only for convenience, it is not necessary for the conclusion.

6.2.3 $L^2(D)$ Is a Hilbert Space

We have used the fact that the integral behaves well under uniform convergence of functions. Now we need to see what happens under pointwise convergence. This is just basic measure theory, but we supply a proof for completeness.

Lemma 6.13 *Let $D \subset \mathbb{C}$ be open.*

(i) Let u and u_n ($n \in \mathbb{N}$) be continuous, nonnegative real-valued functions on D. Assume $u_1 \leq u_2 \leq \cdots$ and $u \leq \sup u_n$ (pointwise, where the supremum can take the value ∞). If there is $M \in \mathbb{R}_+$ with $\int_D u_n \leq M$ for all n, then also $\int_D u \leq M$.

(ii) Let $(f_i)_{i\in\mathbb{N}}$ be a Cauchy sequence in $L^2(D)$ (with respect to the $\|\cdot\|_D$-metric). Suppose (f_i) converges pointwise to some holomorphic function f on D. Then $f \in L^2(D)$, and (f_i) converges to f in the $\|\cdot\|_D$-metric.

Proof. (i) Fix t with $0 < t < 1$. Let $D_n = \{z \in D : u_n(z) > tu(z)\}$. Then the D_n are open subsets of D with $D_1 \subset D_2 \subset \cdots$ and $\bigcup_{n=1}^{\infty} D_n = D$. Then

$$\int_D u = \sup_{n\in\mathbb{N}} \int_{D_n} u \tag{6.1}$$

by the fundamental properties of the Lebesgue integral (See, e.g., [Ru], Th. 1.26). For $z \in D_n$ we have $u(z) \leq t^{-1}u_n(z)$, hence $\int_{D_n} u \leq t^{-1}\int_{D_n} u_n \leq t^{-1}\int_D u_n \leq t^{-1}M$. Then also $\int_D u \leq t^{-1}M$ by (6.1). This proves (i).

(ii) From the definition of Cauchy sequence it follows that there is a subsequence (f_{i_ν}) with $\|f_{i_\nu} - f_{i_{\nu+1}}\|_D \leq 2^{-\nu}$. Fix some $\mu \in \mathbb{N}$. From $f_{i_\mu} - f = \sum_{\nu=\mu}^{\infty}(f_{i_\nu} - f_{i_{\nu+1}})$ we get $|f_{i_\mu} - f| \leq \sum_{\nu=\mu}^{\infty} |f_{i_\nu} - f_{i_{\nu+1}}|$. Consider the partial sums $s_n = \sum_{\nu=\mu}^{n} |f_{i_\nu} - f_{i_{\nu+1}}|$. Set $u = |f_{i_\mu} - f|^2$, $u_n = s_n^2$. Then

$$\left(\int_D u_n\right)^{1/2} = \|s_n\|_D \le \sum_{\nu=\mu}^{n} \|f_{i_\nu} - f_{i_{\nu+1}}\|_D \le \sum_{\nu=\mu}^{n} 2^{-\nu} \le 2^{-\mu+1}.$$

By (i) it follows that $(\int_D u)^{1/2} \le 2^{-\mu+1}$, that is, $\|f_{i_\mu} - f\|_D \le 2^{-\mu+1}$.

Thus $f_{i_\mu} - f \in L^2(D)$, hence $f = f_{i_\mu} - (f_{i_\mu} - f) \in L^2(D)$. Further, the sequence (f_{i_ν}) converges to f in the $\|\cdot\|_D$-metric. Since (f_i) is a Cauchy sequence, (f_i) also converges to f. Hence (ii). □

Proposition 6.14 *Let D be an open subset of* $\mathbb{C}$.

(a) The vector space $L^2(D)$*, equipped with the form* $\langle , \rangle_D$*, is a Hilbert space.*
(b) For each $a \in D$ *and* $n = 0, 1, \ldots$ *the map* $L^2(D) \to \mathbb{C}$, $f \mapsto f^{(n)}(a)$ *(nth derivative at a) is a continuous linear map.*

Proof. Let $f \in L^2(D)$ and $a \in D$. We can choose $0 < r < r'$ such that $D(a, r') \subset D$. Then $\|f\|^2_{D(a,r)} = \sum_{n=0}^{\infty} |c_n|^2 \pi r^{2n+2}(n+1)^{-1}$ by Lemma 6.12, where $c_n = f^{(n)}(a)/n!$. Thus $|c_n|^2 \pi r^{2n+2}(n+1)^{-1} \le \|f\|^2_{D(a,r)} \le \|f\|^2_D$. Hence $|f^{(n)}(a)| \le n!(n+1)^{1/2}\pi^{-1/2}r^{-n-1}\|f\|_D$. This proves (b) (cf. Lemma 6.4).

For the proof of (a), consider a compact subset A of D. Since A is compact, we can choose $r' > 0$ such that $D(a, r') \subset D$ for all $a \in A$. Let again $0 < r < r'$. Then by the preceding paragraph (for $n = 0$) we get

$$\sup_{a \in A} |f(a)| \le \pi^{-1/2} r^{-1} \|f\|_D.$$

Hence if (f_i) is a Cauchy sequence in $L^2(D)$ (relative to the $\|\cdot\|_D$-metric) then the $f_i|_A$ form a Cauchy sequence with respect to the supremum norm.

It follows that (f_i) converges pointwise to some (complex-valued) function f on D, and this convergence is uniform on each compact disc $A \subset D$. Thus f is holomorphic. By the preceding lemma, f lies in $L^2(D)$, and (f_i) converges to f in the $\|\cdot\|_D$-metric. This proves that $L^2(D)$ is complete with respect to the $\|\cdot\|_D$-metric. Hence it is a Hilbert space. □

Remark 6.15 If $D' \subset D \subset \mathbb{C}$ are open, then the restriction map $L^2(D) \to L^2(D')$ is (well defined!) linear and continuous. This follows from Lemma 6.4 (with $C = 1$) and the definitions.

We conclude this section with a lemma for later use.

Lemma 6.16 *Let $D' \subset A \subset D \subset \mathbb{C}$ where D, D' are open and A is compact. Then for each $\epsilon > 0$ there is a closed subspace M of $L^2(D)$ of finite codimension such that*

$$\|f\|_{D'} \le \epsilon \|f\|_D \qquad \textit{for all } f \in M.$$

Proof. Again choose $r' > 0$ such that $D(a, r') \subset D$ for all $a \in A$. Let $0 < r < r'$. Since A is compact, we have $A \subset \cup_{j=1}^{s} D(a_j, r/2)$ for certain $a_1, \ldots, a_s \in A$.

Given $\epsilon > 0$, choose $m \in \mathbb{N}$ with $s2^{-2m-2} \le \epsilon^2$. Let M be the set of all $f \in L^2(D)$ that vanish at each a_j at least of order m. Then M is the intersection of the kernels of the continuous linear maps $L^2(D) \to \mathbb{C}$, $f \mapsto f^{(\mu)}(a_j)$, where $\mu = 0, \ldots, m-1$, $j = 1, \ldots, s$. Each such kernel is a closed subspace of $L^2(D)$ of codimension ≤ 1, hence M is also closed, and of finite codimension.

By Lemma 6.12, for $f \in M$ we have

$$\|f\|^2_{D(a_j,r)} = \sum_{n=m}^{\infty} d_{nj}\, r^{2n+2} \qquad \text{and} \qquad \|f\|^2_{D(a_j,r/2)} = \sum_{n=m}^{\infty} d_{nj}\, r^{2n+2}\, 2^{-2n-2}$$

for certain nonnegative real coefficients d_{nj}. Hence $\|f\|^2_{D(a_j,r/2)} \le 2^{-2m-2} \|f\|^2_{D(a_j,r)} \le 2^{-2m-2} \|f\|^2_D$. Since $D' \subset A \subset \bigcup_{j=1}^{s} D(a_j, r/2)$ we further get $\|f\|^2_{D'} \le \sum_{j=1}^{s} \|f\|^2_{D(a_j,r/2)} \le s2^{-2m-2}\|f\|^2_D \le \epsilon^2 \|f\|^2_D$. □

6.3 Cocycles and Coboundaries

6.3.1 Square Integrable Functions on Coordinate Patches

Let Y be a Riemann surface. Let I be a finite set, and suppose (W_i, z_i) $(i \in I)$ are coordinate maps on Y. For an open set $E \subset W_i$ define $L^2(E, z_i)$ to be the space of all functions f on E with $f \circ z_i^{-1} \in L^2(D)$, where $D = z_i(E)$. We view $L^2(E, z_i)$ as equipped with the hermitian form $\langle f, g\rangle = \langle f \circ z_i^{-1}, g \circ z_i^{-1}\rangle_D$. Thus $L^2(E, z_i)$ is a Hilbert space canonically isomorphic to $L^2(D)$.

The following lemma compares the spaces $L^2(E, z_i)$ and $L^2(E, z_j)$, where $E \subset W_i \cap W_j$ $(i, j \in I)$.

Lemma 6.17 *Suppose E is open and relatively compact in $W_i \cap W_j$. Then there is $\kappa > 0$ such that for each holomorphic function f on E we have*

$$\left\|f \circ z_i^{-1}\right\|_{z_i(E)} \le \kappa \left\|f \circ z_j^{-1}\right\|_{z_j(E)}$$

Proof. Set $D_i = z_i(E)$ and $D_j = z_j(E)$. The transformation theorem for Lebesgue integrals (e.g., [Ru], Th. 8.27) yields

$$\|f \circ z_i^{-1}\|_{D_i}^2 = \int_{D_i} |f \circ z_i^{-1}|^2 = \int_{D_j} |f \circ z_j^{-1}|^2 J$$

where J is the Jacobian of the analytic isomorphism $z_i \circ z_j^{-1} : D_j \to D_i$. (Recall $J = |d(z_i \circ z_j^{-1})/dz|^2$.) Since J is actually defined (and continuous) on $z_j(W_i \cap W_j)$, it is bounded on the relatively compact subset $z_j(E) = D_j$. Now choose a constant $\kappa > 0$ such that $J \leq \kappa^2$ on D_j. Then the claim holds. □

By symmetry of i and j, it follows that $f \circ z_i^{-1}$ is square integrable over $z_i(E)$ if and only if $f \circ z_j^{-1}$ is square integrable over $z_j(E)$. In this case we just say f is square integrable over E.

In other words, $L^2(E, z_i)$ and $L^2(E, z_j)$ coincide as sets, even as topological spaces (by Lemma 6.4), but not (necessarily) as metric spaces. Hence, in situations where only topological aspects are involved, we will just write $L^2(E)$ for any $L^2(E, z_i)$ with E relatively compact in W_i.

6.3.2 Cocycles

Now let $\mathcal{U} = (U_i)_{i \in I}$ be a family of open sets $U_i \subset W_i$, such that U_i is relatively compact in W_i. Define

$$C^0(\mathcal{U}) = \bigoplus_{i \in I} L^2(U_i, z_i) \qquad \text{and} \qquad C^1(\mathcal{U}) = \bigoplus_{(i,j) \in I \times I} L^2(U_i \cap U_j, z_i),$$

the direct sum of Hilbert spaces (see Definition 6.7). We write the elements of $C^0(\mathcal{U})$ and $C^1(\mathcal{U})$ as tuples $(f_i)_{i \in I}$ and $(f_{ij})_{i,j \in I}$, respectively. Fix $i, j, k \in I$ (not necessarily distinct), and consider the map

$$C^1(\mathcal{U}) \to L^2(U_i \cap U_j \cap U_k), \quad (f_{\nu\mu}) \mapsto (f_{ij} - f_{ik} - f_{kj})|_{U_i \cap U_j \cap U_k}.$$

This map is continuous since it is composed of the continuous maps projection $(f_{\nu\mu}) \mapsto f_{ij}$, $(f_{\nu\mu}) \mapsto f_{ik}$, and $(f_{\nu\mu}) \mapsto f_{kj}$, followed by restriction to $U_i \cap U_j \cap U_k$ (Remark 6.15). (Note that we use the above convention about the notation $L^2(E)$.)

Now let i, j, k vary over all triples from I, and let $Z^1(\mathcal{U})$ denote the intersection of the kernels of the corresponding maps. Thus $Z^1(\mathcal{U})$ is a closed subspace of $C^1(\mathcal{U})$, consisting of the $(f_{\nu\mu})$ with

$$f_{ij} = f_{ik} + f_{kj} \qquad \text{for all} \qquad i, j, k \in I.$$

This is called the *cocycle relation*, and the elements of $Z^1(\mathcal{U})$ are called the (square integrable) cocycles. Note that $f_{ii} = 0$ and $f_{ji} = -f_{ij}$.

6.3.3 The Coboundary Map

Define

$$\partial : C^0(\mathcal{U}) \to Z^1(\mathcal{U}), \qquad (g_i)_{i\in I} \mapsto (f_{ij})_{i,j\in I} \qquad \text{where}$$
$$f_{ij} = (g_j - g_i)|_{U_i \cap U_j}.$$

The f_{ij} satisfy the cocycle relation, hence this map is well defined. It is called the *coboundary map*. It is a continuous linear map. To check continuity, it suffices to show that the map $C^0(\mathcal{U}) \to L^2(U_i \cap U_j)$, $(g_\nu) \mapsto (g_j - g_i)|_{U_i \cap U_j}$ is continuous for all i, j. This follows as in the preceding subsection.

6.4 Cocycles on a Disc

The language of cocycles and coboundaries will be used for the patching of locally defined functions. We need more preparation.

6.4.1 Dolbeault's Lemma

This subsection is the (only) place where we need some classical analysis. We write $i = \sqrt{-1}$ (departing from our general convention) to shorten formulas.

Let φ be a (complex-valued) function on an open subset of $\mathbb{C}$. Identifying $\mathbb{C}$ with $\mathbb{R}^2$ as before, via the coordinates $x = \mathrm{Re}(z)$, $y = \mathrm{Im}(z)$, we can consider the partial derivatives $\partial\varphi/\partial x$ and $\partial\varphi/\partial y$. We say φ is *differentiable* (or C^∞) if all iterated partials exist and are continuous. For such φ we define

$$\frac{\partial\varphi}{\partial\bar{z}} = \frac{1}{2}\left(\frac{\partial\varphi}{\partial x} + i\,\frac{\partial\varphi}{\partial y}\right)$$

(one of the partials of the Wirtinger calculus). Recall that a differentiable function φ is holomorphic if and only if $\partial\varphi/\partial\bar{z} = 0$. The latter is called the (homogeneous) Cauchy–Riemann differential equation. The goal of this section is to solve the corresponding inhomogeneous equation on a disc.

Let $\mathbb{D} = \{z \in \mathbb{C} : |z| < 1\}$ be the unit disc.

Lemma 6.18 *Let φ be a differentiable function on $\mathbb{D}$ with compact support (that is, vanishing outside a compact set). Then there is a differentiable function ψ on $\mathbb{D}$ with*

$$\frac{\partial \psi}{\partial \bar{z}} = \varphi.$$

Proof. We can extend φ to a differentiable function on $\mathbb{C}$ vanishing outside $\mathbb{D}$. Then ψ can be written down explicitly as

$$\psi(z) = -\frac{1}{\pi} \int_0^{2\pi} \int_0^2 \varphi(z + re^{i\theta})\, e^{-i\theta}\, dr\, d\theta, \qquad z \in \mathbb{D}.$$

Set $\varphi_1 = \partial\varphi/\partial x$, $\varphi_2 = \partial\varphi/\partial y$. By standard theorems, ψ is differentiable and differentiating commutes with the integral sign, that is,

$$\begin{aligned} -2\pi \frac{\partial \psi}{\partial \bar{z}}(z) &= \int_0^{2\pi} \int_0^2 [\varphi_1(z + re^{i\theta}) + i\, \varphi_2(z + re^{i\theta})] e^{-i\theta}\, dr\, d\theta \\ &= \int_0^{2\pi} \int_0^2 [\varphi_1(z + re^{i\theta}) \cos\theta + \varphi_2(z + re^{i\theta}) \sin\theta]\, dr\, d\theta \\ &\quad + i \int_0^{2\pi} \int_0^2 [-\varphi_1(z + re^{i\theta}) \sin\theta + \varphi_2(z + re^{i\theta}) \cos\theta]\, dr\, d\theta \end{aligned}$$

where we used $e^{-i\theta} = \cos\theta - i \sin\theta$. Using $\mathrm{Re}(z + re^{i\theta}) = \mathrm{Re}(z) + r\cos\theta$ and $\mathrm{Im}(z + re^{i\theta}) = \mathrm{Im}(z) + r\sin\theta$, as well as the chain rule for functions of two variables, we continue as

$$\begin{aligned} -2\pi \frac{\partial \psi}{\partial \bar{z}}(z) &= \int_0^{2\pi} \int_0^2 \frac{d\varphi(z + re^{i\theta})}{dr}\, dr\, d\theta \\ &\quad + i \int_0^{2\pi} \int_0^2 r^{-1} \frac{d\varphi(z + re^{i\theta})}{d\theta}\, dr\, d\theta \\ &= \int_0^{2\pi} [\varphi(z + 2e^{i\theta}) - \varphi(z + 0e^{i\theta})]\, d\theta \\ &\quad + i \int_0^2 r^{-1} [\varphi(z + re^{i2\pi}) - \varphi(z + re^{i0})]\, dr \\ &= \int_0^{2\pi} [-\varphi(z)]\, d\theta = -2\pi\varphi(z), \end{aligned}$$

as desired. (Note that if $z \in \mathbb{D}$ then $z + 2e^{i\theta}$ does not lie in $\mathbb{D}$, hence $\varphi(z + 2e^{i\theta}) = 0$.) □

Lemma 6.19 *Let $\mathbb{D}'$ be an open disc around 0 of radius <1. Then for each differentiable function φ on $\mathbb{D}$ there is a bounded differentiable function ψ on $\mathbb{D}'$ with*

$$\frac{\partial \psi}{\partial \bar{z}} = \varphi$$

on $\mathbb{D}'$.

Proof. Let $\mathbb{D}''$ be another disc around 0, strictly between $\mathbb{D}'$ and $\mathbb{D}$. Then there is a differentiable function χ on $\mathbb{D}$ which is identically 1 on $\mathbb{D}'$ and vanishes outside $\mathbb{D}''$. (Such a function can be constructed using Remark 6.21 below.)

Then the function $\varphi' = \varphi \cdot \chi$ has compact support, hence the preceding lemma yields ψ with $\partial\psi/\partial\bar{z} = \varphi'$. This ψ is bounded on $\mathbb{D}'$ (since continuous on all of $\mathbb{D}$). The restriction of ψ to $\mathbb{D}'$ solves our problem, since φ and φ' coincide on $\mathbb{D}'$. □

6.4.2 Cocycles on a Disc

Again $\mathbb{D}$ denotes the unit disc, and $\mathbb{D}'$ is an open disc around 0 of radius <1.

Proposition 6.20 *Let I be a finite set. Suppose $\mathbb{D} = \bigcup_{i \in I} X_i$, the union of open subsets X_i. Assume there is an associated partition of unity $(\chi_i)_{i \in I}$, that is, the χ_i are differentiable nonnegative real-valued functions on $\mathbb{D}$ with $\sum_{i \in I} \chi_i = 1$, such that χ_i vanishes on an open subset $\tilde{X}_i$ of $\mathbb{D}$ with $\tilde{X}_i \cup X_i = \mathbb{D}$. Suppose further we are given holomorphic functions f_{ij} on $X_i \cap X_j$ $(i, j \in I)$, satisfying the cocycle relation*

$$f_{ij} = f_{ik} + f_{kj} \qquad \text{for all} \qquad i, j, k \in I.$$

Set $X_i' = X_i \cap \mathbb{D}'$. Assume each f_{ij} is square integrable over $X_i' \cap X_j'$. Then there are functions $g_i \in L^2(X_i')$ with

$$f_{ij} = g_i - g_j \qquad \text{on} \qquad X_i' \cap X_j'$$

for all $i, j \in I$.

Proof. Extend each f_{ik} to a (not necessarily continuous) function on $\mathbb{D}$ by setting it zero outside $X_i \cap X_k$. Then the function $\chi_k f_{ik}$ is differentiable on X_i.

Indeed, X_i is the union of its open subsets $X_i \cap \tilde{X}_k$ and $X_i \cap X_k$, and $\chi_k f_{ik}$ is differentiable on both subsets.

Then $\psi_i = \sum_{k \in I} \chi_k f_{ik}$ is a differentiable function on X_i. Since each f_{ik} is square integrable over $X_i' \cap X_k'$, we have $\chi_k f_{ik}$ square integrable over X_i' (since $0 \le \chi_k \le 1$), hence ψ_i is square integrable over X_i'. On $X_i \cap X_j$ we have

$$\psi_i - \psi_j = \sum_{k \in I} \chi_k (f_{ik} - f_{jk}) = \sum_{k \in I} \chi_k f_{ij} = f_{ij} \sum_{k \in I} \chi_k = f_{ij}.$$

This is as desired, only the ψ_i may not be holomorphic. Since f_{ij} is holomorphic on $X_i \cap X_j$, we get $\partial \psi_i / \partial \bar{z} = \partial \psi_j / \partial \bar{z}$ on $X_i \cap X_j$. Hence there is a globally defined (differentiable) function φ on $\mathbb{D}$ with $\varphi = \partial \psi_i / \partial \bar{z}$ on X_i. By the preceding lemma there is a bounded differentiable function ψ on $\mathbb{D}'$ with $\partial \psi / \partial \bar{z} = \varphi$ on $\mathbb{D}'$.

Since ψ_i is square integrable over X_i', the same holds for $g_i := \psi_i - \psi$. Further $\partial g_i / \partial \bar{z} = \partial \psi_i / \partial \bar{z} - \partial \psi / \partial \bar{z} = 0$ on X_i', hence g_i is holomorphic on X_i'. Finally,

$$g_i - g_j = \psi_i - \psi_j = f_{ij}$$

on $X_i' \cap X_j'$. □

Remark 6.21 The following function will be useful in constructing partitions of unity:

$$\chi(z) = \begin{cases} e^{\frac{1}{|z|^2 - R^2}} & \text{for } |z| < R \\ 0 & \text{for } |z| \ge R. \end{cases}$$

This is a differentiable nonnegative real-valued function on $\mathbb{C}$ with $\chi(z) \ne 0$ if and only if $|z| < R$.

6.5 A Finiteness Theorem

Now we embark on the actual proof of Riemann's existence theorem. Let Y be a compact Riemann surface and $a_0 \in Y$. Each point $a \in Y$ has a coordinate neighborhood (W_a, z_a), where z_a maps W_a onto an open disc in $\mathbb{C}$. We can assume $a_0 \notin W_a$ for $a \ne a_0$ (because Y is Hausdorff). Since Y is compact, it is the union of finitely many such W_a, say $Y = \bigcup_{i \in I} W_{a_i}$, where I is a finite index set. Write $W_i = W_{a_i}$ and $z_i = z_{a_i}$ from now on. Translating and dilatating the disc $z_i(W_i)$ we may assume $z_i(W_i) = \mathbb{D}$ (the unit disc). Further, we may assume $I = \{0, \dots, s\}$, such that the given point a_0 lies in W_0. Then $a_0 \notin W_i$ for $i \ne 0$ (by construction).

There is $0 < r < 1$ such that the $U_i = \{a \in W_i : |z_i(a)| < r\}$ still cover Y. Indeed, Y is the union of the open subsets $Y_n = \bigcup_{i \in I} \{a \in W_i : |z_i(a)| < 1 - \frac{1}{n}\}$ for $n = 1, 2, \ldots$. Since Y is compact, it is the union of finitely many such Y_n, that is, equals some Y_n.

Summarizing, the U_i and W_i ($i \in I$) are open subsets of Y with $U_i \subset W_i$ and $Y = \bigcup_{i \in I} U_i$. The coordinate map z_i maps U_i and W_i onto the open disc around 0 of radius r and 1, respectively.

Set $\mathcal{U} = (U_i)_{i \in I}$. We study the Hilbert spaces $Z^1(\mathcal{U})$, $C^0(\mathcal{U})$, etc., from Section 6.3. The main goal of this section is to prove:

Theorem 6.22 *The image of the coboundary map* $\partial : C^0(\mathcal{U}) \to Z^1(\mathcal{U})$ *is of finite codimension in* $Z^1(\mathcal{U})$.

First we show how this implies Riemann's existence theorem.

6.5.1 The Patching Process

Corollary 6.23 *Let* Y *be a compact Riemann surface and* $a_0 \in Y$. *Then there exists a meromorphic function on* Y *that has a pole at* a_0 *and is holomorphic on* $Y \setminus \{a_0\}$.

Proof. We use the notation introduced at the beginning of Section 6.5. Further, let $z = z_0 - z_0(a_0)$, a coordinate map on W_0 with $z(a_0) = 0$. For $\nu = 1, 2, \ldots$ define elements $\xi_\nu \in Z^1(\mathcal{U})$ as follows: Let $\xi_\nu = (f_{ij}^{(\nu)})_{i,j \in I}$ where $f_{ij}^{(\nu)} = 0$ for $i, j \neq 0$; further, $f_{00}^{(\nu)} = 0$ and

$$f_{0j}^{(\nu)} = -f_{j0}^{(\nu)} = z^{-\nu} \qquad \text{on} \qquad U_0 \cap U_j$$

for $j \neq 0$. The function $z^{-\nu}$ is defined and continuous on $W_0 \cap W_j$, since $a_0 \notin W_j$ for $j \neq 0$. Hence $z^{-\nu}$ is bounded on $U_0 \cap U_j$ (since $U_0 \cap U_j$ is relatively compact in $W_0 \cap W_j$). Thus all $f_{ij}^{(\nu)}$ are square integrable over $U_i \cap U_j$. They satisfy the cocycle relation, hence we have actually defined elements $\xi_\nu \in Z^1(\mathcal{U})$, $\nu = 1, 2, \ldots$.

By Theorem 6.22, the ξ_ν are linearly dependent modulo the image of ∂. Hence there are $c_1, \ldots, c_n \in \mathbb{C}$, not all zero, with $\sum_{\nu=1}^n c_\nu \xi_\nu = \partial\eta$ for some $\eta \in C^0(\mathcal{U})$. Let $\eta = (h_i)_{i \in I}$. Now define a function $f : Y \setminus \{a_0\} \to \mathbb{C}$ by

$$f(u) = \begin{cases} h_j(u) & \text{for } u \in U_j,\ j \neq 0 \\ h_0(u) + \sum_{\nu=1}^n c_\nu z^{-\nu} & \text{for } u \in U_0. \end{cases}$$

If this is well defined then clearly f is holomorphic on $Y \setminus \{a_0\}$ and has a pole at a_0 (since not all c_ν are zero). Thus it only remains to check that f is well defined. From the definition of the coboundary map we get on $U_i \cap U_j$

$$h_j - h_i = \sum_{\nu=1}^{n} c_\nu f_{ij}^{(\nu)} = \begin{cases} 0 & \text{for } i, j \neq 0 \\ \sum_{\nu=1}^{n} c_\nu z^{-\nu} & \text{for } i = 0, j \neq 0. \end{cases}$$

This shows that f is well defined. □

Proof of Theorem 6.1 (RET – Analytic Version). Let $a_1, \ldots, a_n$ be distinct points of Y, and $c_1, \ldots, c_n \in \mathbb{C}$. By the previous Corollary, there is a meromorphic function $f^{(i)}$ on Y that has a pole at a_i and is holomorphic on $Y \setminus \{a_i\}$, for each $i = 1, \ldots, n$. Define meromorphic functions

$$g^{(ij)} = \frac{f^{(i)} - f^{(i)}(a_j)}{f^{(i)} - f^{(i)}(a_j) + d_{ij}}$$

for $i, j = 1, \ldots, n, i \neq j$, where the d_{ij} are certain nonzero constants. Then $g^{(ij)}(a_i) = 1$, and $g^{(ij)}(a_j) = 0$. For suitable choice of the d_{ij} we have $g^{(ij)}(a_k) \neq \infty$ for all k. Hence the

$$h^{(i)} = \prod_{j \neq i} g^{(ij)}$$

satisfy $h^{(i)}(a_k) = \delta_{ik}$ (Kronecker delta). It follows that $f = \sum_{i=1}^{n} c_i h^{(i)}$ has the desired property that $f(a_i) = c_i$. □

6.5.2 Restriction $Z^1(\mathcal{V}) \to Z^1(\mathcal{U})$

This subsection contains some more preparations for the proof of Theorem 6.22. Recall the coordinate maps $z_i : W_i \to \mathbb{D}$, mapping U_i to the disc with radius $r < 1$. Now choose r' with $r < r' < 1$, and let $V_i = \{a \in W_i : |z_i(a)| < r'\}$. Then U_i, V_i and W_i $(i \in I)$ are open subsets of Y with $U_i \subset V_i \subset W_i$ and $Y = \bigcup_{i \in I} U_i$. Moreover, U_i (resp., V_i) is relatively compact in V_i (resp., W_i). Set $\mathcal{V} = (V_i)_{i \in I}$.

Combining the restriction maps $L^2(V_i \cap V_j) \to L^2(U_i \cap U_j)$ (cf. Remark 6.15 and the conventions from section 6.3.1) we obtain a continuous linear map $Z^1(\mathcal{V}) \to Z^1(\mathcal{U})$ (sending (g_{ij}) to $(g_{ij}|_{U_i \cap U_j})$). We denote this map by $\theta \mapsto \theta|_{\mathcal{U}}$.

Proposition 6.24 *For each $\xi \in Z^1(\mathcal{U})$ there is $\theta \in Z^1(\mathcal{V})$ and $\eta \in C^0(\mathcal{U})$ with*

$$\xi = \theta|_{\mathcal{U}} + \partial\eta$$

where ∂ is the coboundary map.

For the proof we need

Lemma 6.25 *There is a partition of unity on Y associated with the U_i's, that is, there are differentiable nonnegative real-valued functions χ_i on Y with $\sum_{i\in I} \chi_i = 1$, such that χ_i vanishes on an open subset C_i of Y with $C_i \cup U_i = Y$.*

Proof. As above we see there is r'' with $0 < r'' < r$ such that the $U_i'' = \{a \in U_i : |z_i(a)| < r''\}$ still cover Y. Let $\bar{U}_i'' = \{a \in U_i : |z_i(a)| \leq r''\}$, and $C_i = Y \setminus \bar{U}_i''$. Then C_i is open, and $C_i \cup U_i = Y$.

By Remark 6.21 there are differentiable nonnegative real-valued functions χ_i' on U_i, vanishing exactly on $U_i \setminus U_i''$. Thus we can extend χ_i' to a differentiable function on Y by setting it zero on C_i. Since the U_i'' cover Y, the function $\chi = \sum_{i\in I} \chi_i'$ is strictly positive. Then the $\chi_i = \chi_i'/\chi$ are as desired. □

Proof of the Proposition. Let $\xi = (f_{ij})_{i,j\in I} \in Z^1(\mathcal{U})$. For fixed $\alpha \in I$ the $U_i \cap W_\alpha$ form an open cover of W_α, and there is a partition of unity associated with this cover (restriction to W_α of the χ_i from the Lemma). Since z_α identifies W_α with $\mathbb{D}$, and V_α with the disc $\mathbb{D}'$ of radius $r' < 1$, it follows from Proposition 6.20 that there are $g_{\alpha i} \in L^2(U_i \cap V_\alpha, z_\alpha)$ with

$$f_{ij} = g_{\alpha i} - g_{\alpha j} \qquad \text{on} \qquad U_i \cap U_j \cap V_\alpha \tag{6.2}$$

for all $i, j \in I$. Now we vary α. For all $i, j, \alpha, \beta \in I$ we have on $U_i \cap U_j \cap V_\alpha \cap V_\beta$:

$$g_{\alpha i} - g_{\alpha j} = f_{ij} = g_{\beta i} - g_{\beta j}$$

hence

$$g_{\beta i} - g_{\alpha i} = g_{\beta j} - g_{\alpha j}.$$

Thus the expressions on either side do not depend on i and j, respectively; that is, there is a holomorphic function $F_{\alpha\beta}$ on $V_\alpha \cap V_\beta$ with

$$F_{\alpha\beta} = g_{\beta i} - g_{\alpha i} \qquad \text{on} \qquad U_i \cap V_\alpha \cap V_\beta. \tag{6.3}$$

Since $g_{\beta i}$ and $g_{\alpha i}$ are square integrable over $U_i \cap V_\alpha \cap V_\beta$ (recall our conventions about this notation from Section 6.3.1), the same holds for $F_{\alpha\beta}$. Then $F_{\alpha\beta}$ is also square integrable over $V_\alpha \cap V_\beta$ (since $V_\alpha \cap V_\beta$ is covered by the finitely many pieces $U_i \cap V_\alpha \cap V_\beta$). Thus $F_{\alpha\beta} \in L^2(V_\alpha \cap V_\beta)$. From the definition of the $F_{\alpha\beta}$ it is clear that they satisfy the cocycle relation, hence $\theta = (F_{\alpha\beta})$ is an element of $Z^1(\mathcal{V})$.

Set $h_\alpha := g_{\alpha\alpha} \in L^2(U_\alpha \cap V_\alpha) = L^2(U_\alpha)$. Let $\eta = (h_\alpha) \in C^0(\mathcal{U})$. Replacing α by β in (6.2), and then setting $i = \alpha$, $j = \beta$ we get

$$f_{\alpha\beta} = g_{\beta\alpha} - g_{\beta\beta} \qquad \text{on} \qquad U_\alpha \cap U_\beta.$$

Setting $i = \alpha$ in (6.3) we get

$$F_{\alpha\beta} = g_{\beta\alpha} - g_{\alpha\alpha} \qquad \text{on} \qquad U_\alpha \cap U_\beta.$$

Thus

$$F_{\alpha\beta} - f_{\alpha\beta} = g_{\beta\beta} - g_{\alpha\alpha} = h_\beta - h_\alpha \qquad \text{on} \qquad U_\alpha \cap U_\beta$$

which means that $\theta|_{\mathcal{U}} - \xi = \partial\eta$. □

We denote the norm on the Hilbert spaces $Z^1(\mathcal{U})$ and $C^0(\mathcal{U})$ by $\|\cdot\|_{\mathcal{U}}$, and on $Z^1(\mathcal{V})$ and $C^0(\mathcal{V})$ by $\|\cdot\|_{\mathcal{V}}$. (This distinguishes between $\|\theta\|_{\mathcal{V}}$ and $\|\theta|_{\mathcal{U}}\|_{\mathcal{U}}$, but keeps notation reasonably simple.)

Corollary 6.26 *There is a constant $C > 0$ with the following property: For each $\xi \in Z^1(\mathcal{U})$ there is $\theta \in Z^1(\mathcal{V})$ and $\eta \in C^0(\mathcal{U})$ with*

$$\xi = \theta|_{\mathcal{U}} + \partial\eta$$

and

$$\max(\|\theta\|_{\mathcal{V}}, \|\eta\|_{\mathcal{U}}) \leq C\|\xi\|_{\mathcal{U}}.$$

Proof. Let H_0 be the direct sum of the Hilbert spaces $Z^1(\mathcal{V})$ and $C^0(\mathcal{U})$. The map $\pi_0 : H_0 \to Z^1(\mathcal{U})$, $(\theta, \eta) \mapsto \theta|_{\mathcal{U}} + \partial\eta$ is a continuous linear map by Section 6.3.3 and the remarks at the beginning of this subsection. The map is surjective by the Proposition. Let H be the orthogonal complement in H_0 of the kernel of π_0 (see Proposition 6.6). Then H is a closed subspace of H_0, hence is itself a Hilbert space. The map π_0 restricts to a bijective continuous linear map $\pi : H \to Z^1(\mathcal{U})$.

By Banach's theorem (Theorem 6.9) it follows that $\pi^{-1} : Z^1(\mathcal{U}) \to H$ is continuous. Thus there is $C > 0$ with $\|\pi^{-1}(\xi)\| \le C\|\xi\|_{\mathcal{U}}$ for all $\xi \in Z^1(\mathcal{U})$ (Lemma 6.4). Finally, writing $\pi^{-1}(\xi) = (\theta, \eta)$ we get $\|\pi^{-1}(\xi)\| = \sqrt{\|\theta\|^2_{\mathcal{V}} + \|\eta\|^2_{\mathcal{U}}} \ge \max(\|\theta\|_{\mathcal{V}}, \|\eta\|_{\mathcal{U}})$. This proves the claim. □

6.5.3 *Proof of the Finiteness Theorem*

Let C be the constant from the above Corollary, and let $\epsilon = (2C)^{-1}$. For each $i, j \in I$ we have $U_i \cap U_j$ relatively compact in $V_i \cap V_j$ (since U_i is relatively compact in V_i). Thus by Lemma 6.16 there is a closed subspace M_{ij} of $L^2(V_i \cap V_j, z_i)$ of finite codimension such that for all $g \in M_{ij}$ we have $\|g \circ z_i^{-1}\|_{z_i(U_i \cap U_j)} \le \epsilon \, \|g \circ z_i^{-1}\|_{z_i(V_i \cap V_j)}$. Let M' be the direct sum of the M_{ij}, a closed subspace of $C^1(\mathcal{V})$ of finite codimension. Let $M = M' \cap Z^1(\mathcal{V})$, a closed subspace of $Z^1(\mathcal{V})$ of finite codimension. For each $\vartheta = (g_{ij}) \in M$ we have

$$\|\vartheta|_{\mathcal{U}}\|_{\mathcal{U}} = \sqrt{\sum_{i,j \in I} \left\|g_{ij} \circ z_i^{-1}\right\|^2_{z_i(U_i \cap U_j)}} \le \sqrt{\sum_{i,j \in I} \epsilon^2 \left\|g_{ij} \circ z_i^{-1}\right\|^2_{z_i(V_i \cap V_j)}}$$
$$= \epsilon \, \|\vartheta\|_{\mathcal{V}}.$$

Let F be the orthogonal complement of M in the Hilbert space $Z^1(\mathcal{V})$ (cf. Proposition 6.6). Then F is a finite-dimensional closed subspace of $Z^1(\mathcal{V})$. Let $\pi_M : Z^1(\mathcal{V}) \to M$ and $\pi_F : Z^1(\mathcal{V}) \to F$ be the associated projections. They satisfy $\|\pi_M(\theta)\|_{\mathcal{V}} \le \|\theta\|_{\mathcal{V}}$ and $\|\pi_F(\theta)\|_{\mathcal{V}} \le \|\theta\|_{\mathcal{V}}$ for all $\theta \in Z^1(\mathcal{V})$.

Now fix some $\xi \in Z^1(\mathcal{U})$. We construct sequences $\xi_1, \xi_2, \ldots$ in $Z^1(\mathcal{U})$, $\varphi_1, \varphi_2, \ldots$ in F and $\eta_1, \eta_2, \ldots$ in $C^0(\mathcal{U})$ as follows. Set $\xi_1 = \xi$. Given ξ_n, write it in the form $\xi_n = \theta_n|_{\mathcal{U}} + \partial\eta_n$ as in the above corollary. This defines η_n and $\varphi_n := \pi_F(\theta_n)$. Set $\xi_{n+1} = (\pi_M(\theta_n))|_{\mathcal{U}}$. This concludes the construction of the above sequences. They satisfy

$$\xi_n = \xi_{n+1} + \varphi_n|_{\mathcal{U}} + \partial\eta_n \tag{6.4}$$

for each n.

The corollary yields

$$\max(\|\theta_n\|_{\mathcal{V}}, \|\eta_n\|_{\mathcal{U}}) \le C\|\xi_n\|_{\mathcal{U}}.$$

Thus $\|\xi_{n+1}\|_{\mathcal{U}} = \|(\pi_M(\theta_n))|_{\mathcal{U}}\|_{\mathcal{U}} \le \epsilon\|\pi_M(\theta_n)\|_{\mathcal{V}} \le \epsilon\|\theta_n\|_{\mathcal{V}} \le C\epsilon\|\xi_n\|_{\mathcal{U}} = \frac{1}{2}\|\xi_n\|_{\mathcal{U}}$. Summarizing, $\|\xi_{n+1}\|_{\mathcal{U}} \le \frac{1}{2}\|\xi_n\|_{\mathcal{U}}$. Hence for each $n \in \mathbb{N}$ we get $\|\xi_n\|_{\mathcal{U}} \le 2^{-n+1}c_1$, where $c_1 := \|\xi_1\|_{\mathcal{U}}$. Then $\max(\|\theta_n\|_{\mathcal{V}}, \|\eta_n\|_{\mathcal{U}}) \le 2^{-n+1}c_1C$. Then also $\|\varphi_n\|_{\mathcal{V}} \le 2^{-n+1}c_1C$.

Adding the equations (6.4) for $n = 1, \ldots, N$ yields

$$\xi_1 = \xi_{N+1} + \left(\sum_{n=1}^{N} \varphi_n\right)\Bigg|_{\mathcal{U}} + \partial\left(\sum_{n=1}^{N} \eta_n\right).$$

From the bounds in the preceding paragraph it follows that the partial sums $\sum_{n=1}^{N} \varphi_n$ and $\sum_{n=1}^{N} \eta_n$ form Cauchy sequences, hence they converge to some $\varphi \in F$ and $\eta \in C^0(\mathcal{U})$, respectively. Clearly $\lim_{N\mapsto\infty}\xi_{N+1} = 0$. Combining all this we get

$$\xi = \varphi|_{\mathcal{U}} + \partial\eta.$$

Since $\xi \in Z^1(\mathcal{U})$ was arbitrary, and φ lies in the finite-dimensional space F, this completes the proof of Theorem 6.22. □

Part Two

Further Directions

7
The Descent from $\mathbb{C}$ to $\bar{k}$

In this chapter we prove that a finite extension of $\mathbb{C}(x)$ can be described by an equation whose coefficients are algebraic over the field generated by the branch points. Thus the algebraic version of RET (Theorem 2.13) remains true if $\mathbb{C}$ is replaced by any algebraically closed subfield k. The most interesting case is of course $k = \bar{\mathbb{Q}}$. Applications include the solvability of all embedding problems over $k(x)$, as well as the existence of a unique minimal field of definition for each FG-extension of $\mathbb{C}(x)$ whose Galois group has trivial center.

7.1 Extensions of $\mathbb{C}(x)$ Unramified Outside a Given Finite Set

Lemma 7.1 *Let $s, n \in \mathbb{N}$. There exists a finite group $H = H_{n,s}$ with generators $h_1, \dots, h_s$ satisfying:*

1. *For any group G of order $\leq n$, and $g_1, \dots, g_s \in G$, there is a homomorphism $H \to G$ sending h_i to g_i (for $i = 1, \dots, s$).*
2. *The intersection of all normal subgroups of H of index $\leq n$ is trivial.*
3. *If $h'_1, \dots, h'_s$ are generators of a group H' satisfying (2) then there is a surjective homomorphism $H \to H'$ sending h_i to h'_i.*
4. *If $h'_1, \dots, h'_s$ are any s elements generating H then there is an automorphism of H sending h_i to h'_i.*

Proof. Let $\mathcal{F}_s$ be the free group on s generators. First note that $\mathcal{F}_s$ has only finitely many normal subgroups of index $\leq n$. Indeed, each such subgroup is the kernel of a homomorphism λ from $\mathcal{F}_s$ to a group G of order $\leq n$, and there are only finitely many such G and λ (since λ is determined by its values on the generators).

Now the group $H = H_{n,s}$ can be constructed as the quotient of $\mathcal{F}_s$ by the intersection of all normal subgroups of index $\leq n$. It is then clear that (1), (2), and (3) hold. In particular, H is finite (by the preceding paragraph).

For (4) note that there is a surjective homomorphism $H \to H$ sending h_i to h_i' (by (3)). It is an automorphism since H is finite. □

Remark 7.2 The group $H_{n,s}$ is unique up to isomorphism, by (3). The conjugacy classes of $h_1, \ldots, h_s$, together with the class of $(h_1 \cdots h_s)^{-1}$, form a weakly rigid tuple in $H_{n,s}$ (by (4)).

Suppose $L/\mathbb{C}(x)$ is the composite of the FG-extensions $L_1/\mathbb{C}(x)$ and $L_2/\mathbb{C}(x)$. Then the branch point set of L is the union of that of L_1 and L_2. This follows from Proposition 2.6(d), because an element of $G(L/\mathbb{C}(x))$ is trivial if and only if its restriction to both L_1 and L_2 is trivial.

Now fix a finite subset P of $\mathbb{P}^1$. The above suggests that there is a unique *maximal* extension of $\mathbb{C}(x)$ with branch point set P. Such an object exists, however; it is an *infinite* Galois extension of $\mathbb{C}(x)$. Its Galois group is free profinite of rank $|P| - 1$. We get an approximation in the class of FG-extensions by restricting the degree.

Corollary 7.3 *Let $P \subset \mathbb{P}^1$ finite, and $s = |P| - 1$. For each $n \geq 2$, consider the composite $K_n(P)$ (inside some fixed algebraic closure of $\mathbb{C}(x)$) of all FG-extensions of $\mathbb{C}(x)$ of degree $\leq n$ with branch points contained in P. Then $K_n(P)$ is an FG-extension of $\mathbb{C}(x)$ with Galois group isomorphic to $H_{n,s}$ (and with branch point set P). Conversely, if $K/\mathbb{C}(x)$ is an FG-extension with group isomorphic to $H_{n,s}$ and with branch points contained in P, then $K = K_n(P)$.*

Proof. Consider some FG-extension $K/\mathbb{C}(x)$ with group isomorphic to $H_{n,s}$ and with branch points contained in P. Such K actually exists by RET (Theorem 2.13), applied to the generators $h_1, \ldots, h_{s+1}$ of $H_{n,s}$, where $h_{s+1} = (h_1 \cdots h_s)^{-1}$. (Note that all $h_i \neq 1$ by Lemma 7.1 (1) and (4).)

Condition (2) from the Lemma translates into the fact that K is the composite of the FG-extensions $L/\mathbb{C}(x)$ of degree $\leq n$ contained in K. Those $L/\mathbb{C}(x)$ have their branch points contained in P (see the remarks before the corollary).

Now let $L/\mathbb{C}(x)$ be any FG-extension of degree $\leq n$, with branch points contained in P, and let K' be the composite of L and K (inside the fixed algebraic closure of $\mathbb{C}(x)$). Then K' also has its branch points contained in P, hence by RET the group $H' = G(K'/\mathbb{C}(x))$ can be generated by s elements. Since K is the composite of certain FG-extensions of $\mathbb{C}(x)$ of degree $\leq n$, the same holds for K'. Hence the intersection of all normal subgroups of H' of index $\leq n$ is trivial. Thus it follows from part (3) of the above Lemma that $|H'| \leq |H_{n,s}|$.

Hence $K' = K$, that is, $L \subset K$. It follows that $K = K_n(P)$. This completes the proof. □

Remark 7.4 Let G be a finite group and $P \subset \mathbb{P}^1$ finite. Then the number $\lambda(G, P)$ of FG-extensions of $\mathbb{C}(x)$ with group isomorphic to G and branch points contained in P is finite, and can be computed as follows.

By the corollary and the Galois correspondence, this number equals the number of normal subgroups N of $H_{n,s}$ with $H_{n,s}/N \cong G$. Here $n = |G|$ and $s = |P|-1$. Such N is the kernel of a surjective homomorphism $\psi : H_{n,s} \to G$, and such ψ is determined by the images $g_1, \ldots, g_s$ of $h_1, \ldots, h_s$. Those $g_1, \ldots, g_s$ are generators of G, and any such generators yield some ψ by the defining property of the group $H_{n,s}$. Two such ψ's have the same kernel if and only if there is an automorphism of G mapping the corresponding g_i's into each other. Hence $\lambda(G, P)$ equals the number of Aut(G)-orbits on the set of generating systems of G of length s (cf. Section 5.4).

7.2 Specializing the Coefficients of an Absolutely Irreducible Polynomial

If K is a field extension of k that is finitely generated over k *as a ring extension of* k then K is algebraic over k. This is an easy exercise in basic field theory. (Using Remark 1.2 we can reduce to the case $K = k[t][a_1, \ldots, a_s]$, where t is transcendental over k, and the a_i are algebraic over $k(t)$. Then the a_i are integral over $k[t][f^{-1}]$ for some $f \in k[t]$. Thus K, in particular $k(t)$, would be integral over $k[t][f^{-1}]$. Contradiction.) This proves

Lemma 7.5 *Let R be a finitely generated algebra over the field k (commutative with 1). Then for each maximal ideal M of R the quotient field R/M is algebraic over k. Hence R has a homomorphism to $\bar{k}$ that is the identity on k.*

Let k be a field and $I = (f_1, \ldots, f_m)$ an ideal of the polynomial ring $k[X_1, \ldots, X_n]$. Then I is a proper ideal if and only if the equations $f_1(a_1, \ldots, a_n) = \cdots = f_m(a_1, \ldots, a_n) = 0$ have a common solution in $\bar{k}^n$. This is a special case of Hilbert's Nullstellensatz. (The "if"-direction is clear. For the "only if", embed I into a maximal ideal M of $k[X_1, \ldots, X_n]$. Then the images of $X_1, \ldots, X_n$ in R/M yield a common solution in $\bar{k}^n$ of the above equations, since R/M embeds into $\bar{k}$ by the lemma.)

Lemma 7.6 (Bertini-Noether) *Let F be a field with subring B such that F is the field of fractions of B. Let $f(x, y) \in B[x, y]$ be a polynomial that is irre-*

ducible in $\bar{F}[x, y]$. Then there is $b \neq 0$ in B such that for each homomorphism $\omega : B \to k$ from B to a field k the following holds: If $\omega(b) \neq 0$ then the polynomial f^ω (obtained by applying ω to the coefficients of f) is irreducible in $\bar{k}[x, y]$.

Proof. The (total) degree of a polynomial

$$g(x, y) = \sum c_{ij}\, x^i y^j$$

is the maximal value of $i + j$ such that $c_{ij} \neq 0$. We first prove

Step 1 For any $d_1, d \in \mathbb{N}$ with $d_1 < d$ there is $n \in \mathbb{N}$ and polynomials $\Phi_{ij} \in \mathbb{Z}[X_1, \ldots, X_n]$, $(0 \le i, j \le d)$, with the following property: A polynomial $g(x, y) = \sum c_{ij}\, x^i y^j$ of degree d over a field F has a divisor $g_1 \in \bar{F}[x, y]$ of degree d_1 if and only if the elements $\Phi_{ij} - c_{ij}$, $(0 \le i, j \le d)$, generate a *proper* ideal of the polynomial ring $F[X_1, \ldots, X_n]$.

Proof. The existence of g_1 is equivalent to solvability of $g = g_1 g_2$ in $\bar{F}[x, y]$, with g_1 of degree d_1 and g_2 of degree $d_2 := d - d_1$. Associate variables $X_1, \ldots, X_n$ to the coefficients of g_1 and g_2. From the equation $g = g_1 g_2$ we obtain each coefficient c_{ij} of g as a polynomial Φ_{ij} in the coefficients of g_1 and g_2, where Φ_{ij} is a "universal" polynomial over $\mathbb{Z}$ in the variables $X_1, \ldots, X_n$ (independent of F and g). Thus the existence of g_1 is equivalent to solvability of the equations $\Phi_{ij}(X_1, \ldots, X_n) = c_{ij}$ over $\bar{F}$. The latter is equivalent to the fact that the elements $\Phi_{ij} - c_{ij}$ generate a proper ideal of $F[X_1, \ldots, X_n]$ (see the remarks before the lemma).

Step 2 Now consider some $f \in B[x, y]$ as in the lemma. It suffices to consider homomorphisms ω that do not vanish on the (nonzero) coefficients of f. (Multiply b by the product of those coefficients.) Then f^ω has the same degree d as f.

Since f is irreducible in $\bar{F}[x, y]$, for each d_1 as in Step 1 the corresponding $\Phi_{ij} - c_{ij}$ generate the unit ideal of $F[X_1, \ldots, X_n]$, that is,

$$1 = \sum \Psi_{ij}\,(\Phi_{ij} - c_{ij})$$

for certain $\Psi_{ij} \in F[X_1, \ldots, X_n]$. Multiplying by a suitable nonzero $c \in B$ we get

$$c = \sum \Psi'_{ij}(\Phi_{ij} - c_{ij})$$

with $\Psi'_{ij} \in B[X_1, \ldots, X_n]$. Thus if $\omega(c) \neq 0$ then the $\Phi_{ij} - c^\omega_{ij}$ generate the unit ideal of $k[X_1, \ldots, X_n]$, hence f^ω has no divisor of degree d_1 in

$\bar{k}[X_1, \ldots, X_n]$ (by Step 1). Hence the product of these elements c, as d_1 ranges from 1 to $d-1$, is the desired b. □

Lemma 7.7 *Let F be a field with subring R such that F is the field of fractions of R. Let K/F be an FG-extension, and let $\alpha_1, \ldots, \alpha_s$ be primitive elements for this extension. Assume $f_i(\alpha_i) = 0$ for $i = 1, \ldots, s$, where $f_i(y) \in R[y]$ is a monic polynomial of degree $n = [K : F]$. Then there is $u \neq 0$ in R such that for each homomorphism ω from R to a field F' the following holds: Let f_i' denote the polynomial obtained by applying ω to the coefficients of f_i. If $\omega(u) \neq 0$ and each f_i' is irreducible in $F'[y]$, then the fields $F'[y]/(f_i')$ are mutually F'-isomorphic, and Galois over F' with Galois group isomorphic to $G(K/F)$.*

Proof. Follows from Lemma 1.5: Take A to contain $\alpha_1, \ldots, \alpha_s$ and all their conjugates over F. Taking $\alpha = \alpha_i$ and $f = f_i$ we obtain certain nonzero $u_i \in R$ from that lemma. The product u of those u_i does the job. □

7.3 The Descent from $\mathbb{C}$ to $\bar{k}$

Lemma 7.8 *Consider the field $k((t))$ of formal Laurent series over some infinite field k. Suppose $L/k(t)$ is a Galois extension of finite degree n, with L contained in $k((t))$. Then there is a generator α of L over $k(t)$ with the following properties: $\alpha \in k[[t]]$, and $f(t, \alpha) = 0$ for some polynomial $f(t, y) \in k[t, y]$, monic and of degree n in y, such that the polynomial $f(0, y) \in k[y]$ has n distinct roots in k.*

Proof. For nonzero $\beta \in k((t))$ define its order $\mathrm{ord}(\beta)$ to be that integer N with $\beta = \sum_{\nu=N}^{\infty} a_\nu t^\nu$ and $a_N \neq 0$. Now take β to be a generator of L over $k(t)$. Let $G = G(L/k(t))$. There is $M > 0$ such that each G-conjugate $\neq \beta$ of β differs from β at some t^ν-coefficient with $\nu \leq M$. Let $N' > M$ with $a_{N'} \neq 0$. (Such N' exists unless we are in the trivial case $L = k(t)$.) Then for

$$\gamma := \sum_{\nu=N'}^{\infty} a_\nu t^\nu = \beta - \sum_{\nu=N}^{N'-1} a_\nu t^\nu$$

we have $\mathrm{ord}(\gamma) = N'$ and $\mathrm{ord}(g(\gamma)) \leq M < N'$ for each $g \neq 1$ in G. Thus $\delta = t^{-N'}\gamma$ satisfies $\mathrm{ord}(\delta) = 0$ and $\mathrm{ord}(g(\delta)) < 0$ for each $g \neq 1$ in G. Finally, $\epsilon = \delta^{-1}$ satisfies $\mathrm{ord}(\epsilon) = 0$ and $\mathrm{ord}(g(\epsilon)) > 0$ for each $g \neq 1$ in G. In particular, all $g(\epsilon)$ $(g \in G)$ lie in $k[[t]]$ (and L). Dividing by a constant we may assume $\epsilon(0) = 1$ and $g(\epsilon)(0) = 0$ for $g \neq 1$ in G. Now choose distinct elements b_g

$(g \in G)$ of k, and set

$$\alpha = \sum_{h \in G} b_h \, h(\epsilon).$$

Then all $g(\alpha)$ $(g \in G)$ lie in $L \cap k[[t]]$, and $g(\alpha)(0) = b_{g^{-1}}$. We have $f(t, \alpha) = 0$ for some $f(t, y) \in k[t, y]$, irreducible in y and of y-degree $\leq n$. Then $f(t, g(\alpha)) = 0$, hence $f(0, b_g) = 0$ for all $g \in G$. Since the b_g are pairwise distinct and $f(0, y)$ is of degree $\leq n$, it follows that $f(0, y)$ has degree n, and the roots of $f(0, y)$ are exactly the b_g. In particular, f has y-degree n, hence α generates $L/k(t)$. Finally, we get f to be monic by the usual application of Lemma 1.3. □

Now we can prove the main result of this chapter.

Theorem 7.9 *Let $L/\mathbb{C}(x)$ be an FG-extension with branch points contained in $k \cup \{\infty\}$, where k is an algebraically closed subfield of $\mathbb{C}$. Then L is defined over k.*

Proof. It suffices to prove the claim for some FG-extension $\tilde{L}/\mathbb{C}(x)$ containing L, and having the same branch points. Indeed, if $\tilde{L}$ is defined over k, then so is L, with L_k the fixed field of $G(\tilde{L}/L)$ in $\tilde{L}_k$ (see Lemma 3.1(b)). Thus by Corollary 7.3, we may assume $L = K_n(P)$ for some $n \in \mathbb{N}$ and finite $P \subset \mathbb{P}^1$. By a change of coordinates we may assume $\infty \in P$.

As usual, we consider a generator α for $L = K_n(P)$ over $\mathbb{C}(x)$ such that $f(x, \alpha) = 0$ for some polynomial $f \in \mathbb{C}[x, y]$ that is irreducible and monic as polynomial in y over $\mathbb{C}(x)$. Let $D_f \in \mathbb{C}[x]$ be the y-discriminant of f (i.e., f viewed as polynomial in y). Then the branch points $\neq \infty$ of L are among the roots of D_f (Proposition 2.6).

Now consider the roots q_i $(i = 1, \ldots, m)$ of D_f which are not branch points of L. Then L embeds into $\mathbb{C}((t))$, where $t = x - q_i$ (see Proposition 2.6). By the above Lemma, there is a generator α_i for $L/\mathbb{C}(x)$ such that $f_i(x, \alpha_i) = 0$ for some $f_i \in \mathbb{C}[x, y]$ that is monic and irreducible as polynomial in y over $\mathbb{C}(x)$, and satisfies $D_{f_i}(q_i) \neq 0$. For each $i = 1, \ldots, m$, the branch points $\neq \infty$ of L are among the roots of D_{f_i} (Proposition 2.6).

Set $f_{m+1} = f$ and $\alpha_{m+1} = \alpha$. Then $f_1, \ldots, f_{m+1}$ are irreducible in $\mathbb{C}[x, y]$ (by Lemma 1.4). Further, the common roots of $D_{f_1}, \ldots, D_{f_{m+1}}$ are exactly the branch points $\neq \infty$ of L. Label these branch points as $p_1, \ldots, p_r$. It follows that the g.c.d. of $D_{f_1}, \ldots, D_{f_{m+1}}$ (in $\mathbb{C}[x]$) equals $\prod_{j=1}^{r}(x - p_j)$. Thus there are $\lambda_1, \ldots, \lambda_{m+1} \in \mathbb{C}[x]$ with

$$D_{f_1}\lambda_1 + \cdots + D_{f_{m+1}}\lambda_{m+1} = \prod_{j=1}^{r}(x - p_j). \tag{7.1}$$

There is a finitely generated k-algebra B_1 in $\mathbb{C}$ with $f_i \in B_1[x, y]$ and $\lambda_i \in B_1[x]$ for all i. Let k_1 be the field of fractions of B_1 (in $\mathbb{C}$). Let L_2 be the field obtained by adjoining all conjugates of all α_i (in L) over $\mathbb{C}(x)$ to $k_1(x)$. Then L_2 is finite Galois over $k_1(x)$. Let k_2 be the algebraic closure of k_1 in L_2. Then L is defined over k_2 with $L_{k_2} = L_2$, and $L_2 = k_2(x)(\alpha_i)$ for each i; in addition, k_2 is finite over k_1 (cf. the proof of Lemma 3.1(d)). Write $k_2 = k_1(\gamma)$ for some γ, and set $B = B_1[\gamma]$. Then k_2 is the field of fractions of B. Further, $f_i \in B[x, y]$ is irreducible in $\mathbb{C}[x, y]$ (see above), hence in $\bar{k}_2[x, y]$. By Lemma 7.6 there is $b \neq 0$ in B such that for each homomorphism $\omega : B \to k$ the following holds: If $\omega(b) \neq 0$ then each f_i^ω is irreducible in $k[x, y]$. (Take b as the product of the b_i corresponding to the various f_i.) Then in particular, f_i^ω is irreducible as polynomial in y over $k(x)$.

Now apply Lemma 7.7, taking for K/F the extension $L_2/k_2(x)$, and $R = B[x]$ and $s = m+1$. We obtain $u \neq 0$ in R with the properties from Lemma 7.7. Define $c \in B$ as the product of the above element b with the nonzero coefficients of u (where $u \in R = B[x]$ is viewed as polynomial in x). The ring $B[c^{-1}]$ is a finitely generated algebra over k, hence has a k-algebra homomorphism to k (Lemma 7.5). Its restriction to B yields a homomorphism $\omega : B \to k$ with $\omega(c) \neq 0$. Extend ω to a homomorphism $\omega : R = B[x] \to k[x]$ (that fixes x). Since $\omega(b) \neq 0$, each f_i^ω is irreducible as polynomial in y over $F' := k(x)$. Further $\omega(u) \neq 0$, hence the fields $F'[y]/(f_i^\omega) = k(x)[y]/(f_i^\omega)$ are Galois over $k(x)$ with group isomorphic to $G(L_2/k_2(x)) \cong H_{n,s}$. Further, these fields are all F'-isomorphic to $K' := F'[y]/(f^\omega)$. Thus each f_i^ω is a generating polynomial for $K'/k(x)$, and the g.c.d. of their y-discriminants divides $\prod_{j=1}^{r}(x - p_j)$ by (7.1). (Here we use the hypothesis that all $p_j \in k$.) Hence the branch points $\neq \infty$ of K' are among $p_1, \ldots, p_r$. Thus all branch points of K' lie in P (since $\infty \in P$ by assumption).

By Lemma 3.3 (ii), the field K' is $k(x)$-isomorphic to M_k for some FG-extension $M/\mathbb{C}(x)$ defined over k. Then $G(M/\mathbb{C}(x)) \cong G(K'/k(x)) \cong H_{n,s}$. The branch points of M lie in P as well (Lemma 3.1(e)). It follows by Corollary 7.3 that M is $\mathbb{C}(x)$-isomorphic to $K_n(P)$. Hence $L = K_n(P)$ is defined over k. □

Corollary 7.10 (RET – The Version Over** k**) *Let k be an algebraically closed subfield of $\mathbb{C}$. Let $\mathcal{T} = [G, P, (C_p)_{p \in P}]$ be a ramification type, where*

$P \subset k \cup \{\infty\}$. Label the elements of P as $p_1, \ldots, p_r$ (with $r = |P|$). Then there exists an FG-extension of $k(x)$ of type $\mathcal{T}$ if and only if there exist generators $g_1, \ldots, g_r$ of G with $g_1 \cdot \ldots \cdot g_r = 1$ and $g_i \in C_{p_i}$ for $i = 1, \ldots, r$.

Proof. If the condition holds then RET over $\mathbb{C}$ (Theorem 2.13) yields an FG-extension $L/\mathbb{C}(x)$ of the given type. This L is defined over k by Theorem 7.9, and the corresponding extension $L_k/k(x)$ also has type $\mathcal{T}$ by Lemma 3.1(e). This proves that the condition is sufficient.

Conversely, given an FG-extension of $k(x)$, Lemma 3.3(ii) associates with it an FG-extension of $\mathbb{C}(x)$, and both are of the same type by Lemma 3.1(e). Thus the condition is necessary by RET over $\mathbb{C}$. □

We could formulate RET for any algebraically closed field of characteristic 0. However, since the notion of ramification type requires a choice of a compatible system of roots of 1, we stick to the subfields of $\mathbb{C}$ (where we can use the canonical choice). This is (by far) the most important case anyway.

Everything in Chapter 3 and Section 7.1 that relies on RET can now be done with $\mathbb{C}$ replaced by an algebraically closed subfield. For later reference, we state the analogue of Corollary 7.3.

Corollary 7.11 *Let k be an algebraically closed subfield of $\mathbb{C}$. Let $P \subset k \cup \{\infty\}$ finite, and $s = |P| - 1$. Let $n \geq 2$, and let $K_n(P)$ be the composite (inside some fixed algebraic closure of $k(x)$) of all FG-extensions of $k(x)$ of degree $\leq n$ with branch points contained in P. Then $K_n(P)$ is an FG-extension of $k(x)$ with Galois group isomorphic to $H_{n,s}$ and with branch point set P.*

Over algebraically closed fields k_p of characteristic $p > 0$ the picture is quite different; for example, there are many extensions of $k_p(x)$ with only one branch point (for the suitable definition of branch point in this case). See Chapter 11 for a discussion of Abhyankar's conjecture (recently proved by Raynaud) and more on the characteristic p case.

7.4 The Minimal Field of Definition

By RET over $\bar{\mathbb{Q}}$, for any choice of r branch points in $\bar{\mathbb{Q}}$ and for any choice of generators $g_1, \ldots, g_r$ of a finite group G with product 1 there is an FG-extension $L/\bar{\mathbb{Q}}(x)$ of the corresponding ramification type. This L is defined over some finite number field. By the following proposition, there is actually a unique minimal number field $\kappa_{\min}$ over which L is defined, if $G(L/\bar{\mathbb{Q}}(x))$ has trivial center. The big problem is to compute $\kappa_{\min}$ from the group-theoretic data; especially to find cases with $\kappa_{\min} = \mathbb{Q}$.

More can be said in the case that the branch points are algebraically independent over $\mathbb{Q}$. In this "generic" case, the braid group action gives us additional information about the minimal field of definition (see Section 9.1).

Proposition 7.12 *Let $L/\mathbb{C}(x)$ be an FG-extension, and label its branch points $\neq \infty$ as $p_1, \ldots, p_r$. Let $s_1, \ldots, s_r$ be the elementary symmetric functions in $p_1, \ldots, p_r$ (i.e., $\prod_{i=1}^r (X - p_i) = X^r + s_1 X^{r-1} + \cdots + s_r$). Let ℓ be any subfield of $\mathbb{C}$.*

(a) If L is defined over $\kappa \subset \mathbb{C}$ then κ contains $s_1, \ldots, s_r$.
(b) Now suppose that $G(L/\mathbb{C}(x))$ has trivial center. Then there is a unique minimal subfield κ of $\mathbb{C}$ containing ℓ such that L is defined over κ. This κ is finite over $\ell(s_1, \ldots, s_r)$.

Proof. (a) Set $k = \bar{\kappa}$ and $M = L_k$. Then $p_1, \ldots, p_r$ are also the branch points $\neq \infty$ of $M/k(x)$ (by Lemma 3.1(e)). By Proposition 3.6 there is an α-automorphism of M for each $\alpha \in G(\bar{\kappa}/\kappa)$. Applying the branch cycle argument (Lemma 2.8) to this α-automorphism we see that α permutes $p_1, \ldots, p_r$, hence fixes $s_1, \ldots, s_r$. Thus $s_1, \ldots, s_r \in \kappa$.

(b) Set $\kappa_0 = \ell(s_1, \ldots, s_r)$, and $k = \bar{\kappa}_0$ $(\subset \mathbb{C})$. Then L is defined over k by Theorem 7.9, hence over some FG-extension κ_1 of κ_0. Set $M = L_k$. Let Γ be the group of all $\gamma \in G(\kappa_1/\kappa_0)$ that extend to an α-automorphism λ of M with $\lambda^* = \mathrm{id}$, for some $\alpha \in G(k/\kappa_0)$ (notation as in Section 3.1.2). Define κ as the fixed field of Γ in κ_1.

Each $\alpha \in G(k/\kappa)$ restricts to some $\gamma \in \Gamma$. This γ extends to an α'-automorphism λ' of M with $(\lambda')^* = \mathrm{id}$, for some $\alpha' \in G(k/\kappa)$. Since $\alpha'' := (\alpha')^{-1}\alpha$ lies in $G(k/\kappa_1)$, and M is defined over κ_1, there exists an α''-automorphism λ'' of M with $(\lambda'')^* = \mathrm{id}$. Then $\lambda := \lambda'\lambda''$ is an α-automorphism of M with $\lambda^* = \mathrm{id}$. Hence M is defined over κ by Proposition 3.6. Then also L is defined over κ.

Now assume conversely that L is defined over some field κ' with $\ell \subset \kappa' \subset \mathbb{C}$. We need to show that $\kappa \subset \kappa'$. By (a) we know that $\kappa_0 \subset \kappa'$. Consider some $\alpha' \in G(k'/\kappa')$, where k' is the algebraic closure of κ'. This α' restricts to some $\alpha \in G(k/\kappa_0)$.

Set $L' = L_{k'}$. Then by Proposition 3.6 there is an α'-automorphism λ' of L' with $(\lambda')^* = \mathrm{id}$. This λ' restricts to an α-automorphism λ of M with $\lambda^* = \mathrm{id}$ (using Lemma 3.1(b),(d)). Thus $\alpha|_{\kappa_1}$ lies in Γ. Hence α fixes κ elementwise (by definition of κ). Thus α' fixes κ elementwise, which means that $\kappa \subset \kappa'$. □

Exercise Bound the degree $[\kappa/\kappa_0]$ by a number depending only on the ramification type of L (as in Remark 7.4).

7.5 Embedding Problems over $\bar{k}(x)$

By our latest version of RET, each finite group is a Galois group over $k(x)$, where k is any algebraically closed subfield of $\mathbb{C}$. We can say more than that, however: The Galois extensions of $k(x)$ are arranged in a systematic way, meaning that we can find a Galois realization of a group G compatible with *any* Galois realization of a given factor group of G.

Definition 7.13 *A (finite)* **embedding problem** *over a field K consists of an FG-extension L/K, together with a surjective homomorphism φ from a finite group H to $G(L/K)$. A* **solution** *of the embedding problem is an FG-extension M/K containing L, together with an isomorphism $\psi : H \to G(M/K)$ such that $\varphi = res_{M/L} \circ \psi$. (Here $res_{M/L}$ denotes the restriction homomorphism $G(M/K) \to G(L/K)$.)*

Theorem 7.14 *Suppose k is algebraically closed of characteristic 0. Then all embedding problems over $k(x)$ are solvable.*

Proof. Using Lemma 3.3(i), (ii), and Lemma 3.1(b) one reduces to the case that k is of finite transcendence degree. (The general hypothesis valid in those lemmas is that $k \subset \mathbb{C}$, but this is not necessary for the proof.) Thus we may assume k is a subfield of $\mathbb{C}$.

Consider an FG-extension $L/k(x)$, and a surjective homomorphism $\varphi : H \to G := G(L/k(x))$. Let $P = \{p_1, \dots, p_{s+1}\}$ be a subset of $\mathbb{P}^1_k$ of finite cardinality $s + 1$, such that the branch points of L are $p_1, \dots, p_r$ with $r < s$. Set $\Omega = K_n(P)$, where $n = |H|$ (see Corollary 7.11). Then $L \subset \Omega$.

By RET (Theorem 7.10) the group $G(\Omega/k(x))$ has generators $h_1, \dots, h_{s+1}$ with $h_1 \cdots h_{s+1} = 1$ and $h_i \in C_{p_i}$ (the class associated with p_i). Let h'_i be the restriction image of h_i in G. By Proposition 2.6(d) it follows that $h'_i = 1$ for $i > r$. Now choose $h''_1, \dots, h''_s \in H$ with $\varphi(h''_i) = h'_i$. Thus $h''_{r+1}, \dots, h''_s$ lie in the kernel of φ; we may choose these elements to be generators of $\ker(\varphi)$ (for sufficiently large s).

We know $G(\Omega/k(x)) \cong H_{n,s}$ (Corollary 7.11). Hence by Lemma 7.1 (1) and (4), there is a homomorphism $\lambda : G(\Omega/k(x)) \to H$ with $\lambda(h_i) = h''_i$ for $i = 1, \dots, s$. Clearly, $\varphi \circ \lambda = \mathrm{res}_{\Omega/L}$. Thus $\mathrm{Im}(\lambda)$ is a subgroup of H mapping surjectively to G, that is, $H = \mathrm{Im}(\lambda) \cdot \ker(\varphi)$. But $\ker(\varphi) \subset \mathrm{Im}(\lambda)$ by construction, hence λ is surjective.

Let M be the fixed field of $U := \ker(\lambda)$ in Ω. Then λ induces an isomorphism $\bar{\lambda} : G(M/k(x)) \cong G(\Omega/k(x))/U \to H$. From $\varphi \circ \lambda = \mathrm{res}_{\Omega/L}$ we get

$\varphi \circ \bar{\lambda} = \mathrm{res}_{M/L}$. Thus $\bar{\lambda}^{-1}$ is the desired map $H \to G(M/k(x))$ solving the given embedding problem. □

Actually, the theorem remains true without the restriction to characteristic 0. This is proved in Chapter 11, replacing RET by Harbater's patching method (Theorem 11.32).

Solvability of all embedding problems is a very strong property of a field. For a countable field, it determines the lattice of all separable algebraic extensions up to isomorphism. This is proved in the next chapter.

8
Embedding Problems

A natural question occurring in Galois theory is the following: Given a Galois realization of a factor group $\bar{G}$ of a group G, find a Galois realization of G compatible with that of $\bar{G}$. This is called an embedding problem.

In Section 8.1 we study fields over which all embedding problems are solvable. This is a very strong property of a field. For a countable field, it determines the lattice of all separable algebraic extensions up to isomorphism. As shown in the last chapter, the field $k(x)$ has this property, where k is algebraically closed of characteristic 0. Shafarevich's conjecture expects the same to hold for the cyclotomic field $\mathbb{Q}_{\mathrm{ab}}$.

Actually it suffices to solve embedding problems with characteristic simple kernel. This leads to a basic dichotomy between the abelian and the nonabelian case. Section 8.2 leads up to the theorem that over a hilbertian field, each split abelian embedding problem is solvable. In Section 8.3 we discuss the notion of GAR-, GAL-, and GAP-realization, a way to deal with the case of centerless kernel. If each nonabelian finite simple group had a GAR-realization over $\mathbb{Q}_{\mathrm{ab}}$, then this would imply Shafarevich's conjecture.

From Section 8.2 on we only consider fields of characteristic 0, our case of main interest. The case of positive characteristic can be included with the usual modifications.

8.1 Generalities

8.1.1 Fields over Which All Embedding Problems Are Solvable

See Section 7.5 for the definition of an embedding problem. Recall that the separable closure of a field is obtained by adjoining the roots of all separable polynomials over that field. In characteristic 0, separable closure is the same as algebraic closure.

Proposition 8.1 *Let K, K' be countable fields such that all embedding problems over these fields are solvable. Fix separable closures $\tilde{K}$ and $\tilde{K}'$ of K and K', respectively, and take all separable algebraic extensions inside these closures. Then there is a bijection $M \mapsto M'$ from the set of separable algebraic extensions M of K to the set of separable algebraic extensions M' of K', with the following properties: We have $M_1 \subset M_2$ if and only if $M_1' \subset M_2'$. Thus the bijection also preserves composites and intersections. Further, M is Galois over K if and only if M' is Galois over K'. In this case, $G(M/K) \cong G(M'/K')$, and these isomorphisms can be chosen to be compatible with restriction maps.*

Proof. Since K is countable, there are only countably many separable polynomials f_n ($n \in \mathbb{N}$) over K. Let M_i be the field obtained by adjoining the roots of all f_n with $n \leq i$ to the field K. Then the M_i are FG-extensions of K with $M_1 \subset M_2 \subset \cdots$ and $\tilde{K} = \bigcup_{i=1}^{\infty} M_i$. Similarly, there are FG-extensions L_i of K' with $L_1 \subset L_2 \subset \cdots$ and $\tilde{K}' = \bigcup_{i=1}^{\infty} L_i$.

Set $K_0 = K$, $K_0' = K'$. Set $K_1 = M_1$ and $G_1 = G(K_1/K)$. Solving the embedding problem over K' corresponding to the map $G_1 \to \{1\} = G(K_0'/K')$, we obtain an FG-extension K_1'/K' and an isomorphism $\kappa_1 : G_1 \to G_1' := G(K_1'/K')$.

In the next step, set $K_2' = K_1'L_2$ (composite inside $\tilde{K}'$) and $G_2' = G(K_2'/K')$. Solving the embedding problem over K corresponding to K_1/K and the map $\kappa_1^{-1} \circ \mathrm{res}_{K_2'/K_1'} : G_2' \to G_1 = G(K_1/K)$, we obtain an FG-extension K_2/K and an isomorphism $\kappa_2^{-1} : G_2' \to G_2 := G(K_2/K)$ with

$$\kappa_1^{-1} \circ \mathrm{res}_{K_2'/K_1'} = \mathrm{res}_{K_2/K_1} \circ \kappa_2^{-1}.$$

In the third step, set $K_3 = K_2M_3$ and $G_3 = G(K_3/K)$. Solving the embedding problem over K' corresponding to K_2'/K' and the map $\kappa_2 \circ \mathrm{res}_{K_3/K_2} : G_3 \to G_2' = G(K_2'/K')$, we obtain K_3'/K' and $\kappa_3 : G_3 \to G_3' := G(K_3'/K')$ with $\mathrm{res}_{K_3'/K_2'} \circ \kappa_3 = \kappa_2 \circ \mathrm{res}_{K_3/K_2}$.

Continuing like this (alternately solving embedding problems over K and K') we obtain FG-extensions $K = K_0 \subset K_1 \subset K_2 \subset \cdots$ and $K' = K_0' \subset K_1' \subset K_2' \subset \cdots$ of K and K', respectively, and isomorphisms $\kappa_i : G_i = G(K_i/K) \to G_i' = G(K_i'/K')$ satisfying

$$\mathrm{res}_{K_i'/K_{i-1}'} \circ \kappa_i = \kappa_{i-1} \circ \mathrm{res}_{K_i/K_{i-1}}. \tag{8.1}$$

Further, $\tilde{K} = \bigcup_{i=1}^{\infty} K_i$ and $\tilde{K}' = \bigcup_{i=1}^{\infty} K_i'$ (because this holds for the M_i and L_i).

Any finite separable extension M/K has a primitive element, which lies in some K_i. Then $M \subset K_i$. Define M' to be the fixed field in K_i' of the subgroup of G_i' corresponding to $G(K_i/M)$ under κ_i. This is well defined (i.e., does not depend on the choice of i) by (8.1). If M/K is infinite separable algebraic, it is the union of its finite subextensions. Define M' to be the union of the corresponding finite extensions of K'. It is now a routine check to show that this map $M \mapsto M'$ has the desired properties. We leave the details to the reader. □

Remark 8.2 In the language of infinite Galois theory, the conclusion of the proposition can simply be expressed by saying that the absolute Galois groups of K and K' are isomorphic (as profinite groups). Actually, they are isomorphic to the free profinite group of countable rank (by Iwasawa's criterion; see [FJ], Section 24.1).

By Theorem 7.14, the field $\bar{\mathbb{Q}}(x)$ has the above property. It is a conjecture, usually attributed to Shafarevich, that the same holds for the cyclotomic field $\mathbb{Q}_{ab}$. This conjecture has been one of the guiding problems in Galois theory in recent years. We will see in several steps how this conjecture can be attacked, and what evidence there is for it. In Chapter 11 (see Remark 11.33) we prove the "geometric case" of this conjecture.

8.1.2 Minimal Embedding Problems

Consider an embedding problem, given by an FG-extension L/K and surjection $\varphi : H \to G = G(L/K)$. The kernel U of φ is called the **kernel of the embedding problem**. The embedding problem is called **split** (resp., **abelian**) if H splits as semi-direct product of U and G (resp., if U is abelian).

Now let U_0 be a normal subgroup of H contained in U. This induces a "decomposition" of the given embedding problem into two embedding problems with kernel U_0 and U/U_0, respectively:

Lemma 8.3 *The given embedding problem is solvable if and only if the following two embedding problems are solvable: The first corresponds to the map $\varphi_0 : H/U_0 \to G = G(L/K)$ induced by φ. If M_0/K and $\psi_0 : H/U_0 \to G(M_0/K)$ are a solution to this, then the second embedding problem is given by the map $H \to G(M_0/K)$ that is the natural map $H \to H/U_0$ followed by ψ_0.*

The proof is obvious. Hence it suffices to solve embedding problems that are minimal in the sense that they cannot be decomposed nontrivially in the above way; that is, whose kernel is a *minimal normal subgroup* of H. Then the following simple but crucial lemma applies:

Lemma 8.4 *If U is minimal normal in a finite group H then U is characteristic simple, that is, $U \cong S^m$ is isomorphic to the direct product of $m \geq 1$ copies of a simple group S.*

Proof. Recall that minimal normal means that U is a nontrivial normal subgroup of H that does not properly contain another nontrivial normal subgroup of H. Let U_0 be a minimal normal subgroup of U. The conjugates U_0^h $(h \in H)$ are also minimal normal in U. Thus any two of those conjugates that are not equal have trivial intersection, hence they commute. Further, these U_0^h generate U (because U is minimal normal in H). It follows that there is a homomorphism from the direct product P of the U_0^h $(h \in H)$ onto U, which is the identity on each U_0^h.

By induction we may assume $U_0 \cong S^\ell$ for simple S and $\ell \in \mathbb{N}$. Thus P is of the form $P = S^n$, $n \in \mathbb{N}$. It remains to show that each (nontrivial) quotient of P is again of this form. If S is cyclic (of prime order), then P is a finite vector space, hence the same is true for each quotient. If S is nonabelian, it is an easy exercise to show that each normal subgroup of P is a product of certain factors in the direct product $P = S \times S \times \cdots$. This implies the claim. □

Corollary 8.5 *Consider an embedding problem over a field K, with kernel U. It is solvable if all embedding problems over K with kernel S^m are solvable, where $m \in \mathbb{N}$ and S is any composition factor of U.*

A basic dichotomy emerges between embedding problems with elementary abelian kernel (i.e., the kernel is a vector space over a finite prime field) and those with kernel S^m, where S is nonabelian simple. Actually, this dichotomy is quite sharp. The abelian case is again divided into the split and the nonsplit case. Split abelian embedding problems are always solvable if the ground field K is hilbertian (see the next section). Nonsplit abelian embedding problems lead to questions of an arithmetic flavor, as illustrated by the following:

Example 8.6 Let d be a nonsquare in the field K (of characteristic $\neq 2$), and consider the embedding problem given by $K(\sqrt{d})/K$ and a surjection $H \to G(K(\sqrt{d})/K)$ with H cyclic of order 4. This embedding problem is solvable if and only if d is a sum of two squares in K. (Exercise! Or see [Se1], 2.2.)

On the other hand, group theory begins to play a role for embedding problems with kernel S^m, where S is nonabelian simple. For example, it is known for several classes of simple groups S that such an embedding problem is always solvable if the ground field K is hilbertian (see Section 8.3). We see that the answer seems to depend on the group-theoretic properties of the kernel, and not so much on the arithmetic properties of the given extension L/K.

8.2 Wreath Products and Split Abelian Embedding Problems

Recall that a wreath product is – roughly speaking – a semi-direct product of groups Γ and G where G is the direct product of isomorphic factors $G_1, \ldots, G_m$ and Γ acts on G by permuting those factors. Wreath products are – next to direct products – the simplest (and most common) type of group extension.

This section is based on the following simple idea for realizing wreath products as Galois groups. Starting from a Galois realization of Γ as $G(\ell/k)$, add new variables $x_1, \ldots, x_m$ and let Γ act on them in the same way it acts on $G_1, \ldots, G_m$. Given compatible Galois realizations of the G_i over $\ell(x_i)$, this results in a Galois realization of the corresponding wreath product.

For the rest of Chapter 8 we let k, κ and ℓ denote fields of characteristic 0.

8.2.1 A Rationality Criterion for Function Fields

Lemma 8.7 *Let ℓ/k be finite Galois. Let Γ be a group of automorphisms of the field $L = \ell(x_1, \ldots, x_s)$, where $x_1, \ldots, x_s$ are independent transcendentals over ℓ. Suppose Γ leaves the subfield $\ell \subset L$ invariant and restriction yields an isomorphism from Γ to $G(\ell/k)$. If Γ leaves the ℓ-linear span Ω of $x_1, \ldots, x_s$ invariant, then the fixed field F of Γ is purely transcendental over k:*

$$F = k(t_1, \ldots, t_s)$$

with $t_1, \ldots, t_s$ independent transcendentals over k. The $t_1, \ldots, t_s$ can be chosen to form an ℓ-basis of Ω.

Proof. Let Ω_k be the set of fixed points of Γ in Ω. Clearly, Ω_k is a k-vector space. If $t_1, \ldots, t_h \in \Omega_k$ are linearly independent over k, then they are linearly independent over ℓ: Indeed, if $t_1, \ldots, t_h$ are dependent over ℓ then $\sum_{i=1}^{h} a_i t_i = 0$ with $a_i \in \ell$ and with some $a_i = 1$. Then also $\sum_{i=1}^{h} b_i t_i = 0$ where $b_i = \sum_{\gamma \in \Gamma} \gamma(a_i) \in k$, and some $b_i = |\Gamma| \neq 0$; hence $t_1, \ldots, t_h$ are dependent over k.

Let c be a generator of ℓ over k. Write $\Gamma = \{\gamma_1, \ldots, \gamma_n\}$ with $n = |\Gamma|$. Fix some $u \in \Omega$, and define

$$u_i = \sum_{j=1}^{n} \gamma_j(c)^{i-1} \gamma_j(u).$$

The matrix $(\gamma_j(c)^{i-1})_{i,j=1,\ldots,n}$ is invertible because its determinant is Vandermonde. Hence each $\gamma_j(u)$, in particular u itself, is an ℓ-linear combination of the u_i, which lie in Ω_k. This shows that Ω_k spans Ω over ℓ. In view of the preceding paragraph, it follows that every k-basis $t_1, \ldots, t_s$ of Ω_k is also an ℓ-basis of Ω.

There is an ℓ-linear automorphism ϕ of Ω with $\phi(x_i) = t_i$ for $i = 1, \ldots, s$. This ϕ extends (uniquely) to an automorphism of $L = \ell(x_1, \ldots, x_s)$. It follows that $L = \ell(t_1, \ldots, t_s)$, and $t_1, \ldots, t_s$ are independent transcendentals over ℓ. Since the fixed field F of Γ contains $k(t_1, \ldots, t_s)$, and $[L : k(t_1, \ldots, t_s)] = [\ell(t_1, \ldots, t_s) : k(t_1, \ldots, t_s)] = [\ell : k] = |\Gamma| = [L : F]$, it follows that $F = k(t_1, \ldots, t_s)$. □

8.2.2 The Group-Theoretic Notion of Wreath Product

Definition 8.8 *Let Γ be a group acting on the letters $\{1, \ldots, m\}$, and let G_1 be any group. The* **wreath product** *of G_1 and Γ is the (outer) semi-direct product*

$$H = G_1^m \cdot \Gamma$$

where G_1^m is the direct product of m copies of G_1, and Γ acts in the following way on G_1^m: Each $\gamma \in \Gamma$ sends $(g_1, \ldots, g_m) \in G_1^m$ to $(g_{\gamma^{-1}(1)}, \ldots, g_{\gamma^{-1}(m)})$.

We naturally view G_1^m and Γ as subgroups of H. Then $\gamma \cdot (g_1, \ldots, g_m) \cdot \gamma^{-1} = (g_{\gamma^{-1}(1)}, \ldots, g_{\gamma^{-1}(m)})$ (in the above notation). Further, view G_1 as a subgroup of G_1^m by identifying $g \in G_1$ with the tuple $(g, 1, \ldots, 1)$. Then $\gamma g \gamma^{-1}$ is the tuple $(1, \ldots, 1, g, 1, \ldots 1)$ with g in the position $\gamma(1)$.

Wreath products have the following universal property.

Lemma 8.9 *Let H be the wreath product of G_1 and Γ, where Γ acts transitively on $\{1, \ldots, m\}$. Let Γ_1 be the stabilizer in Γ of the letter 1. Then for any group $\bar{H}$ with subgroups $\bar{G}_1$ and $\bar{\Gamma}$ the following holds: If $G_1 \to \bar{G}_1$ and $\Gamma \to \bar{\Gamma}$, $\gamma \mapsto \bar{\gamma}$, are homomorphisms such that $\bar{G}_1$ centralizes each $\bar{\gamma}\bar{G}_1\bar{\gamma}^{-1}$ with $\gamma \in \Gamma \setminus \Gamma_1$, and each $\bar{\gamma}$ with $\gamma \in \Gamma_1$, then these homomorphisms extend uniquely to a homomorphism $H \to \bar{H}$.*

Proof. Let $\gamma_1, \ldots, \gamma_m \in \Gamma$ with $\gamma_i(1) = i$. Then each element of H can be written uniquely in the form

$$h = (g_1, \ldots, g_m) \cdot \gamma = \gamma_1 g_1 \gamma_1^{-1} \cdots \gamma_m g_m \gamma_m^{-1} \cdot \gamma$$

with $g_1, \ldots, g_m \in G_1$, $\gamma \in \Gamma$. Define its image in $\bar{H}$ by the latter expression where each g_i, γ_i and γ is replaced by its image in $\bar{G}_1$ and $\bar{\Gamma}$, respectively. In order to show that this defines a homomorphism $H \to \bar{H}$, take another element of H and write it as

$$h' = \gamma_1 g_1' \gamma_1^{-1} \cdots \gamma_m g_m' \gamma_m^{-1} \cdot \gamma'$$

with $g_1', \ldots, g_m' \in G_1, \gamma' \in \Gamma$. Then

$$hh' = \left(\gamma_1 g_1 \gamma_1^{-1} \cdot \gamma \gamma_{i_1} g_{i_1}' \gamma_{i_1}^{-1} \gamma^{-1}\right) \cdots \left(\gamma_m g_m \gamma_m^{-1} \cdot \gamma \gamma_{i_m} g_{i_m}' \gamma_{i_m}^{-1} \gamma^{-1}\right) \cdot \gamma \gamma'$$

where $\gamma \gamma_{i_\mu}(1) = \mu$ for $\mu = 1, \ldots, m$. Thus $\gamma \gamma_{i_\mu} \in \gamma_\mu \Gamma_1$. Since Γ_1 centralizes G_1, we have

$$hh' = \gamma_1 (g_1 g_{i_1}') \gamma_1^{-1} \cdots \gamma_m (g_m g_{i_m}') \gamma_m^{-1} \cdot \gamma \gamma'.$$

The hypothesis allows us to do the same regrouping for the images of these elements in $\bar{H}$. This proves the claim. □

8.2.3 Wreath Products as Galois Groups

Lemma 8.10 *Let $x_1, \ldots, x_M, y_1, \ldots, y_m$ be independent transcendentals over ℓ, and set $\mathbf{x} = (x_1, \ldots, x_M)$. Suppose the x's are divided into $\mathbf{x}_1 = (x_1, \ldots, x_{i_1}), \ldots, \mathbf{x}_m = (x_{i_{m-1}+1}, \ldots, x_M)$. For $i = 1, \ldots, m$ let $f_i(\mathbf{x}_i, y_i) \in \ell(\mathbf{x}_i)[y_i]$ be such that $\ell(\mathbf{x}_i)[y_i]/(f_i)$ is a Galois field extension of $\ell(\mathbf{x}_i)$, regular over ℓ. Let G_i be the corresponding Galois group, of order $n_i = deg_y(f_i)$. Consider the field*

$$L_m = \ell(\mathbf{x})(\alpha_1, \ldots, \alpha_m)$$

where α_i is an element (inside some fixed algebraic closure of $\ell(\mathbf{x})$) with $f_i(\mathbf{x}_i, \alpha_i) = 0$.

(a) The field L_m is regular over ℓ, and Galois over $\ell(\mathbf{x})$ with

$$G := G(L_m/\ell(\mathbf{x})) = \prod_{i=1}^{m} G_i$$

where G_i is embedded in G as the subgroup fixing all α_j with $j \neq i$.

(b) The map

$$\ell(\mathbf{x})[y_1, \ldots, y_m]/(f_1(\mathbf{x}_1, y_1), \ldots, f_m(\mathbf{x}_m, y_m)) \to L_m$$

sending the class of y_i to α_i is isomorphic.

Proof. The claim is clear for $m = 1$. Now assume $m > 1$, and the claim holds for $m - 1$. View L_{m-1} as naturally embedded into L_m.

Since the elements of $\mathbf{x}_m$ are independent transcendentals over the algebraic closure A of L_{m-1}, we have $f_m(\mathbf{x}_m, y_m)$ irreducible over $A(\mathbf{x}_m)$ (by Lemma 1.1 (ii)). By the same lemma, it follows that $L_m = L_{m-1}(\mathbf{x}_m, \alpha_m)$ is regular over L_{m-1}. Since L_{m-1} is regular over ℓ (by induction), L_m is also regular over ℓ. Further, the degree of L_m over $K := L_{m-1}(\mathbf{x}_m)$ equals $n_m = \deg_y(f_m)$.

By Lemma 1.1(i), $[K : \ell(\mathbf{x})] = [L_{m-1}(\mathbf{x}_m) : \ell(\mathbf{x}_1, \ldots, \mathbf{x}_{m-1})(\mathbf{x}_m)] = [L_{m-1} : \ell(\mathbf{x}_1, \ldots, \mathbf{x}_{m-1})]$. The latter equals $\prod_{i=1}^{m-1} n_i$ by induction. Thus $[L_m : \ell(\mathbf{x})] = [L_m : K] \cdot [K : \ell(\mathbf{x})] = n_m \cdot \prod_{i=1}^{m-1} n_i = \prod_{i=1}^{m} n_i$.

For $i = 1, \ldots, m$ let $F_i = \ell(\mathbf{x}_i, \alpha_i)$. By hypothesis, F_i is Galois over $\ell(\mathbf{x}_i)$ with group G_i. Thus F_i is generated over $\ell(\mathbf{x}_i)$ by all roots of the polynomial $f_i(\mathbf{x}_i, Y) \in \ell(\mathbf{x}_i)[Y]$. Hence L_m is Galois over $\ell(\mathbf{x})$, and the group $G = G(L_m/\ell(\mathbf{x}))$ leaves F_i invariant. Thus restriction yields a homomorphism $G \to G_i$. Combining these homomorphisms for $i = 1, \ldots, m$ we obtain a map

$$G = G(L_m/\ell(\mathbf{x})) \to \prod_{i=1}^{m} G(F_i/\ell(\mathbf{x}_i)) \cong \prod_{i=1}^{m} G_i.$$

This map is injective, because an element of G that fixes all α_i must be trivial. Then this map is even isomorphic, since $|G| = \prod_{i=1}^{m} n_i = |\prod_{i=1}^{m} G_i|$ by the above. Thus G_i embeds into G as the kernel of the map $G \to \prod_{j \neq i} G_j$, hence as the subgroup fixing all α_j with $j \neq i$. This proves (a).

(b) Let $\bar{y}_i$ be the image of y_i in the ring

$$\ell(\mathbf{x})[y_1, \ldots, y_m]/(f_1(\mathbf{x}_1, y_1), \ldots, f_m(\mathbf{x}_m, y_m)).$$

Each element of this ring can be written as a polynomial in $\bar{y}_1, \ldots, \bar{y}_m$ over $\ell(\mathbf{x})$ of degree $<n_i$ in $\bar{y}_i$. Hence this ring has dimension $\leq \prod_{i=1}^{m} n_i$ over $\ell(\mathbf{x})$. Since $[L_m : \ell(\mathbf{x})] = \prod_{i=1}^{m} n_i$, it follows that the map in (b) above is isomorphic (being surjective). □

The lemma shows that if two finite groups G_1, G_2 occur regularly over ℓ, then the same holds for their direct product. We are going to derive a more general result, where direct products are replaced by wreath products.

Proposition 8.11 *Let ℓ/k be finite Galois with group Γ. Let Γ_1 be a subgroup of Γ, and let ℓ_1 be the fixed field of Γ_1. Let $m = [\Gamma : \Gamma_1]$, and consider a (transitive) permutation representation of Γ on $\{1, \ldots, m\}$ such that Γ_1 is the stabilizer of the letter 1. Form accordingly the wreath product $H = G_1^m \cdot \Gamma$, where G_1 is a* *finite group that occurs regularly over ℓ_1. Let $\varphi : H = G_1^m \cdot \Gamma \to \Gamma$ be the natural projection. Then there is a collection of transcendentals $\mathbf{t} = (t_1, \dots, t_M)$ over ℓ, and an extension field L of $\ell(\mathbf{t})$, regular over ℓ and Galois over $k(\mathbf{t})$, and an isomorphism $\phi : H \to G(L/k(\mathbf{t}))$ with $res_{L/\ell} \circ \phi = \varphi$.*

Proof. By hypothesis, $G_1 = G(K_1/\ell_1(\mathbf{x}_1))$, with K_1 regular over ℓ_1. Here $\mathbf{x}_1$ is some collection of transcendentals over ℓ. Let $\mathbf{x} = (x_1, \dots, x_M)$ be a larger collection of such transcendentals, partitioned into $\mathbf{x}_1, \dots, \mathbf{x}_m$, portions of equal size. Let Γ act on $\{x_1, \dots, x_M\}$ by permuting $\mathbf{x}_1, \dots, \mathbf{x}_m$ according to $\gamma(\mathbf{x}_i) = \mathbf{x}_{\gamma(i)}$ for $\gamma \in \Gamma$. Arrange this such that Γ_1 fixes not only $\mathbf{x}_1$, but each variable in $\mathbf{x}_1$.

Let $y_1, \dots, y_m$ be further variables. Extend the action of Γ from ℓ to $\ell(\mathbf{x})[y_1, \dots, y_m]$ by letting Γ act on $x_1, \dots, x_M$ as described above, and on $y_1, \dots, y_m$ with $\gamma \in \Gamma$ sending y_i to $y_{\gamma(i)}$. Write K_1 in the form $K_1 = \ell_1(\mathbf{x}_1)[y_1]/(f_1)$, with $f_1(\mathbf{x}_1, y_1) \in \ell_1(\mathbf{x}_1)[y_1]$. For $i = 1, \dots, m$ define $f_i(\mathbf{x}_i, y_i) \in \ell(\mathbf{x}_i)[y_i]$ as the image of f_1 under some $\gamma \in \Gamma$ with $\gamma(1) = i$. This does not depend on the choice of γ, because any two such γ differ by an element of Γ_1, and Γ_1 acts trivially on $\ell_1(\mathbf{x}_1)[y_1]$ by our hypotheses. Thus Γ permutes $f_1, \dots, f_m$. It follows that the action of Γ on $\ell(\mathbf{x})[y_1, \dots, y_m]$ induces an action on

$$L = \ell(\mathbf{x})[y_1, \dots, y_m]/(f_1, \dots, f_m).$$

We are now in the situation of Lemma 8.10. By Lemma 8.10(b), we can identify L with the field $L_m = \ell(\mathbf{x})[\alpha_1, \dots, \alpha_m]$, where α_i becomes the image of y_i in L_m. This identification yields an action of Γ on L_m, extending the action on $\ell(\mathbf{x})$ and with $\gamma \in \Gamma$ sending α_i to $\alpha_{\gamma(i)}$. From now on we view Γ as embedded into $\mathrm{Aut}(L)$ via this action.

By Lemma 8.10(a) we have

$$G := G(L/\ell(\mathbf{x})) = \prod_{i=1}^{m} G_i$$

where G_i is the subgroup of G fixing all α_j with $j \neq i$. (For $i = 1$ this is compatible with the above use of G_1.) Since Γ acts on L leaving $\ell(\mathbf{x})$ invariant, Γ normalizes G. Then clearly $\gamma G_i \gamma^{-1} = G_{\gamma(i)}$ for each $\gamma \in \Gamma$. In particular, G_1 centralizes each $\gamma G_1 \gamma^{-1}$ with $\gamma \notin \Gamma_1$.

Γ_1 fixes $\mathbf{x}_1$ and α_1, hence acts trivially in $\ell_1(\mathbf{x}_1)(\alpha_1)$; this field is Galois over $\ell_1(\mathbf{x}_1)$, hence contains all conjugates of α_1 over $\ell(\mathbf{x})$. Thus each $\gamma \in \Gamma_1$ fixes the $g(\alpha_1)$ with $g \in G_1$, and so $\gamma g \gamma^{-1}(\alpha_1) = g(\alpha_1)$. This implies $\gamma g \gamma^{-1} = g$

(because $\gamma g \gamma^{-1} \in G_1$ by the above, and because an element of G_1 fixing α_1 must be trivial). We have proved that Γ_1 centralizes G_1.

In view of the preceding two paragraphs, by Lemma 8.9 there is a unique homomorphism ϕ from the wreath product $H = G_1^m \cdot \Gamma$ to Aut(L), where G_1 and Γ are mapped identically to their canonical images in Aut(L). From the above it is clear that ϕ maps G_1^m onto $\prod_{i=1}^m G_i = G(L/\ell(\mathbf{x}))$. The full image H' of H under this homomorphism is generated by the image of G_1^m and Γ. Hence the fixed field F of H' in L is the intersection of the fixed fields of $G(L/\ell(\mathbf{x}))$ and Γ, hence equals the fixed field of Γ in $\ell(\mathbf{x})$. Since Γ permutes $x_1, \dots, x_M$, it follows from Lemma 8.7 that F is purely transcendental over k, say $F = k(\mathbf{t})$ for independent transcendentals $\mathbf{t} = (t_1, \dots, t_M)$. Thus $H' = G(L/k(\mathbf{t}))$. Clearly H' has the same order as H, hence the map $\phi : H \to H' = G(L/k(\mathbf{t}))$ is an isomorphism. Further, L is regular over ℓ by Lemma 8.10. Finally, from $\phi(G_1^m) = G(L/\ell(\mathbf{x}))$ it follows that $\mathrm{res}_{L/\ell} \circ \phi$ vanishes on G_1^m (as does φ). Since $\mathrm{res}_{L/\ell} \circ \phi$ is the identity on Γ (by construction), we get $\mathrm{res}_{L/\ell} \circ \phi = \varphi$, as desired. □

Addendum *If H has a normal subgroup N with $G_1^m \cdot N = H$, then the group H/N occurs regularly over k.*

Proof. From $G_1^m \cdot N = H$ it follows by the Galois correspondence that

$$\ell(\mathbf{x}) \cap \mathrm{Fix}(N) = k(\mathbf{t})$$

where Fix(N) is the fixed field in L of $\phi(N)$. Since L is regular over ℓ, it follows that Fix(N) is regular over k. Since $H/N \cong G(\mathrm{Fix}(N)/k(\mathbf{t}))$, the claim follows. □

Corollary 8.12 *Let H be the wreath product of G_1 and Γ, where Γ acts transitively on $\{1, \dots, m\}$.*

(a) If both G_1 and Γ occur regularly over the field κ, then so does H.
(b) Let $\varphi : H \to \Gamma$ be the natural projection. Suppose $\Gamma = G(\ell/k)$ for some FG-extension ℓ/k, and G_1 occurs regularly over k. If k is hilbertian then the embedding problem over k given by φ is solvable.

Proof. (a) We can assume $\Gamma = G(\ell/k)$, where $k = \kappa(\mathbf{y})$ for a vector $\mathbf{y}$ of transcendentals over κ, and ℓ regular over κ. Let Γ_1 be the stabilizer in Γ of the letter 1, and let ℓ_1 be the fixed field of Γ_1 in ℓ. If G_1 occurs regularly over κ then also over ℓ_1 by Corollary 1.15.

By the above proposition it follows that $H \cong G(L/k(\mathbf{t})) = G(L/\kappa(\mathbf{y}, \mathbf{t}))$, where L is regular over ℓ, hence over κ, and $\mathbf{t}$ is a vector of transcendentals over $k = \kappa(\mathbf{y})$. Thus $\kappa(\mathbf{y}, \mathbf{t})$ is purely transcendental over κ, and $H \cong G(L/\kappa(\mathbf{y}, \mathbf{t}))$ occurs regularly over κ.

(b) As in the above proposition we obtain the isomorphism $\phi : H \to G(L/k(\mathbf{t}))$. Below we show that if k is hilbertian then there is an FG-extension L'/k containing ℓ, and an isomorphism $\chi : G(L/k(\mathbf{t})) \to G(L'/k)$ with $\mathrm{res}_{L'/\ell} \circ \chi = \mathrm{res}_{L/\ell}$. Thus the extension L'/k, together with $\chi \circ \phi : H \to G(L'/k)$, solves the embedding problem.

It remains to prove the existence of L' and χ. As usual, choose a generator α of $L/k(\mathbf{t})$ with $f(\alpha) = 0$ for monic and irreducible $f(y) = f(\mathbf{t}, y) \in k[\mathbf{t}][y]$. Apply Lemma 1.5 with K/F the extension $L/k(\mathbf{t})$, and $R = k[\mathbf{t}]$, and the set A containing α and a generator for ℓ/k. Let $u \in k[\mathbf{t}]$ be the nonzero element provided by that lemma. Since k is hilbertian we can find $\mathbf{b} = (b_1, \dots, b_M) \in k^M$ with $u(\mathbf{b}) \neq 0$ such that $f(\mathbf{b}, y)$ is irreducible over k (Lemmas 1.4 and 1.10(ii)). Then by Lemma 1.5, the evaluation at $\mathbf{b}$ homomorphism $k[\mathbf{t}] \to k$ extends to $\tilde{\omega} : S \to L'$, where L'/k is finite Galois and S contains ℓ; further, there is an isomorphism $\chi : G(L/k(\mathbf{t})) \to G(L'/k)$ with $\chi(\sigma)(\tilde{\omega}(s)) = \tilde{\omega}(\sigma(s))$ for all $\sigma \in G$, $s \in S$. The latter condition yields the desired relation $\mathrm{res}_{L'/\ell} \circ \chi = \mathrm{res}_{L/\ell}$, after identifying ℓ and $\tilde{\omega}(\ell)$ (via $\tilde{\omega}$). □

The above corollary was the main goal of this section. It gives quite a general way of producing Galois groups. In particular, it includes the case of split abelian embedding problems.

Corollary 8.13

(a) Every finite abelian group occurs regularly over k.

(b) More generally, suppose the finite group E is the semi-direct product of an abelian normal subgroup G_1 and some Γ. If Γ occurs regularly over k then so does E.

(c) If k is hilbertian then every split abelian embedding problem over k is solvable.

Proof. (a) Fix an integer $n \geq 2$. Let ℓ be the field extension of k obtained by adjoining the nth roots of unity. The field $\ell(x^{1/n})$ is Galois over $\ell(x)$, with cyclic Galois group G_1 of order n. Clearly, $\ell(x^{1/n})$ is regular over ℓ, hence G_1 occurs regularly over ℓ.

Let $\Gamma = G(\ell/k)$. Consider the regular permutation representation of Γ (transitive with $\Gamma_1 = \{1\}$). Form accordingly the wreath product H of G_1 and Γ. By

Lemma 8.9 there is a homomorphism $H \to G_1$ that is the identity on G_1 and is trivial on Γ. Thus $G_1^m \cdot N = H$, where N is the kernel of this homomorphism. Now we are in the situation of the above proposition and its addendum. Hence the cyclic group G_1 of order n occurs regularly over k.

Since every finite abelian group is a direct product of cyclic groups, claim (a) follows (cf. the remarks before Proposition 8.11).

(b) Consider again the regular permutation representation of Γ, and form accordingly the wreath product H of G_1 and Γ. By Lemma 8.9 there is a homomorphism $\lambda : H \to E$ that is the identity on G_1 and on Γ. This homomorphism is surjective because its image contains G_1 and Γ. Hence E is a quotient of H.

By (a) we know that G_1 occurs regularly over k. If Γ also occurs regularly over k then so does H (by the previous corollary). Then the quotient E of H also occurs regularly over k.

(c) Consider the embedding problem given by some surjection $\varphi' : E \to \Gamma = G(\ell/k)$ with abelian kernel G_1, where E splits over G_1. Then there is a complement to G_1 in E that we may identify with Γ (via φ'). This allows us to construct a surjection $\lambda : H \to E$ as in the proof of (b). Now $\varphi = \varphi' \circ \lambda : H \to \Gamma$ is the natural projection, hence the corresponding embedding problem is solvable by Corollary 8.12(b). The solution induces one of the embedding problem given by φ'. □

As an application, we see that every (finite) dihedral group occurs regularly over $\mathbb{Q}$. Namely, it is the semi-direct product of a cyclic normal subgroup with the group of order 2 (acting by inversion). More generally, by repeated application of (b) and taking quotients of the groups constructed in this way, it can be shown that every nilpotent group of class 2 occurs regularly over $\mathbb{Q}$ (Thompson). Further, every solvable group with all Sylow subgroups abelian, and all groups of order p^n with p a prime and $n \leq 4$ (Dentzer). See [MM], Ch. 4, 2.3, for those results.

It is still not known, however, whether all p-groups occur regularly over $\mathbb{Q}$ (not even over $\mathbb{Q}_{\text{ab}}$).

8.3 GAR-Realizations and GAL-Realizations

As before, k, κ, and ℓ denote fields of characteristic 0.

8.3.1 Definition and the Main Property of a GAR-Realization

A very useful tool for solving embedding problems with center-free kernel is the notion of GAR-realization, due to Matzat [Ma4]. We shall use a modification of this, which we call a GAL-realization. In Section 8.3.3 we further define GAP-realizations. Here L resp., P, stands for "linear action," resp., "projective action." (GAL-realizations were first called GAT-realizations in [V4], but GAL is better notation in view of the GAP-case.)

Definition 8.14 *Let U be a finite group with trivial center. Let $\mathbf{x} = (x_1, \ldots, x_s)$ be a collection of independent transcendentals over the field k. Suppose U is isomorphic to the Galois group of a (Galois) extension $K/k(\mathbf{x})$, regular over k. This is called a* **GAR-realization (resp., a GAL-realization)** *of U over k if the following conditions* **(GA)** *and* **(R)** *(resp.,* **(GA)** *and* **(L)***) hold:*

(GA) *There is a subfield F of $k(\mathbf{x})$, containing k, such that K is Galois over F with Galois group isomorphic to Aut(U), and under this isomorphism, $G(K/k(\mathbf{x}))$ corresponds to Inn(U).*

(R) *Each subfield R of $\bar{k}(\mathbf{x})$ with $\bar{k}R = \bar{k}(\mathbf{x})$ and $F \subset R$ is purely transcendental over the algebraic closure $\hat{k}$ of k in R.*

(L) *The k-vector space spanned by $x_1, \ldots, x_s$ is invariant under $G(k(\mathbf{x})/F)$.*

The main reason for the above definition is the following.

Proposition 8.15 (Matzat) *Let ℓ/k be finite Galois with group Γ. Let $\varphi : H \to \Gamma$ be a surjective homomorphism of finite groups, with kernel U. Suppose U has a GAR-realization over k. Then there is a collection of transcendentals $\mathbf{t} = (t_1, \ldots, t_s)$ over ℓ, and an extension field L of $\ell(\mathbf{t})$, regular over ℓ and Galois over $k(\mathbf{t})$, and an isomorphism $\phi : H \to G(L/k(\mathbf{t}))$ with $res_{L/\ell} \circ \phi = \varphi$.*

Proof. In the situation of the above definition, fix an isomorphism between $G(K/k(\mathbf{x}))$ and U. This induces an isomorphism between $G(K/F)$ and Aut(U).

Let L (resp., F') be the composite of K (resp., F) and ℓ (inside some algebraic closure of F). Since K is regular over k we have $K \cap \ell(\mathbf{x}) = k(\mathbf{x})$, and L is regular over ℓ (Lemma 1.1). Hence $G(L/k(\mathbf{x})) = G(L/K) \times G(L/\ell(\mathbf{x})) \cong \Gamma \times U$. In particular, $G(L/K)$ is isomorphic to $\Gamma = G(\ell/k)$ via restriction to ℓ. Thus $G(L/K)$ fixes F', and induces a subgroup of Aut(F'/F) of order $[\ell : k]$. Since $[F' : F] \leq [\ell : k]$ it follows that F'/F is Galois, and $G(F'/F) \cong G(L/K)$ via restriction. Thus the identifications from the preceding paragraph induce an isomorphism (Figure 8.1)

$$\phi_1 : G(L/F) = G(L/K) \times G(L/F') \cong \Gamma \times \text{Aut}(U).$$

For $h \in H$ let ι_h be the automorphism of U induced by h; that is, $\iota_h(u) = huh^{-1}$. The homomorphism $\phi_0 : H \to \Gamma \times \text{Aut}(U), h \mapsto (\varphi(h), \iota_h)$ is injective.

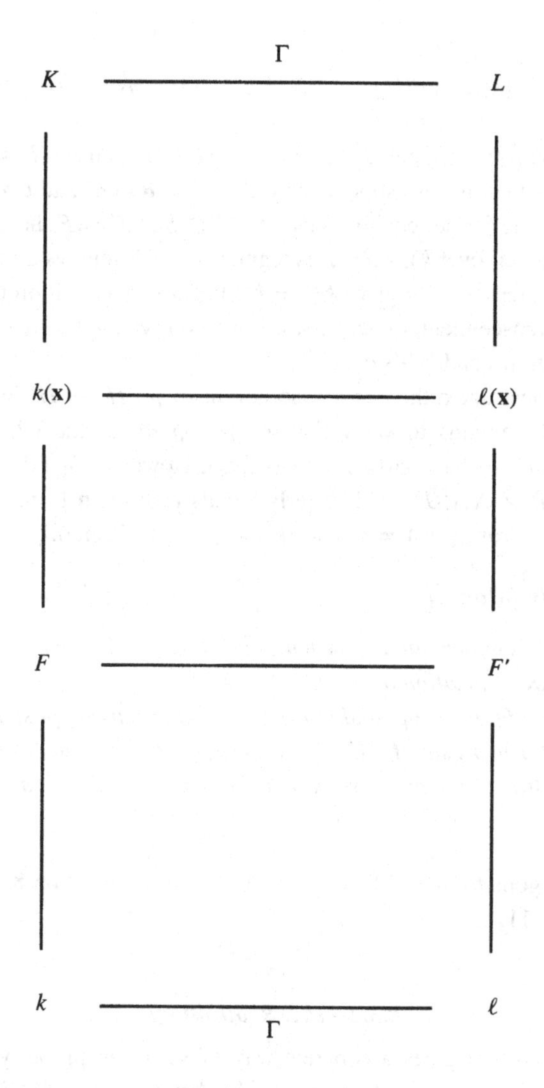

Fig. 8.1.

Indeed, $\ker(\phi_0) = \ker(\varphi) \cap C_H(U) = U \cap C_H(U) = Z(U) = 1$. Define $\phi := \phi_1^{-1}\phi_0$, an injection $H \to G(L/F)$. Let R be the fixed field of $H' := \phi(H)$ in L. Then $F \subset R$.

From the definition of ϕ_0 it is clear that

$$\Gamma \times \mathrm{Aut}(U) = \phi_0(H) \cdot (\{1\} \times \mathrm{Aut}(U))$$

and

$$\{1\} \times \mathrm{Inn}(U) \;=\; \phi_0(H) \cap (\{1\} \times \mathrm{Aut}(U)).$$

Composing with ϕ_1^{-1} we get $G(L/F) = H' \cdot G(L/F')$ and $G(L/\ell(\mathbf{x})) = H' \cap G(L/F')$. The Galois correspondence yields $F = R \cap F'$ and $\ell(\mathbf{x}) = RF'$ (= $RF\ell = R\ell$). The former equality gives $R \cap \ell \subset R \cap F' = F$, hence $R \cap \ell = k$ (since F is regular over k); since L is regular over ℓ it follows that R is regular over k. The latter equality gives $\bar{k}R = \bar{k}(\mathbf{x})$. Thus by condition (R), the field R is purely transcendental over k, say $R = k(\mathbf{t})$, where $\mathbf{t} = (t_1, \ldots, t_s)$ is a collection of transcendentals over k.

We have constructed the desired isomorphism $\phi : H \to H' = G(L/R) = G(L/k(\mathbf{t}))$. It remains to show that $\mathrm{res}_{L/\ell} \circ \phi = \varphi$. Indeed, for $h \in H$ and $h' := \phi(h)$ we have $\phi_1(h') = (\varphi(h), \iota_h)$. However, ϕ_1 followed by projection $\pi_1 : \Gamma \times \mathrm{Aut}(U) \to \Gamma$ clearly equals restriction from L to ℓ. Thus $\mathrm{res}_{L/\ell} \circ \phi(h) = \mathrm{res}_{L/\ell}(h') = \pi_1 \circ \phi_1(h') = \varphi(h)$, as desired. □

Corollary 8.16 (Matzat)

(a) Suppose k is hilbertian. Then each embedding problem over k whose kernel has a GAR-realization over k is solvable.

(b) Let H be a finite group, and U a normal subgroup. Suppose H/U occurs regularly over κ, say, $H/U \cong G(\ell/\kappa(\mathbf{y}))$ with ℓ regular over κ, and $\mathbf{y}$ a vector of transcendentals over κ. If U has a GAR-realization over $k = \kappa(\mathbf{y})$ then H occurs regularly over κ.

Proof. Analogous to that of Corollary 8.12, using Proposition 8.15 instead of Proposition 8.11. □

8.3.2 GAL-Realizations

The GAL-condition gives a concrete way of verifying property (R). (Other methods have been used in the case $s = 1$.) Actually, most GAR-realizations that have been found so far satisfy the stronger GAL-condition. The advantage of GAL-realizations is that they have better invariance properties.

Lemma 8.17

(i) Condition (L) implies (R), therefore each GAL-realization is a GAR-realization.

(ii) If U has a GAL-realization over k, then it has a GAL-realization over each field k′ containing k.

(iii) Suppose U is almost simple (i.e., lies between a nonabelian simple group and its automorphism group). If U has a GAL-realization over k, then the same is true for U^m, the direct product of $m \geq 1$ copies of U.

Proof. (i) Assume the GAL-condition, and let R and $\hat{k}$ be as in condition (R). By the proof of (ii) (see below) we may assume $\hat{k} = k$ (replacing k by $\hat{k}$). Then there is an FG-extension ℓ of k such that $\ell R = \ell(\mathbf{x})$. As in the proof of Proposition 8.15 we get $G(\ell(\mathbf{x})/F) = G(\ell(\mathbf{x})/k(\mathbf{x})) \times G(\ell(\mathbf{x})/F')$, where again $F' = \ell F$. Hence $G(\ell(\mathbf{x})/F')$ is isomorphic to $G(k(\mathbf{x})/F)$ via restriction. Thus by condition (L), the ℓ-vector space spanned by $x_1, \ldots, x_s$ is invariant under $G(\ell(\mathbf{x})/F')$, hence under $G(\ell(\mathbf{x})/F)$. Then this vector space is invariant under $\Gamma := G(\ell(\mathbf{x})/R)$ (since $F \subset R$).

This Γ is a finite group of automorphisms of $\ell(\mathbf{x})$, leaving ℓ invariant. The fixed field of Γ in ℓ equals $R \cap \ell = k$, hence restriction gives a surjection $\Gamma \to G(\ell/k)$. This surjection is injective since $\ell R = \ell(\mathbf{x})$. Hence all hypotheses of Lemma 8.7 are verified, and it follows that $R = \mathrm{Fix}(\Gamma)$ is purely transcendental over k.

(ii) In the given GAL-realization for U over k, choose $x_1, \ldots, x_s$ to be independent transcendentals over k', and write K in the form $K = k(\mathbf{x})[y]/(f)$ as in the proof of Corollary 1.15. By that proof, $K' := k'(\mathbf{x})[y]/(f)$ is (a field) regular over k', and Galois over $k'(\mathbf{x})$. Further, the restriction map $G(K'/k'(\mathbf{x})) \to G(K/k(\mathbf{x}))$ ($\cong U$) is isomorphic.

The k-linear action of the group $Q := G(k(\mathbf{x})/F)$ on $k(\mathbf{x})$ extends uniquely to a k'-linear action on $k'(\mathbf{x})$. The fixed field of Q on $k'(\mathbf{x})$ contains $F' := Fk'$ (composite inside K'). Hence $[k'(\mathbf{x}) : F'] \geq |Q| = [k(\mathbf{x}) : F]$. Since $[K' : k'(\mathbf{x})] = [K : k(\mathbf{x})]$, it follows that $[K' : F'] \geq [K : F]$.

Let y_1 be a generator of K over F, and $y_1, \ldots, y_t$ the conjugates of y_1 over F. Then K' is generated by $y_1, \ldots, y_t$ over F', hence K' is Galois over F', and the group $G(K'/F')$ embeds into $G(K/F)$ via restriction. Since $[K' : F'] \geq [K : F]$ it follows that the restriction map $G(K'/F') \to G(K/F)$ is isomorphic. Thus the fields K', $k'(\mathbf{x})$, and F' yield a GA-realization for U over k'. This is even a GAL-realization, because $G(K'/F')$ fixes the k-span of $x_1, \ldots, x_s$, hence also the k'-span of those elements.

(iii) ***Step 1*** The automorphism group of U^m

The hypothesis implies $Z(U) = 1$. Let $A = \mathrm{Aut}(U)$, and view U as a subgroup of A naturally. Consider the wreath product $\tilde{A} = A^m \cdot S_m$ of A with the symmetric group S_m (in its natural permutation representation). The group U^m embeds naturally as a normal subgroup of $\tilde{A}$, and the centralizer of U^m in $\tilde{A}$ is trivial. Hence $\tilde{A}$ embeds into $\mathrm{Aut}(U^m)$.

By hypothesis, U has a unique minimal normal subgroup W, and W is nonabelian simple. Each automorphism of U^m acts on $W^m \leq U^m$, hence permutes the m maximal normal subgroups of W^m. The centralizers in U^m of these maximal normal subgroups are just the m factors $\cong U$ of U^m. Thus $\text{Aut}(U^m)$ permutes these factors, and the kernel of this permutation action is clearly A^m. Thus $\text{Aut}(U^m)$ has order $\leq |A^m| \cdot |S_m| = |\tilde{A}|$. This proves that actually $\text{Aut}(U^m) = \tilde{A}$.

Step 2 The GAL-rcalization for U^m

Now consider a GAL-realization for U over k. In the notation of Definition 8.14, identify U with $G(K/k(\mathbf{x}))$. Write K as $K = k(\mathbf{x})(\alpha_1)$ where α_1 has minimum polynomial $f(\mathbf{x}, Y) \in k[\mathbf{x}, Y]$ over $k(\mathbf{x})$. Set $\mathbf{x}_1 = \mathbf{x} = (x_1, \ldots, x_s), \ldots, \mathbf{x}_m = (x_{M-s+1}, \ldots, x_M)$, and $\tilde{\mathbf{x}} = (x_1, \ldots, x_M)$, where $x_1, \ldots, x_M$ are independent transcendentals over k. Consider the field $\tilde{K} = k(\tilde{\mathbf{x}})(\alpha_1, \ldots, \alpha_m)$, where $f(\mathbf{x}_i, \alpha_i) = 0$ for $i = 1, \ldots, m$. Then $\tilde{K}$ is regular over k, and Galois over $k(\tilde{\mathbf{x}})$ with group canonically isomorphic to U^m (see Lemma 8.10). As in the proof of Proposition 8.11, one obtains an action of S_m on $\tilde{K}$ by k-algebra automorphisms, permuting $\mathbf{x}_1, \ldots, \mathbf{x}_m$ and $\alpha_1, \ldots, \alpha_m$.

View $K = k(\mathbf{x}_1)(\alpha_1)$ as naturally embedded into $\tilde{K}$. Let F be the subfield of $k(\mathbf{x}_1)$ given by the (GA)-condition, and $A := G(K/F)$. Then A induces $\text{Aut}(U)$ via conjugation action, and leaves the k-span of the variables in the vector $\mathbf{x}_1$ invariant. We claim that the action of A extends to $\tilde{K}$, fixing each variable in $\mathbf{x}_2, \ldots, \mathbf{x}_m$, and each of $\alpha_2, \ldots, \alpha_m$. This follows as in the proof of Proposition 8.11, because $\tilde{K} = K(\mathbf{x}_2, \ldots, \mathbf{x}_m)[\alpha_2, \ldots, \alpha_m]$ with $\mathbf{x}_2, \ldots, \mathbf{x}_m$ consisting of independent transcendentals over K, and $f(\mathbf{x}_i, \alpha_i) = 0$ with f having coefficients in $k \subset F = \text{Fix}(A)$.

View A and S_m as subgroups of $\text{Aut}(\tilde{K})$ via the preceding two paragraphs. One checks that the conditions of Lemma 8.9 hold (as in Proposition 8.11), hence we get a homomorphism ϕ from the wreath product $\tilde{A} = A^m \cdot S_m$ to $\text{Aut}(\tilde{K})$ that is the identity on A and S_m. This ϕ maps U^m isomorphically onto $G(\tilde{K}/k(\tilde{\mathbf{x}}))$ (by the first paragraph of Step 2). Thus the kernel of ϕ is a normal subgroup of $\tilde{A}$ that intersects U^m trivially. Hence $\ker(\phi)$ centralizes U^m. Since U^m has trivial centralizer in $\tilde{A}$, it follows that $\ker(\phi) = 1$. Hence ϕ yields the desired embedding of $\tilde{A} = \text{Aut}(U^m)$ into $\text{Aut}(\tilde{K})$.

Let $\tilde{F}$ be the fixed field of $\tilde{A}$ in $\tilde{K}$. Then the fields $\tilde{F} \subset k(\tilde{\mathbf{x}}) \subset \tilde{K}$ satisfy the (GA)-condition for U^m (in place of U). Clearly, A as well as S_m leaves the k-span of $x_1, \ldots, x_M$ invariant. Hence the same holds for their wreath product $\tilde{A}$. Thus we have found the desired GAL-realization for U^m. □

Corollary 8.18

(a) Suppose H is a finite group all of whose composition factors (are non-abelian and) have a GAL-realization over κ. Then H occurs regularly over κ.

(b) Suppose U is almost simple. If U has a GAL-realization over k then U^m has a GAR-realization over k', for each extension field k' of k, and for each $m \geq 1$. Hence if k' is hilbertian then all embedding problems over k' with kernel U^m are solvable.

Proof. (a) Choose a minimal normal subgroup of H. We know it is of the form U^m, with U simple. We may assume by induction that H/U^m occurs regularly over κ. By hypothesis, U has a GAL-realization over κ. Using (i), (ii), and (iii) of the above lemma, it follows from Corollary 8.16(b) that H occurs regularly over κ.

(b) Follows from the above lemma and Corollary 8.16 (a). □

Remark 8.19 From the GAR-condition for almost simple U over k, one can directly derive the solvability of embedding problems over (hilbertian) k with kernel U^m. This is Matzat's "embedding theorem for characteristic simple kernel" (see [MM], Ch. IV, Th. 3.5).

For a group U (with trivial center and) with no outer automorphisms, the GAL-condition just means that U occurs regularly over k. For example, this applies to the Monster M (see [At]). Hence M has a GAL-realization over $\mathbb{Q}$ by Example 3.23.

The next case to consider is when $[\mathrm{Aut}(U) : \mathrm{Inn}(U)] = 2$. This includes the alternating groups A_n, $n > 6$, and the groups $\mathrm{PSL}_2(p)$ for p an odd prime.

Lemma 8.20 *Let U be a finite group with trivial center, naturally embedded in $H := Aut(U)$.*

(a) If $[H : U] = 2$ and H admits a rigid triple of rational conjugacy classes, then U has a GAL-realization over $\mathbb{Q}$.

(b) If H/U is cyclic and H admits a rigid triple of conjugacy classes C_1, C_2, C_3 with $C_3 \subset U$ then U has a GAL-realization over $\mathbb{Q}_{\mathrm{ab}}$.

Proof. Set $k = \mathbb{Q}$ in case (a), and $k = \mathbb{Q}_{\mathrm{ab}}$ in case (b). Choose branch points $p_1 = 0$, $p_2 = \infty$, $p_3 = 1$, and associate the given classes with these points to obtain a rigid and k-rational type $\mathcal{T}$ (as in Corollaries 3.13 and 3.14). By RET over $\bar{\mathbb{Q}}$ (Theorem 7.10) and Theorem 3.8 there is an FG-extension $L/\bar{\mathbb{Q}}(x)$ of type $\mathcal{T}$ defined over k. We can identify H with $G(L/\bar{\mathbb{Q}}(x))$, as well as with $G(K/k(x))$, where $K := L_k$.

The fixed field K^U of U in K is a cyclic extension of $k(x)$ of degree $n = [H:U]$. We may assume $n > 1$. Below we show that $K^U = k(z)$ with $z^n \in k(x)$. Thus $U = G(K/k(z))$ occurs regularly over k, and condition (GA) holds with $F := k(x)$. Further, (L) holds because the group $G(k(z)/F)$ permutes the ζz, where ζ runs over the nth roots of 1 (in k). Hence U has a GAL-realization over k.

It remains to show that $K^U = k(z)$ with $z^n \in k(x)$. We generalize the proof of Lemma 3.20, where this was done for $n = 2$. Since k contains the nth roots of 1 (in either case (a) or (b)), it follows by Kummer theory (see, e.g., [Jac], II, Section 8.9) that $K^U = k(x)(f^{1/n})$ with $f = f(x) \in k(x)$. We can adjust f by nth powers to assume that $f(x) \in k[x]$ has no irreducible factor of multiplicity $\geq n$.

We have $N := \bar{\mathbb{Q}}(x)(f^{1/n}) = L^U$ by Lemma 3.1(a). In either case (a) or (b) we can label the given classes such that $C_3 \subset U$. Then by Lemma 2.6(d), the branch points of $N/\bar{\mathbb{Q}}(x)$ are $p_1 = 0$ and $p_2 = \infty$. By Example 2.9 it follows that f is of the form $f(x) = c\,x^m$ with $1 \leq m \leq n-1$ relatively prime to n, and $c \neq 0$ in k. Thus $ma + nb = 1$ for certain integers a, b. Set $z := x^b(f^{1/n})^a$. Then $z^n = c^a x \in k(x)$. Further, $K^U = k(x)(f^{1/n}) = k(z)$ (since $z^m = x^{bm}(f^{1/n})^{(1-nb)} = c^{-b} f^{1/n}$). This completes the proof. □

Let us consider the example of the alternating group A_n. For $n > 6$, its automorphism group is S_n. Using the rigid triple in S_n from Lemma 3.19 we see that A_n has a GAL-realization over $\mathbb{Q}$ for $n > 6$. Similarly, one gets a GAL-realization over $\mathbb{Q}$ for the sporadic group M_{12} (see Example 3.24).

It is not difficult to construct a rigid triple in the group $\mathrm{PGL}_2(p)$ for any odd prime p (see [MM2]). Since $\mathrm{PGL}_2(p)$ is the automorphism group of $\mathrm{PSL}_2(p)$, it follows that $\mathrm{PSL}_2(p)$ has a GAL-realization over $\mathbb{Q}_{ab}$. For $p \not\equiv \pm 1 \bmod 24$ one even gets a GAL-realization over $\mathbb{Q}$; see [MM2].

Corollary 8.16(a) yields

Example 8.21 Let U be a group that has a GAR-realization over $\mathbb{Q}$, and has index two in its automorphism group H. Examples are the groups A_n ($n > 6$), $\mathrm{PSL}_2(p)$ (p odd and $p \not\equiv \pm 1 \bmod 24$), $M_{12}, \ldots$. Then each quadratic number field can be embedded in a Galois extension of $\mathbb{Q}$ with group H.

It is not known whether all simple groups have a GAR-realization (or even GAL-realization) over $\mathbb{Q}$, or at least over $\mathbb{Q}_{ab}$. If true, that would be a very strong result. In the next section we discuss some implications, and survey the known results on GAR-realizations.

On the other hand, it is easy to give examples of nonsimple (yet almost simple) groups U that cannot have a GAR-realization over $\mathbb{Q}$. For $\mathbb{Q}_{\text{ab}}$, I do not know such an example.

Example 8.22 (A group with no GAR-realization over $\mathbb{Q}$) Let S be nonabelian simple, and H a subgroup of $\text{Aut}(S)$ containing $\text{Inn}(S) \cong S$ with H/S cyclic of order 4. Let U be the group properly between S and H. If U has a GAR-realization over $\mathbb{Q}$ then each quadratic number field could be embedded in a Galois extension of $\mathbb{Q}$ with group H (Corollary 8.16(a)). Hence each quadratic number field could be embedded in a $\mathbb{Z}/4\mathbb{Z}$-extension of $\mathbb{Q}$, contradicting Example 8.6.

8.3.3 Digression: Fields of Cohomological Dimension 1 and the Shafarevich Conjecture

In this subsection we collect some conjectures, remarks, and open problems that expand on the subject of Chapter 8. Proofs are only sketched.

A surjection of finite groups $\varphi : H \to G$ is called Frattini if no proper subgroup of H maps surjectively to G; equivalently, if $\ker(\phi)$ lies in the Frattini subgroup of H. (Recall that the Frattini subgroup of a finite group is the intersection of all maximal subgroups. It is always nilpotent.)

Definition 8.23

(a) An embedding problem, given by the FG-extension L/K and surjection $\varphi : H \to G(L/K)$, is called **Frattini** *if the map φ is Frattini.*
(b) The field K is called **of cohomological dimension 1**, *written as $cdim(K) = 1$, if K is not separably closed, and each Frattini embedding problem over K is solvable.*

The latter is equivalent to the usual (cohomological) definition by [Se2], I.3.4. or [Shatz], Th. 15 and Prop. 16. (Note that if $\varphi : H \to G$ is a surjection of finite groups, and $H_1 \leq H$ minimal with $\varphi(H_1) = G$ then the map $H_1 \to G$ is Frattini. Hence the solvability of all Frattini embedding problems over K is equivalent to the fact that $G_K := G(\tilde{K}/K)$ has the lifting property from [Shatz], p. 66, for all extensions of finite groups. The latter property means G_K is "projective" in the sense of [FJ], 20.4. Here $\tilde{K}$ is a separable closure of K.)

The field $\mathbb{Q}$ is not of cohomological dimension 1, as shown by Example 8.6. The field $\mathbb{Q}_{\text{ab}}$, and each proper subfield of $\bar{\mathbb{Q}}$ containing $\mathbb{Q}_{\text{ab}}$, however, are of cohomological dimension 1 (see [Se2], II.3.3.) This is a basic result in Galois theory. Its proof uses cohomological methods and the local-global principle for the Brauer group, hence does not fit into the framework of this book.

Lemma 8.24 *Suppose $cdim(K) = 1$, and let U be a finite group. If all* **split** *embedding problems over K with kernel U are solvable, then all embedding problems over K with kernel U are solvable.*

Proof. Consider an embedding problem given by the surjection $\varphi : H \to G = G(L/K)$, with kernel U. Let H_1 be a subgroup of H with $\varphi(H_1) = G$, but $\varphi(H_2) \neq G$ for all proper subgroups H_2 of H_1. Then the restriction of φ yields a Frattini map $\varphi_1 : H_1 \to G$.

Form the outer semi-direct product $\tilde{H} = U \cdot H_1$, where H_1 acts on U via conjugation inside H. Then the map $\lambda : \tilde{H} \to H$, $(u, h_1) \mapsto u \cdot h_1$ is a surjective homomorphism. Let $\lambda_1 : \tilde{H} \to H_1$ be the projection $(u, h_1) \mapsto h_1$. Then $\tilde{\varphi} = \varphi_1 \circ \lambda_1 : \tilde{H} \to G$ is the composition of the Frattini map φ_1 and the map λ_1 which gives a split extension of H_1 by U. Correspondingly, the embedding problem over K given by $\tilde{\varphi}$ decomposes into a Frattini embedding problem and a split embedding problem with kernel U (in the sense of Section 8.1.2). The latter two are solvable by hypothesis, hence by Lemma 8.3 also the embedding problem given by $\tilde{\varphi}$ is solvable.

But $\tilde{\varphi}$ also decomposes as $\tilde{\varphi} = \varphi \circ \lambda$. Thus solvability of the embedding problem given by $\tilde{\varphi}$ implies solvability of that given by φ. □

For technical reasons, we now again restrict attention to fields k of characteristic 0 (since we developed the notion of hilbertian fields only in this case).

Corollary 8.25 *If k is hilbertian of cohomological dimension 1 then all embedding problems over k with solvable kernel are solvable.*

Proof. All split abelian embedding problems over k are solvable by Corollary 8.13. Hence by the preceding lemma, all abelian embedding problems over k are solvable. Inductively, we also get those with solvable kernel (Corollary 8.5). □

By Corollary 1.28 this applies to $k = \mathbb{Q}_{ab}$. Thus each finite solvable group is a Galois group over $\mathbb{Q}_{ab}$. Moreover, the lattice of all solvable extensions of $\mathbb{Q}_{ab}$ is known (analogously to Section 8.1): The composite of all finite solvable extensions of $\mathbb{Q}_{ab}$ is infinite Galois over $\mathbb{Q}_{ab}$, with Galois group the free prosolvable profinite group of countable rank. Shafarevich's conjecture expects the same to hold without the restriction to the solvable case.

Conjecture I *(Shafarevich) All embedding problems over $\mathbb{Q}_{\mathrm{ab}}$ are solvable.*

The following, even stronger conjecture was formulated in [FV2].

Conjecture II *If k is hilbertian of cohomological dimension 1 then all embedding problems over k are solvable.*

This was proved in [FV2] under the stronger hypothesis that k is hilbertian and pseudo-algebraically closed, that is, every polynomial in $k[x, y]$ that is irreducible in $\bar{k}[x, y]$ has infinitely many zeroes in k^2. (See Section 10.3.3; this has recently been extended to fields of positive characteristic by Pop [P2].) It is known that $\mathbb{Q}_{\mathrm{ab}}$ is not pseudo-algebraically closed (a result of Frey, see [FJ], Cor. 10.15). However, for $\mathbb{Q}_{\mathrm{solv}}$ (see Example 1.29) this is an open problem. By a result of Pop [P1], the field k obtained by adjoining $\sqrt{-1}$ to the field of all totally real algebraic numbers is pseudo-algebraically closed. This k is hilbertian by Weissauer's theorem (Thm. 1.26), hence satisfies the conclusion of Conjecture II by [FV2].

In dealing with fields of cohomological dimension 1, the following weakening of the GAL-property seems useful. For this we need the natural action of the projective linear group $\mathrm{PGL}_{s+1}(k)$ on the rational function field $k(x_1, \dots, x_s)$. This action is induced from the natural action of $\mathrm{PGL}_{s+1}(k)$ on projective s-space (whose function field is $k(x_1, \dots, x_s)$). In more concrete terms, consider the natural action of $\mathrm{GL}_{s+1}(k)$ on the rational function field $k(z_1, \dots, z_{s+1})$ (by linear substitutions of the variables). This action leaves the subfield $k(x_1, \dots, x_s)$ invariant, where $x_1 = z_1/z_{s+1}, \dots, x_s = z_s/z_{s+1}$. Here the scalars act trivially, hence we get an induced action of $\mathrm{PGL}_{s+1}(k)$ on $k(x_1, \dots, x_s)$.

Definition 8.26 *In the situation of Definition 8.14, we say U has a* **GAP-realization** *over k if condition (GA) and the following condition (P) hold:* **(P)** *The group $G(k(\mathbf{x})/F)$ acts on the field $k(\mathbf{x})$ as a subgroup of $PGL_{s+1}(k)$.*

Note that in the case $s = 1$, condition (P) automatically holds since $\mathrm{Aut}(k(x)) \cong \mathrm{PGL}_2(k)$.

Clearly, the GAL-property implies the GAP-property. If $\mathrm{cdim}(k) = 1$ then the GAP-property still implies GAR. This follows from the fact (see [Se3], Ch. X, §5 or [Wa], p. 146) that the Galois cohomology set $H^1(k, \mathrm{PGL}_{s+1})$ vanishes if $\mathrm{cdim}(k) = 1$. (Similarly, the vanishing of $H^1(k, \mathrm{GL}_s)$ shows the implication GAL $\to$ GAR; see [V1], proof of Lemma 1.5.) We remind the reader that (as shown in the above references) $H^1(k, \mathrm{PGL}_{s+1})$ embeds naturally into

$H^2(k, \mathrm{GL}_1)$ (the Brauer group of k), and the latter vanishes if $\mathrm{cdim}(k) = 1$. This embedding comes from the coboundary map associated with the surjection $\mathrm{GL}_{s+1} \to \mathrm{PGL}_{s+1}$ (with kernel $\cong \mathrm{GL}_1$).

Lemma 8.27

(i) If $\mathrm{cdim}(k) = 1$ then each GAP-realization over k is a GAR-realization.
(ii) If U has a GAP-realization over k, then it has a GAP-realization over each extension field of k.

Proof. The proof of (ii) is exactly as in the GAL-case (Lemma 8.17). The proof of (i) is also similar; however, we need to replace Lemma 8.7 (which reflects the fact that $H^1(k, \mathrm{GL}_s) = 0$) by the fact that $H^1(k, \mathrm{PGL}_{s+1}) = 0$. We only give a brief sketch. We use the notation from Definition 8.14. As for the implication GAL $\to$ GAR, we may assume $k = \hat{k}$. Then

$$G(\bar{k}(\mathbf{x})/F) \;=\; G(\bar{k}(\mathbf{x})/k(\mathbf{x})) \times Q \;=\; \Gamma \cdot Q$$

(the latter a semi-direct product) where $\Gamma = G(\bar{k}(\mathbf{x})/R)$ and $Q = G(\bar{k}(\mathbf{x})/\bar{k}F) \cong G(k(\mathbf{x})/F)$ via restriction. Thus Q embeds into $\mathrm{PGL}_{s+1}(k)$, hence into $\mathrm{PGL}_{s+1}(\bar{k})$, by condition (P). View all those groups as subgroups of $\mathrm{Aut}(\bar{k}(\mathbf{x})/k)$ (in the natural way). Further, identify $G(\bar{k}(\mathbf{x})/k(\mathbf{x}))$ with $G_k = G(\bar{k}/k)$ naturally. Thus

$$G_k \cdot \mathrm{PGL}_{s+1}(\bar{k}) \;=\; \Gamma \cdot \mathrm{PGL}_{s+1}(\bar{k}),$$

that is, Γ is a complement to $\mathrm{PGL}_{s+1}(\bar{k})$ in the semi-direct product $G_k \cdot \mathrm{PGL}_{s+1}(\bar{k})$ (where G_k acts via matrix coefficients on $\mathrm{PGL}_{s+1}(\bar{k})$). These complements are classified (up to conjugation by $\mathrm{PGL}_{s+1}(\bar{k})$) by the set $H^1(G_k, \mathrm{PGL}_{s+1}(\bar{k})) = H^1(k, \mathrm{PGL}_{s+1})$. Vanishing of this cohomology means that Γ is conjugate to G_k under an element of $\mathrm{PGL}_{s+1}(\bar{k}) \subset \mathrm{Aut}(\bar{k}(\mathbf{x})/\bar{k})$. Thus $R = \mathrm{Fix}(\Gamma)$ is conjugate to $k(\mathbf{x}) = \mathrm{Fix}(G_k)$ under an element of $\mathrm{Aut}(\bar{k}(\mathbf{x})/\bar{k})$. Hence R is purely transcendental over k. □

Using this lemma, Corollary 8.5, Lemma 8.25, Corollary 8.18, and Remark 8.19, we can summarize as

Proposition 8.28 *Conjecture II holds under the additional assumption that each nonabelian finite simple group has a GAR-realization over k. For this it suffices that each such group has a GAL-realization, or at least a GAP-realization, over k.*

So far, several classes of simple groups have been shown to have a GAR-realization over $\mathbb{Q}$, or at least over $\mathbb{Q}_{ab}$. These all actually satisfy the GAP-property. Most of these GAR-realizations are in the case $s = 1$ (arising as in Lemma 8.20, and partially requiring deep results from group theory):

Theorem 8.29 *([MM], Th. IV.4.6) The following simple groups possess GAR-realizations in one variable over $\mathbb{Q}_{ab}$:*

(a) All nonabelian simple alternating groups, and all sporadic simple groups.
(b) The groups of Lie type $G(p)$ for primes $p > 2$ with the possible exceptions ${}^3D_4(p)$, $E_7(3)$, $E_8(3)$, and $E_8(5)$.
(c) The groups $Sp_{2n}(2)$, $\Omega^+_{2n}(2)$, $\Omega^-_{2n}(2)$.

Most of these results are due to Belyi and Malle; see [MM] for precise references. The picture over $\mathbb{Q}$ is less complete; see [MM, Th. IV.4.3].

For the Lie type groups, one notes the restriction to groups $G(p)$ over prime fields. If q is any prime power then the automorphisms of the field $\mathbb{F}_q$ act as outer automorphisms on $G(q)$. Since the GA?-conditions depend heavily on the automorphism group of U, it was an open question whether those conditions could hold in the presence of field automorphisms. This question was answered in [V4] in the positive. We describe this construction in Chapter 9. It is based on using moduli spaces for covers of the Riemann sphere. (These examples can also be described in the language of [MM], see Kap. IV, 4.3.) In particular, we construct GAL-realizations over $\mathbb{Q}$ for certain simple groups U for which $\mathrm{Out}(U)$ does not embed into $\mathrm{PGL}_2(\mathbb{Q}_{ab})$ (see Remark 9.22). Thus these groups have a GAR-realization in several variables over $\mathbb{Q}_{ab}$, but cannot have such a realization in one variable.

8.3.4 GAL-Realizations over $\bar{k}$

The degree of difficulty in constructing GA?-realizations over $\mathbb{Q}$ and $\mathbb{Q}_{ab}$ is illustrated by a question of Jarden, who suggested looking first at the case $k = \bar{\mathbb{Q}}$. Each U occurs regularly over $\bar{\mathbb{Q}}$ (by RET over $\bar{\mathbb{Q}}$), so the first part of Definition 8.14 presents no problem. Condition (R) becomes trivial over $\bar{\mathbb{Q}}$, but it is not clear how to accommodate the automorphisms of U, that is, how to satisfy (GA). The case $s = 1$ yields the following partial result.

Remark 8.30 Every center-free group U with $\mathrm{Out}(U)$ cyclic (resp., dihedral or isomorphic to A_4, S_4, or A_5) has a GAL-realization (resp., a GAP-realization) over $\bar{\mathbb{Q}}$.

Proof. Under the respective conditions on $\mathrm{Out}(U)$, this group embeds into $\mathrm{GL}_1(\bar{\mathbb{Q}}) \cong \bar{\mathbb{Q}}^*$ (resp., into $\mathrm{PGL}_2(\bar{\mathbb{Q}})$). This yields an action of $\mathrm{Out}(U)$ on the field $\bar{\mathbb{Q}}(x)$, satisfying condition (L) resp., (P). Let F be the fixed field of $\mathrm{Out}(U)$. Then $F = \bar{\mathbb{Q}}(t)$ by Lüroth's theorem (for some t). By Theorem 7.14, the embedding problem given by the natural map $\mathrm{Aut}(U) \to \mathrm{Out}(U) \cong G(\bar{\mathbb{Q}}(x)/\bar{\mathbb{Q}}(t))$ is solvable. The solution yields an extension K of F with Galois group $\mathrm{Aut}(U)$, satisfying the (GA)-condition. Condition (L) (resp., (P)) holds by construction. □

The condition of cyclic $\mathrm{Out}(U)$ is fulfilled by all alternating and sporadic simple groups (except A_6), and by 7 of the 16 families of simple groups of Lie type. The condition of dihedral $\mathrm{Out}(U)$ holds in some further cases. There are families of simple groups, however, whose $\mathrm{Out}(U)$ does not embed into $\mathrm{PGL}_2(\bar{\mathbb{Q}})$. Hence the case $s = 1$ does not suffice to settle Jarden's question. It gives solid evidence for a positive answer, however.

Problem (Jarden) *Does every (nonabelian) finite simple group have a GAL-realization over* $\bar{\mathbb{Q}}$*?*

A positive answer would imply that each embedding problem over a hilbertian field k' containing $\bar{\mathbb{Q}}$ is solvable, provided the kernel has only nonabelian composition factors. This would shed new light on the Galois theory of the fields $\bar{\mathbb{Q}}(x_1, \ldots, x_n)$. (It is known that not all abelian embedding problems over such a field are solvable if $n > 1$; namely, its cohomological dimension is > 1.)

Finally, if each nonabelian simple group had at least a GAP-realization over $\bar{\mathbb{Q}}$, then Conjecture II would hold in the case $\bar{\mathbb{Q}} \subset k$ (by Lemma 8.27(ii) and Proposition 8.28).

Exercise 8.31 (a) Suppose k is hilbertian and $\bar{\mathbb{Q}} \subset k$. Prove that each embedding problem over k given by a surjection $H \to G(\ell/k)$ is solvable if ℓ/k is **cyclic.** (Hint: Replacing H by some wreath product with a simple group, we may assume $C_H(U) = 1$, where U is the kernel of the embedding problem. Thus we may view H as a subgroup of $\mathrm{Aut}(U)$. Then Remark 8.30 shows that U has a GAL_H-realization over $\bar{\mathbb{Q}}$, meaning that the GAL-condition holds with H replacing the full automorphism group of U. Again this implies that U has a GAL_H-realization over k, hence a GAR_H-realization over k (similarly defined). From this we derive solvability of the given embedding problem, just as in the ordinary GAR-case.)

(b) Show that $k = \bar{\mathbb{Q}}(x_1, \ldots, x_n)$ is hilbertian. (Use the trick in the Application in Section 1.3.2.)

9
Braiding Action and Weak Rigidity

Weakly rigid (Galois) extensions of $\mathbb{C}(x)$ are those that are uniquely determined by their ramification type $\mathcal{T}$. If such an extension $L/\mathbb{C}(x)$ is even rigid then its minimal field of definition equals a certain field $\kappa_{\mathcal{T}}$ that is quite explicitly given in terms of the ramification type $\mathcal{T}$. Thus $G = G(L/\mathbb{C}(x))$ occurs as a Galois group over $\kappa_{\mathcal{T}}(x)$, and this yields the rigidity criteria from Chapter 3.

In the weakly rigid case, we still get a Galois realization over the field $\kappa_{\mathcal{T}}(x)$; not for G, however, but for a certain group $G_{\mathcal{T}}$ of automorphisms of G (that contains the inner automorphisms); see Proposition 9.2. This raises the question of computing $G_{\mathcal{T}}$, or equivalently, the image of $G_{\mathcal{T}}$ in $\mathrm{Out}(G)$. In general, this is a very difficult question, essentially equivalent to the regular version of the Inverse Galois Problem (since every finite extension of $\mathbb{C}(x)$ embeds into a weakly rigid one). A partial answer can be given in the "generic" case that the branch points of $L/\mathbb{C}(x)$ are algebraically independent. Then the image of $G_{\mathcal{T}}$ in $\mathrm{Out}(G)$ has a normal subgroup Δ that can be described purely combinatorially in terms of the ramification type $\mathcal{T}$. This description involves the braiding action on generating systems of G (Theorem 9.5).

The proof of this theorem is given in Chapter 10. It requires topological and analytic methods, as for Riemann's existence theorem. The hypothesis of generic branch points means geometrically that we are looking at families of covers of the Riemann sphere (instead of just a single cover as for RET). Thus we are naturally led to a study of moduli spaces for covers of the Riemann sphere.

If Δ is self-normalizing in $\mathrm{Out}(G)$ then it equals the image of $G_{\mathcal{T}}$, hence Δ occurs as a Galois group over the field $\kappa_{\mathcal{T}}(x)$. The resulting criterion for the realization of groups as Galois groups (including GAL-realizations) is given in Section 9.2. We call it the Outer Rigidity Criterion, since its most interesting applications are for realizations of the group Δ (a group of outer automorphisms of the given group G). For a particular choice of G (semi-direct product of a finite vector space with a group of scalars), this leads to Galois realizations, even GAL-realizations over $\mathbb{Q}$, of certain groups $\mathrm{PGL}_n(q)$ and $\mathrm{PU}_n(q)$ (see Section 9.4).

In Chapters 9 and 10 we mostly assume that $Z(G) = 1$, so that we can view G as a subgroup of $\mathrm{Aut}(G)$. This assumption is not necessary for many of the results, however, as indicated in the references. The main results of this chapter appear in the paper [V4], but we follow a somewhat different approach here.

9.1 Certain Galois Groups Associated with a Weakly Rigid Ramification Type

Let $L/\mathbb{C}(x)$ be an FG-extension of ramification type $\mathcal{T}$ (see Definition 2.12). If L is defined over a subfield κ of $\mathbb{C}$ then $\mathcal{T}$ is κ-rational (see Definition 3.7 and the remarks following it). Thus the minimal field of definition of L contains the minimal field κ such that $\mathcal{T}$ is κ-rational (if such minimal fields exist). If $\mathcal{T}$ is rigid then those two fields coincide (Theorem 3.8 and Proposition 7.12). Now we are going to study the interplay between those two fields in the weakly rigid case. Another field comes into play, related to the notion of weakly κ-rational type (introduced in Remark 3.9(b)). To fix notation, we recall the definition.

Let $\mathcal{T} = [G, P, \mathbf{C}]$ be a ramification type. Let $n = |G|$, and let ζ_n be a primitive nth root of 1 (in $\mathbb{C}$). Let κ be a subfield of $\mathbb{C}$. Then $\mathcal{T}$ is called **weakly κ-rational** iff $P \subset \bar{\kappa} \cup \{\infty\}$, and for each $\alpha \in G(\bar{\kappa}/\kappa)$ there is $f \in \mathrm{Aut}(G)$ such that for all $p \in P$ we have $\alpha(p) \in P$ and

$$C_{\alpha(p)} = f\left(C_p^m\right) \tag{9.1}$$

where m is an integer such that $\alpha^{-1}(\zeta_n) = \zeta_n^m$. (Recall our convention that $\alpha(\infty) = \infty$.) Further, $\mathcal{T}$ is called **κ-rational** if we can always take $f = \mathrm{id}$.

We fix a subfield ℓ of $\mathbb{C}$ (our base field, later to be taken as $\mathbb{Q}$ or $\mathbb{Q}_{\mathrm{ab}}$). First we state the analogue of Proposition 7.12.

Proposition 9.1 *Let $\mathcal{T} = [G, P, \mathbf{C}]$ be a ramification type, where $\mathbf{C} = (C_p)_{p\in P}$. Label the elements of P as $p_1, \ldots, p_r$, and assume they are $\neq \infty$. Let $\mathrm{Aut}_{\mathbf{C}}(G)$ (resp., $\mathrm{Aut}^{\mathbf{C}}(G)$) be the group of all automorphisms f of G that fix each C_p (resp., for which $f(C_{p_1}), \ldots, f(C_{p_r})$ is a permutation of $C_{p_1}, \ldots, C_{p_r}$). Let $s_1, \ldots, s_r$ be the elementary symmetric functions in $p_1, \ldots, p_r$ (i.e., $\prod_{i=1}^r (X - p_i) = X^r + s_1X^{r-1} + \cdots + s_r$). Let $n = |G|$, and let ζ_n be a primitive nth root of 1 (in $\mathbb{C}$).*

(i) Consider subfields κ of $\mathbb{C}$ containing ℓ such that $\mathcal{T}$ is κ-rational (resp., weakly κ-rational). There is a unique minimal such field κ, which we denote

by $\kappa_{\mathcal{T}}$ (resp., $\kappa^{\mathcal{T}}$). Then $\kappa_{\mathcal{T}}$ is Galois over $\kappa^{\mathcal{T}}$, and

$$\ell(s_1, \ldots, s_r) \subset \kappa^{\mathcal{T}} \subset \kappa_{\mathcal{T}} \subset \ell(\zeta_n)(p_1, \ldots, p_r).$$

(ii) *Now suppose that $p_1, \ldots, p_r$ are algebraically independent over ℓ, and the tuple $(C_{p_1}, \ldots, C_{p_r})$ is ℓ-rational (Definition 3.15). Then $\kappa_{\mathcal{T}}$ is purely transcendental over ℓ, say $\kappa_{\mathcal{T}} = \ell(t_1, \ldots, t_r)$. The elements $t_1, \ldots, t_r$ can be chosen so that their ℓ-linear span is invariant under $G(\kappa_{\mathcal{T}}/\kappa^{\mathcal{T}})$. The latter group is isomorphic to $Aut^{\mathbf{C}}(G)/Aut_{\mathbf{C}}(G)$.*

Proof. (i) For κ as in (i) the group $G(\bar{\kappa}/\kappa)$ permutes $p_1, \ldots, p_r$, hence fixes $s_1, \ldots, s_r$. Thus $\kappa_0 := \ell(s_1, \ldots, s_r) \subset \kappa$. Set $\ell' = \ell(\zeta_n)$ and $\kappa_1' = \ell'(p_1, \ldots, p_r)$. Then κ_1'/κ_0 is finite Galois, and its Galois group permutes $p_1, \ldots, p_r$. Let Λ be the group of all $\alpha \in G(\kappa_1'/\kappa_0)$ for which there is $f \in \mathrm{Aut}(G)$ satisfying (9.1) for each $p \in P$. Let Γ be the subgroup consisting of all such α for which we can take $f = \mathrm{id}$. We note that Γ is normal in Λ.

If $\mathcal{T}$ is κ-rational then each $\alpha \in G(\bar{\kappa}/\kappa)$ restricts to an element of Γ, hence κ contains $(\kappa_1')^{\Gamma}$, the fixed field of Γ. Thus the minimal field $\kappa_{\mathcal{T}}$ exists, and equals $(\kappa_1')^{\Gamma}$. Analogously, we get $\kappa^{\mathcal{T}} = (\kappa_1')^{\Lambda}$. The rest of (i) follows since Γ is normal in Λ.

(ii) Assume the hypothesis of (ii). Set $\kappa_0' := \ell'(s_1, \ldots, s_r)$ and $\kappa_{\mathcal{T}}' := \kappa_{\mathcal{T}}(\zeta_n) = \kappa_{\mathcal{T}}\ell'$. Let $P = \cup_{\nu=1}^{s} P_\nu$ be the partitioning of P into sets P_ν that are maximal with respect to the property that all C_p with $p \in P_\nu$ are equal.

Claim 1 The group $\Gamma' := G(\kappa_1'/\kappa_{\mathcal{T}}')$ equals $\{\alpha \in G(\kappa_1'/\kappa_0') : \alpha(P_\nu) = P_\nu$ for all $\nu\}$.

Proof. We have $\Gamma' = G(\kappa_1'/\kappa_{\mathcal{T}}\kappa_0') = G(\kappa_1'/\kappa_{\mathcal{T}}) \cap G(\kappa_1'/\kappa_0') = \{\alpha \in G(\kappa_1'/\kappa_0') : C_{\alpha(p)} = C_p$ for all $p \in P\}$. This proves Claim 1.

Let $r_\nu = |P_\nu|$, and define elements $q_{\nu\mu}$ by

$$\prod_{p \in P_\nu} (X - p) = X^{r_\nu} + q_{\nu 1}X^{r_\nu - 1} + \cdots + q_{\nu r_\nu}.$$

These $q_{\nu\mu}$ are fixed by Γ' (by Claim 1), hence lie in $\kappa_{\mathcal{T}}'$. Given $\alpha \in \Lambda$, the $C_{\alpha(p)}$ with $p \in P_\nu$ are all equal by (9.1). Hence $\alpha(P_\nu) = P_{\nu'}$ for some ν'. Thus $\alpha(q_{\nu\mu}) = q_{\nu'\mu}$. Hence Λ permutes the $q_{\nu\mu}$.

Since $p_1, \ldots, p_r$ are algebraically independent over ℓ, the same holds for the $q_{\nu\mu}$ ($\nu = 1, \ldots, s$, $\mu = 1, \ldots, r_\nu$), and we have

Claim 2 $\kappa_{\mathcal{T}}' = \ell'(q_{\nu\mu} : \text{all } \nu, \mu)$.

Proof. The group $G(\kappa_1'/\kappa_0') = G(\ell'(p_1, \ldots, p_r)/\ell'(s_1, \ldots, s_r))$ is isomorphic to the symmetric group S_r, permuting $p_1, \ldots, p_r$ (see Example 1.17).

Under this isomorphism, Γ' corresponds to a conjugate of the subgroup $S_{r_1} \times \cdots \times S_{r_s}$ (by Claim 1). As in Example 1.17 one shows that the fixed field of this group equals $\ell'(q_{\nu\mu} : \text{all } \nu, \mu)$. Since $\kappa'_T = (\kappa'_1)^{\Gamma'}$, Claim 2 follows.

Claim 3 $G(\kappa_T/\kappa^T) \cong \mathrm{Aut}^{\mathbf{C}}(G)/\mathrm{Aut}_{\mathbf{C}}(G)$.

Proof. With each $\alpha \in \Lambda$ there is associated some $f \in \mathrm{Aut}(G)$ satisfying (9.1). This f is not unique, but if f' is another choice then $f^{-1}f' \in \mathrm{Aut}_{\mathbf{C}}(G)$ (by (9.1)). Thus f is unique modulo $\mathrm{Aut}_{\mathbf{C}}(G)$. Further, f lies in $\mathrm{Aut}^{\mathbf{C}}(G)$. Indeed, (9.1) yields $f(C_p) = C_{\alpha(p)}^{-m}$, and the $C_{\alpha(p_1)}^{-m}, \dots, C_{\alpha(p_r)}^{-m}$ form a permutation of $C_{p_1}, \dots, C_{p_r}$ by ℓ-rationality of $\mathbf{C}$.

Thus we get a homomorphism $\Lambda \to \mathrm{Aut}^{\mathbf{C}}(G)/\mathrm{Aut}_{\mathbf{C}}(G)$, sending α to the class of the associated f. The kernel of this homomorphism equals Γ. Also, the homomorphism is surjective: If $g \in \mathrm{Aut}^{\mathbf{C}}(G)$ then by definition of $\mathrm{Aut}^{\mathbf{C}}(G)$ there is some $\alpha \in G(\kappa'_1/\kappa'_0) \cong S_r$ with $C_{\alpha(p)} = g(C_p)$ for all $p \in P$. This α lies in Λ, and is associated with the given g.

It follows that $\Lambda/\Gamma \cong \mathrm{Aut}^{\mathbf{C}}(G)/\mathrm{Aut}_{\mathbf{C}}(G)$. But also $\Lambda/\Gamma \cong G(\kappa_T/\kappa^T)$. Hence Claim 3.

Claim 4 Restriction $\Gamma \to G(\ell'/\ell)$ is surjective.

Proof. Since $\kappa'_1 = \kappa_1\kappa'_0$ and $\kappa_1 \cap \kappa'_0 = \kappa_0$ we have

$$G(\kappa'_1/\kappa_0) = G(\kappa'_1/\kappa_1) \times G(\kappa'_1/\kappa'_0).$$

Let $\alpha \in G(\ell'/\ell)$. Then $\alpha^{-1}(\zeta_n) = \zeta_n^m$ for some integer m. By the hypothesis of ℓ-rationality, $C_{p_1}^m, \dots, C_{p_r}^m$ is a permutation of $C_{p_1}, \dots, C_{p_r}$. Hence there is $\beta \in G(\kappa'_1/\kappa'_0) \cong S_r$ with $C_{\beta(p_i)} = C_{p_i}^m$ for $i = 1, \dots, r$. Extend α to an element of $G(\ell'(p_1, \dots, p_r)/\ell(p_1, \dots, p_r)) = G(\kappa'_1/\kappa_1)$ and set $\gamma = \alpha\beta$. Then $C_{\gamma(p_i)} = C_{\beta(p_i)} = C_{p_i}^m$ where $\gamma^{-1}(\zeta_n) = \alpha^{-1}(\zeta_n) = \zeta_n^m$. Hence $\gamma \in \Gamma$. Also, $\gamma|_{\ell'} = \alpha$. This proves Claim 4.

Conclusion The kernel of the map in Claim 4 is $G(\kappa'_1/\ell'\kappa_T) = G(\kappa'_1/\kappa'_T)$. Hence $\bar{\Gamma} := G(\kappa'_T/\kappa_T)$ is isomorphic to $G(\ell'/\ell)$ via restriction. This $\bar{\Gamma}$ is a group of automorphisms of the rational function field $\kappa'_T = \ell'(q_{\nu\mu})$ (Claim 2), and it permutes the transcendentals $q_{\nu\mu}$ (since Λ does). Thus Lemma 8.7 applies, showing that $\kappa_T = (\kappa'_T)^{\bar{\Gamma}}$ is purely transcendental over ℓ, say $\kappa_T = \ell(t_1, \dots, t_r)$. More precisely, the $t_1, \dots, t_r$ can be chosen such that their ℓ'-linear span equals the ℓ'-span S' of the $q_{\nu\mu}$. The ℓ-span of $t_1, \dots, t_r$ equals $S' \cap \kappa_T$, hence is invariant under Λ (since S' is). This proves (ii). □

Proposition 9.2 *Let $L/\mathbb{C}(x)$ be an FG-extension of weakly rigid type $\mathcal{T}$. Let $G = G(L/\mathbb{C}(x))$, let $P = \{p_1, \dots, p_r\}$ be the set of branch points of L, and $\mathbf{C} = (C_p)_{p \in P}$ the family of associated conjugacy classes of G. Assume $Z(G) = 1$ and $\infty \notin P$. Let κ be the minimal field of definition of L containing ℓ (see*

Proposition 7.12). Set $M = L_\kappa$*. Identify* $G = G(L/\mathbb{C}(x))$ *canonically with* $G(M/\kappa(x))$ *(Lemma 3.1(b)). Let* $\kappa_{\mathcal{T}}$ *and* $\kappa^{\mathcal{T}}$ *be as in the preceding proposition. Then we have:*

(a) For any two of the following fields, the larger is Galois over the smaller.

$$\kappa^{\mathcal{T}}(x) \subset \kappa_{\mathcal{T}}(x) \subset \kappa(x) \subset M.$$

This gives a series of normal subgroups

$$G = G(M/\kappa(x)) \leq G_{\mathcal{T}} := G(M/\kappa_{\mathcal{T}}(x)) \leq G^{\mathcal{T}} := G(M/\kappa^{\mathcal{T}}(x)).$$

(b) Via conjugation action on G*, the group* $G_{\mathcal{T}}$ *embeds into* $Aut_{\mathbf{C}}(G)$*. This induces an embedding of* $G(\kappa/\kappa_{\mathcal{T}}) \cong G(\kappa(x)/\kappa_{\mathcal{T}}(x))$ *into* $Out_{\mathbf{C}}(G) = Aut_{\mathbf{C}}(G)/Inn(G)$*. Let* $\hat{\Delta} \leq Out_{\mathbf{C}}(G)$ *denote the image of this embedding.*
(c) Via conjugation action on G*, the group* $G^{\mathcal{T}}$ *embeds into* $Aut(G)$*. It embeds into* $Aut^{\mathbf{C}}(G)$ *if* $(C_{p_1}, \ldots, C_{p_r})$ *is* ℓ*-rational. This induces an embedding of* $G(\kappa/\kappa^{\mathcal{T}})$ *into* $Out^{\mathbf{C}}(G) = Aut^{\mathbf{C}}(G)/Inn(G)$*. The inverse image of* $Out_{\mathbf{C}}(G)$ *in* $G(\kappa/\kappa^{\mathcal{T}})$ *equals* $G(\kappa/\kappa_{\mathcal{T}})$*.*
(d) The field $k_{\mathcal{T}} := \kappa_{\mathcal{T}}\bar{\ell}$ *(resp.,* $k := \kappa\bar{\ell}$*) is the minimal field containing* $\bar{\ell}$ *such that* $\mathcal{T}$ *is* $k_{\mathcal{T}}$*-rational (resp., such that* L *is defined over* k*). As in (b) we get a canonical embedding of* $G(k/k_{\mathcal{T}})$ *into* $Out_{\mathbf{C}}(G)$*. The image* Δ *of this embedding is a normal subgroup of* $\hat{\Delta}$*. If* $\Delta = \hat{\Delta}$ *then* $\kappa \cap k_{\mathcal{T}} = \kappa_{\mathcal{T}}$*.*

Proof. (a) We have already noted that κ contains $\kappa_{\mathcal{T}}$ (i.e., $\mathcal{T}$ is κ-rational), since L is defined over κ. Further, κ is finite over $\ell(s_1, \ldots, s_r)$ by Proposition 7.12, hence finite over $\kappa_{\mathcal{T}}$. (Notation as in the preceding proposition.)

For simplicity we use that every element α of $G(\overline{\kappa^{\mathcal{T}}}/\kappa^{\mathcal{T}})$ extends to an automorphism α of $\mathbb{C}$ that we still denote by α. (A consequence of Zorn's Lemma; we could avoid it by using the field $\bar{\kappa}$ instead of $\mathbb{C}$.) By Lemma 3.5 there is an α-isomorphism $\lambda : L \to L'$, for some FG-extension $L'/\mathbb{C}(x)$. Let $G' = G(L'/\mathbb{C}(x))$, and let C'_p be the class of G' associated with any $p \in P$.

Since $\mathcal{T}$ is weakly $\kappa^{\mathcal{T}}$-rational, we have $\alpha(P) = P$, and there is some $f \in Aut(G)$ satisfying (9.1) for all $p \in P$. Coupled with Lemma 2.8, this yields that for all $p, q \in P$ with $q = \alpha(p)$ we have

$$C'_q = \lambda^*(C_p^m) = \lambda^*(f^{-1}(C_q)) \tag{9.2}$$

where m is as in (9.1). Hence L and L' are of the same type $\mathcal{T}$ (via the isomorphism $\lambda^* f^{-1} : G \to G'$). Since $\mathcal{T}$ is weakly rigid it follows that L' is

$\mathbb{C}(x)$-isomorphic to L (Theorem 2.17). Thus we may assume $L = L'$ (composing λ with such a $\mathbb{C}(x)$-isomorphism). Then λ becomes an α-automorphism of L which is the identity on ℓ. It leaves the minimal field κ of definition over ℓ invariant (by uniqueness of κ), hence also the field $M = L_\kappa$ (by uniqueness of L_κ, see Lemma 3.1(d)). We have proved that each $\alpha \in G(\overline{\kappa^T}/\kappa^T)$ leaves κ invariant, hence κ/κ^T is Galois. Further, each element of $G(\kappa/\kappa^T) \cong G(\kappa(x)/\kappa^T(x))$ extends to an automorphism of M. Hence M is Galois over $\kappa^T(x)$. This proves (a). (Note that κ_T/κ^T is Galois by the preceding proposition.)

(b) Let C be the centralizer of G in G^T. We have $G \cap C \leq Z(G) = 1$, hence the group $H = \langle G, C \rangle$ is the direct product of G and C. It follows as in the proof of Proposition 3.6 that L is defined over κ', where $\kappa'(x) = M^H$, with $L_{\kappa'} = M^C$. Hence $\kappa' = \kappa$ and $M^C = M$ by minimality of κ. Thus $C = 1$, which means that G^T embeds into Aut(G) via conjugation action on G. Thus also G_T embeds this way into Aut(G).

Now we use the set-up of part (a) of the proof, with α the identity on κ_T, and with $L = L'$. Then (9.2) holds with $f = \mathrm{id}$ (since T is κ_T-rational). It follows that λ^* fixes all C_q, hence $\lambda^* \in \mathrm{Aut}_{\mathbf{C}}(G)$. This λ^* is the image in Aut(G) of $\lambda|_M \in G_T$. It follows that G_T embeds into $\mathrm{Aut}_{\mathbf{C}}(G)$ (since those $\lambda|_M$ generate G_T modulo G). The rest of (b) is clear.

(c) Let α, m and λ be as in the proof of (a). Then $\lambda^*(C_p) = C_{\alpha(p)}^{-m}$ for all $p \in P$, by Lemma 2.8. If $(C_{p_1}, \ldots, C_{p_r})$ is ℓ-rational then $C_{\alpha(p_1)}^{-m}, \ldots, C_{\alpha(p_r)}^{-m}$ is a permutation of $C_{p_1}, \ldots, C_{p_r}$. Hence $\lambda^* \in \mathrm{Aut}^{\mathbf{C}}(G)$. As for (b), this implies that G^T embeds into $\mathrm{Aut}^{\mathbf{C}}(G)$ (via conjugation action). The last assertion in (c) follows from the fact that if $\lambda^* \in \mathrm{Aut}_{\mathbf{C}}(G)$ then $C_{\alpha(p)} = C_p^m$ for all $p \in P$, hence α is the identity on κ_T (by the proof of Proposition 9.1(i)).

(d) The first assertion is clear. So is the second. Now consider the canonical embeddings

$$\phi : G(k/k_T) \to \mathrm{Out}_{\mathbf{C}}(G) \qquad \text{and} \qquad \hat{\phi} : G(\kappa/\kappa_T) \to \mathrm{Out}_{\mathbf{C}}(G)$$

(see (b)). Since $k = \kappa\bar{\ell}$ and $k_T = \kappa_T\bar{\ell}$, restriction yields an injection $\rho : G(k/k_T) \to G(\kappa/\kappa_T)$. We have $\hat{\phi} \circ \rho = \phi$. Indeed, if α is an element of Aut$(\mathbb{C}/k_T)$ and λ an α-automorphism of L then ϕ (resp., $\hat{\phi}$) maps the element $\alpha|_k \in G(k/k_T)$ (resp., the element $\alpha|_\kappa \in G(\kappa/\kappa_T)$) to λ^* mod Inn(G) (see the proof of (b)). Since these elements correspond under ρ, we get $\hat{\phi} \circ \rho = \phi$.

The image of ρ has fixed field $\kappa \cap k_T$, hence equals $G(\kappa/\kappa \cap k_T)$. This group is normal in $G(\kappa/\kappa_T)$ since $\kappa \cap k_T$ is Galois over κ_T. Hence the group $\Delta = \mathrm{Im}(\phi) = \mathrm{Im}(\hat{\phi} \circ \rho) = \hat{\phi}(\mathrm{Im}(\rho))$ is normal in $\hat{\Delta} = \hat{\phi}(G(\kappa/\kappa_T))$. Thus if $\Delta = \hat{\Delta}$ then $\rho : G(k/k_T) \to G(\kappa/\kappa_T)$ is an isomorphism, and thus $\kappa \cap k_T = \kappa_T$. □

Addendum *Generalizing (d), assume A is a normal subgroup of $\mathrm{Out}_{\mathbf{C}}(G)$ such that $\Delta A/A = \hat{\Delta} A/A$. Let k_A (resp., κ_A) be the field such that $G(k/k_A)$ (resp., $G(\kappa/\kappa_A)$) is the inverse image of A in $G(k/k_T)$ (resp., in $G(\kappa/\kappa_T)$) under the embedding from (d) (resp., (b)). Then $\kappa_A \cap k_T = \kappa_T$.*

Proof. Notation as in the proof of (d). Let $\alpha \in G(k/k_T)$. Since $\hat{\phi} \circ \rho = \phi$ we have $\alpha \in G(k/k_A)$ if and only if $\rho(\alpha) \in G(\kappa/\kappa_A)$. Thus $G(k/k_A)$ acts trivially on κ_A, which means that $\kappa_A \subset k_A$. Hence $\kappa_A k_T \subset k_A$. On the other hand, if $\alpha \in G(k/\kappa_A k_T)$ then $\rho(\alpha) \in G(\kappa/\kappa_A)$, hence $\alpha \in G(k/k_A)$. Thus $\kappa_A k_T = k_A$.

Hence restriction induces an injection $\bar{\rho}$ from $G(k_A/k_T)$ into $G(\kappa_A/\kappa_T)$. The former group is isomorphic to $\Delta A/A$, the latter to $\hat{\Delta} A/A$. Hence if $\Delta A/A = \hat{\Delta} A/A$ then $\bar{\rho}$ is an isomorphism. This means that $\kappa_A \cap k_T = \kappa_T$ (as in the proof of (d)). □

Consider an FG-extension $L/\mathbb{C}(x)$ of weakly rigid type $\mathcal{T}$, as in the above proposition. Let κ again be its minimal field of definition over ℓ. The group $G(\kappa/\kappa_T)$ gives us information on the "size" of κ, since the field κ_T is fairly well understood (see Proposition 9.1). This group $G(\kappa/\kappa_T)$ contains Δ as a normal subgroup, and equals Δ if Δ is self-normalizing in $\mathrm{Out}_{\mathbf{C}}(G)$. Thus Δ holds important information on the minimal field of definition of $L/\mathbb{C}(x)$. In the case of generic branch points $p_1, \ldots, p_r$ (and $\mathbf{C}$ being ℓ-rational), the field κ_T is purely transcendental over ℓ, hence if Δ is self-normalizing in $\mathrm{Out}_{\mathbf{C}}(G)$ then Δ occurs as a Galois group over $\ell(x_1, \ldots, x_r)$. Thus also from a point of view of realizing groups as Galois groups, it is important to know the group Δ.

Since there are no further obvious conditions distinguishing Δ as a subgroup of $\mathrm{Out}_{\mathbf{C}}(G)$, one could expect that in the case of generic branch points, "usually" Δ equals $\mathrm{Out}_{\mathbf{C}}(G)$. This is not true, however, as shown by the examples in Section 9.4. Actually, the description of Δ is quite complicated, and requires the notion of braid orbits. As for RET, it is again the theory of the fundamental group that allows us to pin down the group Δ. Topological and analytic methods apply since Δ is defined as a Galois group between function fields over the algebraically closed field $\bar{\ell}$, hence in a geometric situation. As for RET, no purely algebraic proof is known.

9.2 Combinatorial Computation of Δ via Braid Group Action and the Resulting Outer Rigidity Criterion

Let $\mathbf{C} = (C_1, \ldots, C_r)$ be a tuple of conjugacy classes of a finite group G. Define $\mathrm{Ni}(\mathbf{C})$ to be the set of all $(g_1, \ldots, g_r) \in G^r$ with the following properties: $g_1 \cdots g_r = 1$, the group G is generated by $g_1, \ldots, g_r$, and there is a permutation $\pi \in S_r$ with $g_{\pi(i)} \in C_i$ for all i. (The notation Ni(C) refers to the name "Nielsen class" which M. Fried coined for this set. Allowing all those permutations makes things independent of the choice of ordering of the classes in **C**.) Let $\mathcal{E}(\mathbf{C})$ be the subset consisting of all $(g_1, \ldots, g_r)$ with $g_i \in C_i$ for all i.

Definition 9.3 *A* **braid orbit** *on Ni(***C***) is a minimal (nonempty) subset closed under the* **braid operations**

$$Q_i\colon \qquad (g_1, \ldots, g_r) \mapsto \left(g_1, \ldots, g_{i-1},\ g_i g_{i+1} g_i^{-1},\ g_i\ , \ldots, g_r\right)$$

and their inverses

$$Q_i'\colon \qquad (g_1, \ldots, g_r) \mapsto \left(g_1, \ldots, g_{i-1},\ g_{i+1},\ g_{i+1}^{-1} g_i g_{i+1}\ , \ldots, g_r\right)$$

for $i = 1, \ldots, r-1$. (In the latter tuple, g_{i+1} is in the ith position, and $g_{i+1}^{-1} g_i g_{i+1}$ in the $(i+1)$st position. All other positions are unchanged.)

The braid operations Q_i are bijections of Ni(**C**) onto itself. Thus a braid orbit is an orbit of the permutation group on Ni(**C**) generated by the Q_i's. Since Ni(**C**) is a finite set, it suffices to require that the braid orbits be closed under either the Q_i or their inverses Q_i'.

For $\pi \in S_r$ let $\mathbf{C}^\pi$ be the tuple $(C_{\pi(1)}, \ldots, C_{\pi(r)})$. This defines a right action of S_r on the set of tuples. The operation Q_i maps $\mathcal{E}(\mathbf{C}^\pi)$ onto $\mathcal{E}(\mathbf{C}^{\pi\tau})$, where τ is the transposition $(i, i+1)$. Since these transpositions generate S_r, it follows that each braid orbit on Ni(**C**) intersects $\mathcal{E}(\mathbf{C})$ nontrivially.

The group $\mathrm{Aut}_{\mathbf{C}}(G)$ acts on Ni(**C**) by the rule that $f \in \mathrm{Aut}_{\mathbf{C}}(G)$ maps $(g_1, \ldots, g_r)$ to $(f(g_1), \ldots, f(g_r))$. This action commutes with the operations Q_i, hence $\mathrm{Aut}_{\mathbf{C}}(G)$ permutes the braid orbits. By the following lemma (which appears in [BF1]), Inn(G) leaves each braid orbit invariant, hence we get an induced action of $\mathrm{Out}_{\mathbf{C}}(G)$ on the set of braid orbits.

Lemma 9.4 *Inn(G) leaves each braid orbit invariant.*

Proof. For this proof, we use the notation $h^g := g^{-1}hg$ for $g, h \in G$. Let B be the braid orbit of the tuple $\mathbf{g} = (g_1, \ldots, g_r)$. The operations

$$\begin{aligned} \mathbf{g} &\mapsto \left(g_2, g_1^{g_2}, g_3, \ldots\right) \mapsto \left(g_2, g_3, g_1^{g_2 g_3}, g_4, \ldots\right) \mapsto \cdots \\ &\mapsto \left(g_2, \ldots, g_{r-1}, g_1^{g_2 \cdots g_{r-1}}\right) = (g_2, \ldots, g_{r-1}, g_1) \end{aligned}$$

show that B contains $(g_2, \ldots, g_{r-1}, g_1)$. Hence B contains each tuple obtained by a cyclic permutation of the entries of $\mathbf{g}$. Further, the operations

$$\mathbf{g} \mapsto \left(\ldots, g_{r-2}, g_r, g_{r-1}^{g_r}\right) \mapsto \left(\ldots, g_{r-3}, g_r, g_{r-2}^{g_r}, g_{r-1}^{g_r}\right) \mapsto \cdots$$
$$\mapsto \left(g_r, g_1^{g_r}, \ldots, g_{r-1}^{g_r}\right) = \left(g_r^{g_r}, g_1^{g_r}, \ldots, g_{r-1}^{g_r}\right),$$

coupled with a cyclic permutation, show that the inner automorphism $h \mapsto h^{g_r}$ maps $\mathbf{g}$ into B. Since this automorphism permutes the braid orbits, it leaves B invariant. Again using cyclic permutations, it follows that each of the inner automorphisms $h \mapsto h^{g_i}$ $(i = 1, \ldots, r)$ leaves B invariant. Since the g_i generate G, the lemma follows. □

Now assume that $\mathbf{C}$ is weakly rigid. Then by definition, $\mathcal{E}(\mathbf{C})$ is nonempty, and $\mathrm{Aut}_{\mathbf{C}}(G)$ acts transitively on this set. Since each braid orbit intersects $\mathcal{E}(\mathbf{C})$ nontrivially, it follows that $\mathrm{Aut}_{\mathbf{C}}(G)$ permutes the braid orbits transitively. Hence $\mathrm{Out}_{\mathbf{C}}(G)$ also permutes the braid orbits transitively. Thus the stabilizers of the braid orbits form a class of conjugate subgroups of $\mathrm{Out}_{\mathbf{C}}(G)$. It is those subgroups that yield the group Δ from Proposition 9.2.

Theorem 9.5 *Let $L/\mathbb{C}(x)$ be a finite Galois extension of weakly rigid type $\mathcal{T}$. Suppose its branch points $p_1, \ldots, p_r$ are (in $\mathbb{C}$ and) algebraically independent over some algebraically closed subfield $\bar{\ell}$ of $\mathbb{C}$. Let $\mathbf{C} = (C_{p_1}, \ldots, C_{p_r})$ be the tuple of the conjugacy classes of $G := G(L/\mathbb{C}(x))$ associated with $p_1, \ldots, p_r$. Assume $Z(G) = 1$. Let $k_{\mathcal{T}}$ (resp., k) be the minimal field containing $\bar{\ell}$ such that $\mathcal{T}$ is $k_{\mathcal{T}}$-rational (resp., such that L is defined over k). Then Proposition 9.2(d) yields an embedding of $G(k/k_{\mathcal{T}})$ into $\mathrm{Out}_{\mathbf{C}}(G)$. Let Δ be the image of this embedding. Then Δ is the stabilizer in $\mathrm{Out}_{\mathbf{C}}(G)$ of a braid orbit on $\mathrm{Ni}(\mathbf{C})$.*

The proof is given in the next chapter. As a corollary, we can now state the Outer Rigidity Criterion (see [V4], Th. 1.4). It gives group-theoretic conditions on a finite group G that produce Galois realizations for a certain subgroup of $\mathrm{Out}(G)$. Here a subgroup is called self-normalizing if it equals its own normalizer.

Corollary 9.6 (Outer Rigidity Criterion) *Let $\ell \subset \mathbb{C}$ be of finite transcendence degree over $\mathbb{Q}$. Let $\mathbf{C} = (C_1, \ldots, C_r)$ be a weakly rigid and ℓ-rational tuple of nontrivial conjugacy classes of a finite group G with $Z(G) = 1$. Let Δ be the stabilizer in $\mathrm{Out}_{\mathbf{C}}(G)$ of a braid orbit on $\mathrm{Ni}(\mathbf{C})$. Then the following holds:*

1. *If Δ is self-normalizing in $\mathrm{Out}_{\mathbf{C}}(G)$ then Δ occurs regularly over ℓ, and the full inverse image of Δ in $\mathrm{Aut}_{\mathbf{C}}(G)$ also occurs regularly over ℓ.*

2. *Let A be a normal subgroup of $Out^{\mathbf{C}}(G)$ contained in $Out_{\mathbf{C}}(G)$. Assume that $\Delta_A := \Delta A/A$ is self-normalizing in $Out_{\mathbf{C}}(G)/A$, and has trivial centralizer in $Out^{\mathbf{C}}(G)/A$. If the order of $Out(\Delta_A)$ is $\leq [Out^{\mathbf{C}}(G) : Out_{\mathbf{C}}(G)]$, then Δ_A has a GAL-realization over ℓ.*

Proof. Let G, $\mathbf{C}$, and ℓ be as in the corollary. Choose $p_1, \ldots, p_r$ in $\mathbb{C}$, algebraically independent over ℓ, and set $P = \{p_1, \ldots, p_r\}$. Let $\mathcal{T} = [G, P, \mathbf{C}]$ be the associated weakly rigid type. By Theorem 2.17 there is an FG-extension $L/\mathbb{C}(x)$ of this type, unique up to $\mathbb{C}(x)$-isomorphism. Thus there is an isomorphism $\iota : G \to G(L/\mathbb{C}(x))$, mapping $C_1, \ldots, C_r$ to the classes associated with $p_1, \ldots, p_r$. Identify G with $G(L/\mathbb{C}(x))$ via ι. This puts us into the set-up of Proposition 9.2. Let κ, $\kappa_{\mathcal{T}}$, $\kappa^{\mathcal{T}}$, k, $k_{\mathcal{T}}$, Δ, and $\hat{\Delta}$ be as in that proposition. By the theorem, this use of Δ is compatible with the present use – Δ being the stabilizer in $\text{Out}_{\mathbf{C}}(G)$ of a braid orbit on $\text{Ni}(\mathbf{C})$. (Since these stabilizers are all conjugate in $\text{Out}_{\mathbf{C}}(G)$, it suffices to prove the claim for one of them.)

(1). Assume Δ is self-normalizing in $\text{Out}_{\mathbf{C}}(G)$. Then by Proposition 9.2(d) we have $\Delta = \hat{\Delta}$, hence $G(\kappa/\kappa_{\mathcal{T}}) \cong G(k/k_{\mathcal{T}}) \cong \Delta$ and $\kappa \cap k_{\mathcal{T}} = \kappa_{\mathcal{T}}$. Further, by Proposition 9.1 the field $\kappa_{\mathcal{T}}$ is purely transcendental over ℓ, say $\kappa_{\mathcal{T}} = \ell(x_1, \ldots, x_r)$. In particular, $\kappa_{\mathcal{T}}$ is regular over ℓ, hence the same holds for κ since $\kappa \cap \bar{\ell}\kappa_{\mathcal{T}} = \kappa \cap k_{\mathcal{T}} = \kappa_{\mathcal{T}}$. Thus $\Delta \cong G(\kappa/\kappa_{\mathcal{T}}) = G(\kappa/\ell(x_1, \ldots, x_r))$ occurs regularly over ℓ.

The inverse image of Δ in $\text{Aut}_{\mathbf{C}}(G)$ is isomorphic to $G_{\mathcal{T}}$ by Proposition 9.2(b). We have $G_{\mathcal{T}} = G(M/\kappa_{\mathcal{T}}(x)) = G(M/\ell(x_1, \ldots, x_r, x))$. Here M is regular over κ, hence over ℓ (since κ is regular over ℓ). This proves (1).

(2). Assume the hypothesis of (2), and let κ_A be as in Proposition 9.2 Addendum. We claim that the fields

$$\kappa^{\mathcal{T}} \subset \kappa_{\mathcal{T}} = \ell(x_1, \ldots, x_r) \subset \kappa_A$$

form a GAL-realization for Δ_A over ℓ (see Definition 8.14). Proposition 9.1(ii) yields condition (L), plus the isomorphism $G(\kappa_{\mathcal{T}}/\kappa^{\mathcal{T}}) \cong \text{Aut}^{\mathbf{C}}(G)/\text{Aut}_{\mathbf{C}}(G) \cong \text{Out}^{\mathbf{C}}(G)/\text{Out}_{\mathbf{C}}(G)$. Since Δ_A is self-normalizing in $\text{Out}_{\mathbf{C}}(G)/A$ we have $\Delta_A = \Delta A/A = \hat{\Delta}A/A$. Thus the Addendum to Proposition 9.2 yields $\kappa_A \cap k_{\mathcal{T}} = \kappa_{\mathcal{T}}$. Hence κ_A is regular over ℓ. (Same argument as in the proof of (1).)

By Proposition 9.2(b),(c), the group $G(\kappa/\kappa^{\mathcal{T}})$ embeds into $\text{Out}^{\mathbf{C}}(G)$, and this restricts to an isomorphism of $G(\kappa/\kappa_{\mathcal{T}})$ with $\hat{\Delta} \leq \text{Out}_{\mathbf{C}}(G)$. Further, $G(\kappa/\kappa_{\mathcal{T}})$ is the full inverse image of $\text{Out}_{\mathbf{C}}(G)$ in $G(\kappa/\kappa^{\mathcal{T}})$. Since $A \subset \text{Out}_{\mathbf{C}}(G)$, the inverse image of A in $G(\kappa/\kappa^{\mathcal{T}})$ equals the inverse image of A in $G(\kappa/\kappa_{\mathcal{T}})$ (which equals $G(\kappa/\kappa_A)$). Thus κ_A is Galois over $\kappa^{\mathcal{T}}$ (since A is normal in $\text{Out}^{\mathbf{C}}(G)$). Further, the above induces an embedding of $G(\kappa_A/\kappa^{\mathcal{T}})$ into $\text{Out}^{\mathbf{C}}(G)/A$, and this restricts to an isomorphism of $G(\kappa_A/\kappa_{\mathcal{T}})$ with $\hat{\Delta}A/A \leq \text{Out}_{\mathbf{C}}(G)/A$. As noted above, we have $\hat{\Delta}A/A = \Delta_A$. Since Δ_A has trivial centralizer in $\text{Out}^{\mathbf{C}}(G)/A$, it follows that $G(\kappa_A/\kappa_{\mathcal{T}})$ has trivial centralizer in $G(\kappa_A/\kappa^{\mathcal{T}})$, hence $G(\kappa_A/\kappa^{\mathcal{T}})$ embeds into the automorphism group of $G(\kappa_A/\kappa_{\mathcal{T}})$ via conjugation action. On the other hand, the previous paragraph together with the hypothesis of (2) implies that $G(\kappa_{\mathcal{T}}/\kappa^{\mathcal{T}})$ has order greater than or equal to $\text{Out}(\Delta_A)$. It follows that $G(\kappa_A/\kappa^{\mathcal{T}})$ induces the full automorphism group of $G(\kappa_A/\kappa_{\mathcal{T}})$. This yields the (GA)-condition, completing the proof of (2). □

Remark 9.7

(a) If $\mathbf{C}$ is even rigid then $\text{Out}_{\mathbf{C}}(G) = \Delta = 1$, hence (1) realizes $G = \text{Inn}(G)$ regularly over ℓ. Thus the usual rigidity criterion Theorem 3.17 is a special case of the Outer Rigidity Criterion.

(b) If in (2) we drop the hypothesis on the centralizer of Δ_A and on $\text{Out}(\Delta_A)$ then $\Delta_A \cong G(\kappa_A/\ell(x_1, \ldots, x_r))$ still occurs regularly over ℓ (by the proof of (2)).

(c) By Theorem 9.5, the braid orbits are associated with the case of generic branch points. The reason for this is that the fundamental group of affine space minus the discriminant locus has certain generators that induce the braid operations Q_i (see the next chapter). For nongeneric branch points, one must use the fundamental group of the corresponding subvariety of affine space minus the discriminant locus. If $\kappa_{\mathcal{T}}$ is again purely transcendental, one obtains analogues of the Outer Rigidity Criterion where the braid orbits are replaced by possibly smaller subsets of $\text{Ni}(\mathbf{C})$. The following case has been worked out in [SV2]: The branch points are of the form $p_1, \ldots, p_{r/2}, -p_1, \ldots, -p_{r/2}$, where $p_1, \ldots, p_{r/2}$ are algebraically independent.

9.3 Construction of Weakly Rigid Tuples

To apply the Outer Rigidity Criterion, we need to find examples where weak rigidity holds. The following lemma shows that such examples exist in abundance (contrary to the rigid case). Generalizing the construction in Section 7.1, we get:

Lemma 9.8 *Let $\mathbf{C} = (C_1, \ldots, C_r)$ be a tuple of nontrivial conjugacy classes of a finite group G, with $\mathcal{E}(\mathbf{C})$ nonempty. Then there exist a finite group G^**

and a tuple $\mathbf{C}^* = (C_1^*, \ldots, C_r^*)$ *of conjugacy classes of* G^* *such that* $\mathcal{E}(\mathbf{C}^*)$ *is nonempty and the following holds:*

1. *The number of surjective homomorphisms* $f : G^* \to G$ *with* $f(\mathbf{C}^*) = \mathbf{C}$ *equals* $|\mathcal{E}(\mathbf{C})|$.
2. *The intersection of the kernels of all homomorphisms* f *from (1) is trivial.*

The pair $(G^*, \mathbf{C}^*)$ *is determined by* $(G, \mathbf{C})$ *up to isomorphism, and* $\mathbf{C}^*$ *is weakly rigid.*

Proof. Let $\mathcal{F}$ be the group defined by generators $u_1, \ldots, u_r$ subject to the single relation $u_1 \cdots u_r = 1$. Let $\mathcal{N}$ be the intersection of the kernels of all surjective homomorphisms $\tilde{f} : \mathcal{F} \to G$ with $\tilde{f}(u_i) \in C_i$. Then $\mathcal{N}$ is a normal subgroup of finite index in $\mathcal{F}$ (cf. the proof of Lemma 7.1).

Suppose $(G^*, \mathbf{C}^*)$ satisfies (1) and (2), and $\mathcal{E}(\mathbf{C}^*)$ is nonempty. Pick $(g_1^*, \ldots, g_r^*)$ in this set. Then there is a homomorphism $\psi : \mathcal{F} \to G^*$ with $\psi(u_i) = g_i^*$. The map $f \mapsto f \circ \psi$ is an injection from the set of homomorphisms $f : G^* \to G$ as in (1) into the set of homomorphisms $\tilde{f} : \mathcal{F} \to G$ as in the preceding paragraph. The latter set has cardinality $\leq |\mathcal{E}(\mathbf{C})|$, since $\tilde{f}$ is determined by its values on $u_1, \ldots, u_r$, and these values yield an element of $\mathcal{E}(\mathbf{C})$. Thus (1) implies that $f \mapsto f \circ \psi$ is a bijection between the above sets. Hence $\ker(\psi) \subset \mathcal{N}$. By (2), the image of $\mathcal{N}$ in G^* is trivial, hence $\mathcal{N} = \ker(\psi)$. Thus ψ induces an isomorphism $\bar{\psi} : \mathcal{F}/\mathcal{N} \to G^*$, mapping u_i to g_i^*. This proves the uniqueness of the pair $(G^*, \mathbf{C}^*)$. Further, it proves that $\mathbf{C}^*$ is weakly rigid: For any $(h_1^*, \ldots, h_r^*)$ in $\mathcal{E}(\mathbf{C}^*)$, we get as above an isomorphism $\bar{\phi} : \mathcal{F}/\mathcal{N} \to G^*$ mapping u_i to h_i^*; then $\bar{\phi}\bar{\psi}^{-1}$ is an automorphism of G^* mapping g_i^* to h_i^*.

Conversely, if we define G^* as $\mathcal{F}/\mathcal{N}$, and C_i^* as the class of $u_i\mathcal{N}$, then clearly (1) and (2) hold (using the freeness property of $\mathcal{F}$). □

Remark 9.9 *Let* $L/\mathbb{C}(x)$ *be a finite Galois extension of ramification type* $[G, P, \mathbf{C}]$. *Let* K *be the composite of all* $L'/\mathbb{C}(x)$ *of the same ramification type. Then* K *has type* $[G^*, P, \mathbf{C}^*]$; *in particular,* K *is of weakly rigid type. (Proof analogous to that of Corollary 7.3.)*

Example 9.10

(a) If $\mathbf{C}$ is weakly rigid then $G^* = G$. This holds in particular if G is abelian.

(b) If G is nonabelian simple, then G^* is the direct product of n copies of G, where n is the number of $\mathrm{Aut}_{\mathbf{C}}(G)$-orbits on $\mathcal{E}(\mathbf{C})$. Further, $\mathrm{Aut}_{\mathbf{C}^*}(G^*)$ is the wreath product of S_n (in its natural permutation representation) with $\mathrm{Aut}_{\mathbf{C}}(G)$.

(c) Suppose G is the one-dimensional affine group over a finite field $\mathbb{F}_q$ (semi-direct product of the additive and the multiplicative group of $\mathbb{F}_q$, where the multiplicative group $\mathbb{F}_q^\times$ acts on the additive group through left multiplication). Further, assume $r \geq 3$, and the classes $C_1, \ldots, C_r$ are represented by nontrivial elements $\zeta_1, \ldots, \zeta_r$ of $\mathbb{F}_q^\times$ that generate the field $\mathbb{F}_q$. Then G^* is the semi-direct product of $\mathbb{F}_q^n$ with $\mathbb{F}_q^\times$, where $n = r - 2$. Further, $\mathrm{Aut}_{\mathbf{C}^*}(G^*)$ is the semi-direct product of $\mathbb{F}_q^n$ with $\mathrm{GL}_n(q)$.

Proofs are left as an exercise. In case (b) (as in most other cases) we expect the Outer Rigidity Criterion to give only Galois realizations of S_n and its wreath products. Case (c) is more interesting; it yields Galois realizations of $\mathrm{GL}_n(q)$ and some interesting subgroups. This is worked out in the next section. Before we get to this, some further general observations about the interrelation between G and G^*.

Definition 9.11 *For $r \geq 2$, define the (abstract) Artin braid group $\mathcal{B}_r$ as the group generated by elements $\tilde{Q}_1, \ldots, \tilde{Q}_{r-1}$ subject to the relations*

$$\tilde{Q}_i\tilde{Q}_{i+1}\tilde{Q}_i = \tilde{Q}_{i+1}\tilde{Q}_i\tilde{Q}_{i+1} \quad \text{for } i = 1, \ldots, r-2 \qquad \text{and}$$
$$\tilde{Q}_i\tilde{Q}_j = \tilde{Q}_j\tilde{Q}_i \qquad \text{for } |i-j| > 1.$$

Since the transpositions $(i, i+1)$ in the symmetric group S_r satisfy the same relations, we get a surjective homomorphism $\kappa : \mathcal{B}_r \to S_r$, $\tilde{Q}_i \mapsto (i, i+1)$. The kernel of κ is called the pure braid group, and denoted by $\mathcal{B}^{(r)}$.

One checks that the braid operations Q_i, as well as the Q_i', from the previous section satisfy those relations. Thus sending $\tilde{Q}_i$ to Q_i' yields a right action of $\mathcal{B}_r$ on $\mathrm{Ni}(\mathbf{C})$, for any r-tuple $\mathbf{C}$ of classes of a finite group G. We denote this action as $(g_1, \ldots, g_r) \mapsto (g_1, \ldots, g_r)^Q$, for $Q \in \mathcal{B}_r$. For technical reasons, we prefer to have $\mathcal{B}_r$ act from the right, just for the remainder of Chapter 9. In Chapter 10, we have the concrete version of $\mathcal{B}_r$ (fundamental group of affine space minus the discriminant locus) act from the left. This helps to emphasize the difference in the set-up. In Chapter 9, the braid group itself does not play an essential role. We could also replace it by the free group on $\tilde{Q}_1, \ldots, \tilde{Q}_{r-1}$; we are mainly interested in the finite group of permutations on $\mathrm{Ni}(\mathbf{C})$ generated by the braid operations.

We have seen that Q_i' maps $\mathcal{E}(\mathbf{C})$ to $\mathcal{E}(\mathbf{C}^{(i,i+1)})$. Hence each $Q \in \mathcal{B}_r$ maps $\mathcal{E}(\mathbf{C})$ to $\mathcal{E}(\mathbf{C}^{\kappa(Q)})$. In particular, the stabilizer $\mathcal{B}_r(\mathbf{C})$ in $\mathcal{B}_r$ of the set $\mathcal{E}(\mathbf{C})$ consists of all $Q \in \mathcal{B}_r$ with $\mathbf{C}^{\kappa(Q)} = \mathbf{C}$.

Now let G, G^* be as in the above lemma. Fix some $\mathbf{g}^* = (g_1^*, \dots, g_r^*) \in \mathcal{E}(\mathbf{C}^*)$. Let M be the set of homomorphisms f from (1). The group $\mathrm{Aut}_{\mathbf{C}^*}(G^*)$ acts on M by the rule that $\alpha \in \mathrm{Aut}_{\mathbf{C}^*}(G^*)$ sends f to $f \circ \alpha^{-1}$. This action is faithful: If $f \circ \alpha^{-1} = f$ for all $f \in M$ then $\alpha^{-1}(g_i^*) \cdot (g_i^*)^{-1}$ lies in the kernel of each $f \in M$, hence is trivial by (2); thus $\alpha = 1$.

By (1) the map $M \to \mathcal{E}(\mathbf{C})$, $f \mapsto f(\mathbf{g}^*) := (f(g_1^*), \dots, f(g_r^*))$ is bijective. Via this bijection, the above action of $\mathrm{Aut}_{\mathbf{C}^*}(G^*)$ on M yields an action on $\mathcal{E}(\mathbf{C})$. Let λ be the corresponding (injective) homomorphism from $\mathrm{Aut}_{\mathbf{C}^*}(G^*)$ to the symmetric group on $\mathcal{E}(\mathbf{C})$.

Lemma 9.12 *Let Δ^* be the stabilizer in $\mathrm{Aut}_{\mathbf{C}^*}(G^*)$ of the braid orbit of $\mathbf{g}^*$. Then $\lambda(\Delta^*)$ equals the permutation group on $\mathcal{E}(\mathbf{C})$ induced by the action of $\mathcal{B}_r(\mathbf{C}^*)$.*

Proof. For each $\alpha \in \Delta^*$ there is $Q \in \mathcal{B}_r$ with $\alpha^{-1}(\mathbf{g}^*) = (\mathbf{g}^*)^Q$. Since α^{-1} preserves the set $\mathcal{E}(\mathbf{C}^*)$, we must have $Q \in \mathcal{B}_r(\mathbf{C}^*)$. Conversely, each such Q occurs for some $\alpha \in \Delta^*$ (since $\mathbf{C}^*$ is weakly rigid).

Let $\mathbf{g} \in \mathcal{E}(\mathbf{C})$. It corresponds to that $f \in M$ with $f(\mathbf{g}^*) = \mathbf{g}$. Then $\lambda(\alpha)$ maps $\mathbf{g}$ to $f \circ \alpha^{-1}(\mathbf{g}^*) = f((\mathbf{g}^*)^Q) = f(\mathbf{g}^*)^Q = \mathbf{g}^Q$. Thus $\lambda(\alpha)$ is the permutation induced by $Q \in \mathcal{B}_r(\mathbf{C}^*)$. This proves the claim. □

Sending α to Q (in the above notation) yields an isomorphism between Δ^* and the permutation group on $\mathcal{E}(\mathbf{C}^*)$ induced by $\mathcal{B}_r(\mathbf{C}^*)$ (which acts freely on $\mathcal{E}(\mathbf{C}^*)$). Note that $\mathcal{B}_r(\mathbf{C}^*) \subset \mathcal{B}_r(\mathbf{C})$ (since there is a homomorphism $f : G^* \to G$ with $f(\mathbf{C}^*) = \mathbf{C}$). Whether equality holds in general is not clear. It certainly holds in the above examples.

Corollary 9.13 *Let $\mathbf{C}$ be an r-tuple of nontrivial conjugacy classes of a finite group G, where $Z(G) = 1$ and $\mathcal{E}(\mathbf{C})$ is nonempty. If the permutation group Δ' on $\mathcal{E}(\mathbf{C})$ induced by the pure braid group $\mathcal{B}^{(r)}$ is self-normalizing in the full symmetric group on $\mathcal{E}(\mathbf{C})$ then Δ' occurs regularly over $\mathbb{Q}_{\mathrm{ab}}$; it occurs regularly over $\mathbb{Q}$ if the tuple $\mathbf{C}^*$ is rational.*

Proof. Note $\mathcal{B}^{(r)}$ is normal in $\mathcal{B}_r$, hence in $\mathcal{B}_r(\mathbf{C}^*)$. Hence if Δ' is self-normalizing in $\mathrm{Sym}(\mathcal{E}(\mathbf{C}))$ then Δ' equals the permutation group induced by $\mathcal{B}_r(\mathbf{C}^*)$. Thus $\Delta' = \lambda(\Delta^*)$ by the lemma. Thus $\lambda(\Delta^*)$ is self-normalizing in $\lambda(\mathrm{Aut}_{\mathbf{C}^*}(G^*))$, and so Δ^* is self-normalizing in $\mathrm{Aut}_{\mathbf{C}^*}(G^*)$. Then the image Δ of Δ^* in $\mathrm{Out}_{\mathbf{C}^*}(G^*)$ is also self-normalizing (since $\mathrm{Inn}(G^*) \subset \Delta^*$ by Lemma 9.4). In addition, $Z(G^*) = 1$ (since every central element of G^* lies in the kernel of each $f \in M$, hence is trivial by (2)). Now the claim follows from the Outer Rigidity Criterion (since $\mathbf{C}^*$ is weakly rigid). □

A version of this corollary appears in [MM], Ch. III, Th. 10.2.

9.4 An Application of the Outer Rigidity Criterion

In this section we study the case of weak rigidity found in Example 9.10. We do not need the original group $\mathbb{F}_q \cdot \mathbb{F}_q^\times$ anymore, so we choose our notation as follows.

Fix an integer $n \geq 2$ and set $r = n + 2$. Let q be a power of the prime p, and let $\mathbb{F}_q$ be the finite field with q elements. Let Z be a subgroup of the multiplicative group $\mathbb{F}_q^\times$, generated by elements $\zeta_1, \dots, \zeta_r$ with $\zeta_1 \dots \zeta_r = 1$ and $\zeta_i \neq 1$ for all i. Assume further $\mathbb{F}_q = \mathbb{F}_p(\zeta_1, \dots, \zeta_r)$. Set $\zeta = (\zeta_1, \dots, \zeta_r)$. For $\pi \in S_r$ we write $\zeta^\pi = (\zeta_{\pi(1)}, \dots, \zeta_{\pi(r)})$.

Let $V = \mathbb{F}_q^n$ be the $\mathbb{F}_q$-vector space of column vectors of length n over $\mathbb{F}_q$. Let $G = V \cdot Z$ be the semi-direct product of (the additive group of) V with Z, where Z acts on V via scalar multiplication. We write the elements of G as pairs $[v, z]$ with $v \in V$, $z \in Z$. The multiplication of these pairs is explicitly given as $[v, z] \cdot [v', z'] = [v + zv', zz']$. We have $[v, 1] \cdot [0, z] \cdot [v, 1]^{-1} = [(1 - z)v, z]$. Hence if $z \neq 1$ then the conjugacy class of G containing $[0, z]$ consists of all $[v, z]$, $v \in V$. For $i = 1, \dots, r$, let C_i be the conjugacy class of G consisting of all $[v, \zeta_i]$, $v \in V$. Set $\mathbf{C}_\zeta = (C_1, \dots, C_r)$, and $\mathcal{E}(\zeta) = \mathcal{E}(\mathbf{C}_\zeta)$, $\mathrm{Ni}(\zeta) = \mathrm{Ni}(\mathbf{C}_\zeta)$.

Lemma 9.14 *Let V act on $\mathcal{E}(\zeta)$ via conjugating the entries of the tuples in $\mathcal{E}(\zeta)$. Then each V-orbit on $\mathcal{E}(\zeta)$ contains exactly one element of the form*

$$([0, \zeta_1],\ [v_1, \zeta_2], \dots, [v_{n+1}, \zeta_r]), \qquad v_i \in V \tag{9.3}$$

Define $\lambda_1, \dots, \lambda_n \in Z$ by setting $\lambda_i = \zeta_{i+1}^{-1} \cdots \zeta_{n+1}^{-1}$. Then an element of the form (9.3) *lies in $\mathcal{E}(\zeta)$ if and only if $v_1, \dots, v_n$ is an $\mathbb{F}_q$-basis of V, and $v_{n+1} = -\lambda_1 v_1 - \cdots - \lambda_n v_n$.*

Proof. The above formula $[v, 1] \cdot [0, \zeta_1] \cdot [v, 1]^{-1} = [(1 - \zeta_1)v, \zeta_1]$ shows that for any $v_0 \in V$ there is a unique $v \in V$ with $[v, 1]^{-1} \cdot [v_0, \zeta_1] \cdot [v, 1] = [0, \zeta_1]$. This proves the first assertion.

The tuple (9.3) lies in $\mathcal{E}(\zeta)$ if and only if the entries of the tuple generate G, and their product (in the given order) is 1. The latter condition means that $v_{n+1} = -\lambda_1 v_1 - \cdots - \lambda_n v_n$ (by a straightforward computation). Thus it remains to show that the entries of the tuple (9.3) generate G if and only if $v_1, \dots, v_n$ is an $\mathbb{F}_q$-basis of V. For the "only if," assume $v_1, \dots, v_n$ do not form a basis. Then they lie in a proper $\mathbb{F}_q$-subspace W of V, hence the entries of (9.3) lie in the proper subgroup $W \cdot Z$ of G.

Let U be the intersection of V with the subgroup H of G generated by $[0, \zeta_1], [v_1, \zeta_2], \ldots, [v_{n+1}, \zeta_r]$. Conjugation action on V of these elements is just scalar multiplication by $\zeta_1, \ldots, \zeta_r$. Thus U is invariant under scalar multiplication by $\zeta_1, \ldots, \zeta_r$. The ring of endomorphisms of U generated by these multiplications is all of $\mathbb{F}_q$, since $\mathbb{F}_q = \mathbb{F}_p[\zeta_1, \ldots, \zeta_r]$. Hence U is an $\mathbb{F}_q$-subspace of V. It contains the commutator $[0, \zeta_1][v_i, \zeta_{i+1}][0, \zeta_1]^{-1}[v_i, \zeta_{i+1}]^{-1} = (\zeta_1 - 1)v_i$ for $i = 1, \ldots, n$. Hence $v_1, \ldots, v_n \in U$. Thus if $v_1, \ldots, v_n$ form a basis of V then $U = V$, hence $V \subset H$. But H maps surjectively to Z since $Z = \langle \zeta_1, \ldots, \zeta_r \rangle$. Hence if $V \subset H$ then $H = G$. This completes the proof. □

9.4.1 Braiding Action through the Matrices $\Phi(Q, \zeta)$

The braiding action of $\mathcal{B}_r$ on $\mathrm{Ni}(\zeta)$ commutes with the action of V, hence induces an action of $\mathcal{B}_r$ on the set $\mathrm{Ni}^{(V)}(\zeta)$ of V-orbits on $\mathrm{Ni}(\zeta)$. Let $\mathcal{E}^{(V)}(\zeta)$ be the set of V-orbits on $\mathcal{E}(\zeta)$. For each matrix $B \in \mathrm{GL}_n(q)$ let $[[B, \zeta]]$ denote the V-orbit of the tuple (9.3) from the lemma, where $v_1, \ldots, v_n$ are the column vectors of B, and v_{n+1} is given in terms of $v_1, \ldots, v_n$ and ζ as in the lemma. Then the lemma shows that the map $B \mapsto [[B, \zeta]]$ is a bijection between $\mathrm{GL}_n(q)$ and $\mathcal{E}^{(V)}(\zeta)$. In terms of the matrices B, we can now explicitly describe the $\mathcal{B}_r$-action on $\mathrm{Ni}^{(V)}(\zeta)$. Let $e_1, \ldots, e_n$ be the standard basis of $V = \mathbb{F}_q^n$ (i.e., e_1 is the vector with entries $1, 0, \ldots, 0$, etc.). Straightforward computations yield:

Lemma 9.15 *For $i = 1, \ldots, r-1$ and $B \in GL_n(q)$ the element $\tilde{Q}_i$ sends $[[B, \zeta]]$ to*

$$[[B \cdot \Phi_i(\zeta),\ \zeta^{(i,i+1)}]]$$

where $(i, i+1)$ is the transposition switching i and $i+1$, and $\Phi_i(\zeta) \in GL_n(q)$ is the following matrix:

(a) For $i = 2, \ldots, n$, the matrix $\Phi_i(\zeta)$ has jth column e_j for $j \notin \{i, i-1\}$, has $(i-1)$st column e_i and ith column $\zeta_{i+1}^{-1}e_{i-1} + \zeta_{i+1}^{-1}(\zeta_i - 1)e_i$.
(b) The matrix $\Phi_1(\zeta)$ has first column $\zeta_2^{-1}(\zeta_2 - 1)^{-1}(1 - \zeta_1)e_1$ and jth column $(\zeta_2 - 1)^{-1}(1 - \zeta_{j+1})e_1 + e_j$ for $j = 2, \ldots, n$.
(c) The matrix $\Phi_{n+1}(\zeta)$ has jth column e_j for $j = 1, \ldots, n-1$, and nth column $-\lambda_1 e_1 - \cdots - \lambda_n e_n$ with $\lambda_1, \ldots, \lambda_n$ as in Lemma 9.14.

Since the $\tilde{Q}_i$'s generate $\mathcal{B}_r$, it follows that for each $Q \in \mathcal{B}_r$ there is unique $\Phi(Q, \zeta) \in \mathrm{GL}_n(q)$ such that

$$[[B, \zeta]]^Q = [[B \cdot \Phi(Q, \zeta),\ \zeta^{\kappa(Q)}]]$$

for all $B \in \mathrm{GL}_n(q)$. (Note that since $\mathrm{Ni}(\zeta)$ is a finite set, each $\tilde{Q}_i^{-1}$ acts as some positive power of $\tilde{Q}_i$ on this set; we could of course also check directly that each $\tilde{Q}_i^{-1}$ acts as claimed). Here $\kappa : \mathcal{B}_r \to S_r$ is the homomorphism sending $\tilde{Q}_i$ to the transposition $(i, i+1)$ (see the previous section.) We get the multiplication rule:

$$\Phi(QQ', \zeta) = \Phi(Q, \zeta) \ \Phi(Q', \zeta^{\kappa(Q)}).$$

Let $\mathcal{B}_r(\zeta)$ be the group of all $Q \in \mathcal{B}_r$ with $\zeta^{\kappa(Q)} = \zeta$. The group $\mathcal{B}_r(\zeta)$ is the stabilizer in $\mathcal{B}_r$ of the set $\mathcal{E}(\zeta)$, and it contains the pure braid group $\mathcal{B}^{(r)} = \ker(\kappa)$. The map $\Phi_\zeta : \mathcal{B}_r(\zeta) \to \mathrm{GL}_n(q)$ sending Q to $\Phi(Q, \zeta)$ is a homomorphism (by the above multiplication rule). The image of this homomorphism is a subgroup of $\mathrm{GL}_n(q)$ that we denote by Δ_ζ.

Remark 9.16 If $\zeta_1 = \cdots = \zeta_r$ then $\mathcal{B}_r(\zeta) = \mathcal{B}_r$, hence in this case Φ_ζ gives a linear representation of the full braid group $\mathcal{B}_r$. The matrices representing the generators $\tilde{Q}_i$ are the $\Phi_i(\zeta)$ from the above lemma. From this we see that Φ_ζ is essentially the classical **Burau representation** of the braid group (see [Bir], p. 118), suitably normalized and with coefficients in the finite field $\mathbb{F}_q$. If the ζ_i's are not all equal then Φ_ζ yields a representation of the pure braid group that is related to the **Gassner representation**, see [Bir], p. 119.

In the case $\zeta_1 = \cdots = \zeta_r$, the matrices $\Phi_i(\zeta)$ not only satisfy the braid relations (from Definition 9.3), but also the quadratic relations of the Hecke algebra of type A_ℓ (see [CR], section 67).

The group Δ_ζ determines the braid orbits on $\mathrm{Ni}^{(V)}(\zeta)$: They correspond to the right cosets of Δ_ζ in $\mathrm{GL}_n(q)$. In particular, $\mathcal{B}_r$ acts transitively on $\mathrm{Ni}^{(V)}(\zeta)$ if and only if $\Delta_\zeta = \mathrm{GL}_n(q)$. In general, the structure of the braid orbits varies quite a bit, as shown by the following classification of the groups Δ_ζ. Here and in the following we view $V = \mathbb{F}_q^n$ as $\mathbb{F}_q$-vector space of column vectors, on which the matrix group $\mathrm{GL}_n(q)$ acts by left multiplication.

Theorem 9.17 *Let $\zeta_1, \ldots, \zeta_r$ be generators of the finite field $\mathbb{F}_q$ satisfying $\zeta_1 \cdots \zeta_r = 1$ and $0 \neq \zeta_i \neq 1$ for all i. Set $\zeta = (\zeta_1, \ldots, \zeta_r)$ and $n = r - 2$. Suppose $n \geq 2$. Let Δ_ζ be the image of the homomorphism $\Phi_\zeta : \mathcal{B}_r(\zeta) \to GL_n(q)$. Then Δ_ζ acts absolutely irreducibly on $V = \mathbb{F}_q^n$. Furthermore:*

1. *Δ_ζ leaves a nonzero bilinear form on V invariant if and only if $q = p$ is a prime, n is even, and $\zeta = (-1, \ldots, -1)$. In this case,*

$$\Delta_\zeta = \mathrm{Sp}_n(p).$$

2. *Δ_ζ leaves a nonzero hermitian form on V invariant if and only if $q = q_0^2$ is a square and all ζ_i have norm 1 over $\mathbb{F}_{q_0}$. In this case,*

$$\mathrm{SU}_n(q) \le \Delta_\zeta \le \mathrm{U}_n(q)$$

with possible exceptions (E1)–(E4) below.

3. *If ζ is not as in (1) or (2), and if $n > 2$, then*

$$\mathrm{SL}_n(q) \le \Delta_\zeta \le \mathrm{GL}_n(q)$$

with exceptions (E3) and (E4).

Let $\bar{\Delta}_\zeta$ denote the image of Δ_ζ in $PGL_n(q)$.

4. *If $n = 2$ then $\bar{\Delta}_\zeta$ is (conjugate to) $PSL_2(q_0)$ or $PGL_2(q_0)$, $q \in \{q_0, q_0^2\}$, with exceptions (E1) and (E2).*

The exceptional cases are as follows:

(E1) *$n = 2$ and $\zeta = (t, t, -t^{-1}, -t^{-1})$ (up to permutation) with $t^4 \neq 1$. In this case, $\bar{\Delta}_\zeta$ is dihedral of order $2m$, with m prime to q.*

(E2) *$n = 2$, and $\bar{\Delta}_\zeta \cong A_4$, S_4 or A_5. If $p > 5$ then $\zeta_i\zeta_j \neq 1$ for all $i \neq j$.*

(E3) *$n = 3$, $p > 3$ and $\zeta = (-\epsilon, -\epsilon, -\epsilon, -\epsilon, \epsilon^{-1})$ with $\epsilon^3 = 1$ (up to permutation). In this case, $\bar{\Delta}_\zeta \cong PU_3(4) \cong \mathbb{F}_3^2 \cdot SL_2(3)$.*

(E4) *$n = 4$, $p > 3$ and $\zeta = (-\epsilon, -\epsilon, -\epsilon, -\epsilon, -\epsilon, -\epsilon)$ with $\epsilon^3 = 1$, $\epsilon \neq 1$. In this case, $\bar{\Delta}_\zeta \cong PSU_4(4) \cong PSp_4(3)$.*

Here $\mathrm{Sp}_n(q)$ (resp., $\mathrm{U}_n(q)$) denotes the invariance group in $\mathrm{GL}_n(q)$ of a nondegenerate symplectic (resp., hermitian) form on V. And $\mathrm{SU}_n(q)$ is the intersection of $\mathrm{U}_n(q)$ and $\mathrm{SL}_n(q)$.

Remark 9.18

(a) The case $n = 2$:
If $\zeta = (t, t, t, t)$ with $t^4 = 1$, but $t^2 \neq 1$, then case (E2) occurs with type S_4. If $\zeta = (s, s, s, -1)$ with $s^3 = -1$, but $s \neq -1$, then case (E2) occurs with type A_4. Further, to compare (1), (2) with (4), note the isomorphisms $\mathrm{SU}_2(q_0^2) \cong \mathrm{SL}_2(q_0)$ and $\mathrm{Sp}_2(q) \cong \mathrm{SL}_2(q)$.

(b) The groups in (E3) and (E4) were classically studied in low-dimensional linear group theory (see [Mi], and the remarks in [Wag]; the group in (E4) belongs to the 27 lines on a cubic surface). Their Galois-theoretic significance has not yet been explored.

The proof of the theorem occupies most of the paper [V2]. We do not include it here, since this would lead us too far away from the main theme. The proof uses a theorem of Wagner [Wag] that classifies primitive linear groups generated by noninvolutory homologies (= elements fixing a hyperplane pointwise, and a point outside). Another ingredient is the construction of a dual pairing between the representations Φ_ζ and $\Phi_{\zeta^{-1}}$. The parallel case of the symplectic braid group is treated in [MSV].

9.4.2 Galois Realizations for $PGL_n(q)$ and $PU_n(q)$

Recall that a semi-linear automorphism f of the $\mathbb{F}_q$-vector space V is an automorphism of the additive group of V for which there exists some $\alpha \in \text{Aut}(\mathbb{F}_q)$ with $f(tv) = \alpha(t) f(v)$ for all $v \in V$ and $t \in \mathbb{F}_q$. Let $\Gamma L_n(q)$ denote the group of all semi-linear automorphisms of V. It contains $\text{GL}_n(q)$ as the (normal) subgroup defined by the condition $\alpha = \text{id}$. Embed Z into $\text{GL}_n(q)$ as the group of scalar matrices. Forming semi-direct products with V, we obtain the inclusions

$$G = V \cdot Z < V \cdot \text{GL}_n(q) < V \cdot \Gamma L_n(q).$$

The group G is normal and has trivial centralizer in both overgroups. Hence they embed into $\text{Aut}(G)$ via conjugation.

Lemma 9.19

(a) Conjugation action on G identifies $V \cdot GL_n(q)$ with $Aut_{\mathbf{C}}(G)$, and $V \cdot \Gamma L_n(q)$ with $Aut(G)$. Here $\mathbf{C} = \mathbf{C}_\zeta$ is the tuple of conjugacy classes of G associated with ζ.

(b) The tuple $\mathbf{C} = \mathbf{C}_\zeta$ is weakly rigid.

(c) The group $V \cdot \Delta_\zeta$ is the stabilizer in $Aut_{\mathbf{C}}(G)$ of a braid orbit on $Ni(\zeta)$.

Proof. First note that each element $[0, A] \in V \cdot GL_n(q)$ maps the tuple (9.3) from Lemma 9.14 to the tuple

$$([0, \zeta_1],\ [Av_1, \zeta_2], \ldots, [Av_{n+1}, \zeta_r]).$$

Hence the image of $V \cdot \text{GL}_n(q)$ in $\text{Aut}(G)$ is actually contained in $\text{Aut}_{\mathbf{C}}(G)$. Further, $V \cdot \text{GL}_n(q)$ acts transitively on $\mathcal{E}(\zeta)$. Indeed, by Lemma 9.14 each element of $\mathcal{E}(\zeta)$ is V-conjugate to an element of the form (9.3), and any two elements of $\mathcal{E}(\zeta)$ of the form (9.3) are $\text{GL}_n(q)$-conjugate (since any two $\mathbb{F}_q$-bases of V are $\text{GL}_n(q)$-conjugate).

Since $V \cdot \text{GL}_n(q)$ embeds into $\text{Aut}_{\mathbf{C}}(G)$, $\text{Aut}_{\mathbf{C}}(G)$ also acts transitively on $\mathcal{E}(\zeta)$. This proves (b). Further, since $\text{Aut}_{\mathbf{C}}(G)$ acts freely on $\mathcal{E}(\zeta)$ (i.e., with trivial stabilizers), it follows that $V \cdot \text{GL}_n(q)$ induces all of $\text{Aut}_{\mathbf{C}}(G)$. This proves the first part of (a).

Since $\mathrm{Inn}(G)$ is normal in $\mathrm{Aut}(G)$, the group of automorphisms of V induced by $\mathrm{Aut}(G)$ normalizes the group induced by $\mathrm{Inn}(G)$. This group induced by $\mathrm{Inn}(G)$ consists of the scalar multiplications by elements of Z. These scalar multiplications generate the ring R of endomorphisms of V consisting of all scalar multiplications (as in the proof of Lemma 9.14). Thus $\mathrm{Aut}(G)$ normalizes $R \cong \mathbb{F}_q$, hence acts through elements of $\mathrm{Aut}(\mathbb{F}_q)$ on $G/V \cong Z$. It follows that the orbit of any element of $\mathcal{E}(\zeta)$ under $\mathrm{Aut}(G)$ is contained in the union $\mathcal{U}$ of all $\mathcal{E}(\alpha(\zeta))$, $\alpha \in \mathrm{Aut}(\mathbb{F}_q)$. On the other hand, the subgroup $V \cdot \Gamma L_n(q)$ of $\mathrm{Aut}(G)$ acts transitively on $\mathcal{U}$, since $V \cdot \mathrm{GL}_n(q)$ acts transitively on $\mathcal{E}(\zeta)$. As above, this implies that $\mathrm{Aut}(G) = V \cdot \Gamma L_n(q)$. This completes the proof of (a).

(c) Both $\mathcal{B}_r(\zeta)$ and $V \cdot \mathrm{GL}_n(q)$ act naturally on the set $\mathcal{E}^{(V)}(\zeta)$ (the latter group because it contains V as a normal subgroup). By Section 9.4.1, we can write the elements of $\mathcal{E}^{(V)}(\zeta)$ in the form $[[B, \zeta]]$ with $B \in \mathrm{GL}_n(q)$. Then $Q \in \mathcal{B}_r(\zeta)$ sends $[[B, \zeta]]$ to $[[B \cdot \Phi_\zeta(Q), \zeta]]$ (see Section 9.4.1), and $[v, A] \in V \cdot \mathrm{GL}_n(q)$ sends $[[B, \zeta]]$ to $[[A \cdot B, \zeta]]$ (see the first sentence of this proof).

Let I_n be the $n \times n$ identity matrix. It follows that the $\mathcal{B}_r(\zeta)$-orbit $\mathcal{O}$ of $[[I_n, \zeta]]$ in the set $\mathcal{E}^{(V)}(\zeta)$ consists of all $[[D, \zeta]]$ with $D \in \Delta_\zeta$. Hence the stabilizer of $\mathcal{O}$ in $\mathrm{Aut}_{\mathbf{C}}(G) = V \cdot \mathrm{GL}_n(q)$ equals $V \cdot \Delta_\zeta$. The full $\mathcal{B}_r$-orbit of $[[I_n, \zeta]]$ (in the set $\mathrm{Ni}^{(V)}(\zeta)$) has the same stabilizer in $\mathrm{Aut}_{\mathbf{C}}(G)$ (since any $Q \in \mathcal{B}_r$ fixing $\mathcal{E}^{(V)}(\zeta)$ must lie in $\mathcal{B}_r(\zeta)$). Together with Lemma 9.4, this proves (c). □

For the weakly rigid tuple $\mathbf{C} = \mathbf{C}_\zeta$ we have now computed the stabilizer in $\mathrm{Out}_{\mathbf{C}}(G)$ of a certain braid orbit on $\mathrm{Ni}(\mathbf{C})$: The natural isomorphism $\mathrm{Out}_{\mathbf{C}}(G) \cong \mathrm{GL}_n(q)/Z$ identifies this stabilizer with $\Delta_\zeta Z/Z$. Thus if $\mathbf{C}$ is rational and $\Delta_\zeta Z/Z$ is self-normalizing in $\mathrm{GL}_n(q)/Z$ then part (1) of the Outer Rigidity Criterion shows that $\Delta_\zeta Z$ occurs regularly over $\mathbb{Q}$. In particular, if $\mathbf{C}$ is rational and $\Delta_\zeta = \mathrm{GL}_n(q)$ then $\mathrm{GL}_n(q)$ occurs regularly over $\mathbb{Q}$.

Recall that $\mathbf{C} = (C_1, \ldots, C_r)$ is called rational if for every integer m prime to the order of G the classes $C_1^m, \ldots, C_r^m$ form a permutation of $C_1, \ldots, C_r$. Define $\zeta = (\zeta_1, \ldots, \zeta_r)$ to be rational if for every integer m prime to $q-1$ the elements $\zeta_1^m, \ldots, \zeta_r^m$ form a permutation of $\zeta_1, \ldots, \zeta_r$. Then ζ is rational if and only if $\mathbf{C} = \mathbf{C}_\zeta$ is rational. For rational ζ we get $\mathrm{Aut}^{\mathbf{C}}(G) = \mathrm{Aut}(G) \cong V \cdot \Gamma L_n(q)$ in Lemma 9.19(a). Indeed, each element of $\Gamma L_n(q)$ permutes $\zeta_1, \ldots, \zeta_r$ since it acts on Z by raising each element to some (fixed) power of p.

Unfortunately, not for every choice of n and q is there a rational ζ satisfying our hypotheses (specified at the beginning of Section 9.4). This forces us to impose certain restrictions on n and q. Let φ denote Euler's φ-function.

Lemma 9.20 *Let $n \geq 4$ be an even integer.*

(i) Assume $q > 4$ is either odd or a power of 4. If $n \geq \varphi(q-1)$ then there is rational ζ with $\Delta_\zeta = GL_n(q)$.

(ii) Let $q = p^{2s}$ with p a prime, s a positive integer. Assume either q or s is odd. If $n \geq \varphi(\sqrt{q}+1)$ then there is rational ζ with $\Delta_\zeta = U_n(q)$.

The given conditions on n and q are by no means necessary for the conclusion, but it is too complicated to find necessary and sufficient conditions. The lemma follows essentially from Theorem 9.17. Additionally, one uses the fact that the element $\Phi_\zeta(\tilde{Q}_i^2)$ of Δ_ζ has determinant $\zeta_i^{-1}\zeta_{i+1}^{-1}$ (for $i = 1, \ldots, r-1$). The detailed proof is given in Lemmas 9 and 10 of [V2].

Finally, we summarize the resulting Galois realizations (see [V1] and [V4]). Here $\mathrm{PU}_n(q)$ is the image of $\mathrm{U}_n(q)$ in $\mathrm{PGL}_n(q)$.

Theorem 9.21 *If n and q are as in (i) (resp., (ii)) of the above lemma then the groups $GL_n(q)$ and $PGL_n(q)$ (resp., $PU_n(q)$) occur regularly over $\mathbb{Q}$. For $PGL_n(q)$ and $PU_n(q)$ we even get GAL-realizations over $\mathbb{Q}$.*

Sketch of Proof. The case of $\mathrm{GL}_n(q)$ follows from part (1) of the Outer Rigidity Criterion, using Lemma 9.19 and Lemma 9.20(i). If $\mathrm{GL}_n(q)$ occurs regularly over $\mathbb{Q}$, then the same holds for its quotient $\mathrm{PGL}_n(q)$.

The same method does not work for $\mathrm{Sp}_n(p)$ and $\mathrm{U}_n(q)$ (the other cases in Theorem 9.17), since they are not self-normalizing in $\mathrm{GL}_n(q)$. Actually, this prevents us from getting any Galois realizations over $\mathbb{Q}$ for either $\mathrm{Sp}_n(p)$ or $\mathrm{PSp}_n(p)$. However, $\mathrm{PU}_n(q)$ is self-normalizing in $\mathrm{PGL}_n(q)$. (For this and further facts about classical groups needed below, we refer the reader to [D].) Assume now that ζ is rational and $\Delta_\zeta = \mathrm{U}_n(q)$. We are going to apply part (2) of the Outer Rigidity Criterion, taking A to be the center of $\mathrm{Out}_\mathrm{C}(G) \cong \mathrm{GL}_n(q)/Z$. Then $\Delta_A \cong \mathrm{PU}_n(q)$ is self-normalizing in $\mathrm{Out}_\mathrm{C}(G)/A \cong \mathrm{PGL}_n(q)$ and has trivial centralizer in $\mathrm{Out}^\mathrm{C}(G)/A \cong P\Gamma L_n(q)$ (see Lemma 9.19 and the remarks following it). Further,

$$\begin{aligned}|\mathrm{Out}(\Delta_A)| &= |\mathrm{Out}(\mathrm{PU}_n(q))| = |\mathrm{Aut}(\mathbb{F}_q)| = [\Gamma L_n(q)/Z : \mathrm{GL}_n(q)/Z] \\ &= \left[\mathrm{Out}^\mathrm{C}(G) : \mathrm{Out}_\mathrm{C}(G)\right].\end{aligned}$$

Thus $\mathrm{PU}_n(q)$ has a GAL-realization over $\mathbb{Q}$.

A GAL-realization for $\mathrm{PGL}_n(q)$ ($n > 2$) cannot be found analogously, since

$$|\mathrm{Out}(\mathrm{PGL}_n(q))| = 2\,|\mathrm{Aut}(\mathbb{F}_q)|.$$

The problem is that the inverse transpose automorphism of $\mathrm{PGL}_n(q)$ does not act on the module V, hence does not act on G. Therefore we need to replace G by another group $\tilde{G}$.

Assume now that ζ is rational and $\Delta_\zeta = \mathrm{GL}_n(q)$. Let $\tilde{V} = V \oplus V$, and let Z act on $\tilde{V}$ such that $z \in Z$ sends (u, v) to $(zu, z^{-1}v)$. Set $\tilde{G} = \tilde{V} \cdot Z$ (semi-direct product). For $i = 1, \ldots, r$, let $\tilde{C}_i$ be the conjugacy class of $\tilde{G}$ containing ζ_i, and set $\tilde{\mathbf{C}} = (\tilde{C}_1, \ldots, \tilde{C}_r)$. Then $\tilde{\mathbf{C}}$ is weakly rigid, and $\mathrm{Aut}_{\tilde{\mathbf{C}}}(\tilde{G}) \cong \tilde{V} \cdot \mathrm{GL}_n(q)^2$. Here $\mathrm{GL}_n(q)^2$ is the direct product of two copies of $\mathrm{GL}_n(q)$, acting in the natural way on $\tilde{V} = V \oplus V$. Further, $\mathrm{Aut}^{\tilde{\mathbf{C}}}(\tilde{G})$ contains $\tilde{V} \cdot \{(f, h) \in \Gamma L_n(q)^2 : fh^{-1} \in \mathrm{GL}_n(q)\}$ as a subgroup of index 2; this subgroup is complemented by the subgroup of order 2 generated by the automorphism $[(u, v), z] \mapsto [(v, u), z^{-1}]$. Hence

$$|\mathrm{Out}(\mathrm{PGL}_n(q))| = 2\, |\mathrm{Aut}(\mathbb{F}_q)| = \left[\mathrm{Out}^{\tilde{\mathbf{C}}}(\tilde{G}) : \mathrm{Out}_{\tilde{\mathbf{C}}}(\tilde{G})\right].$$

Now part (2) of the Outer Rigidity Criterion applies, with $A = (\mathbb{F}_q^\times)^2/Z$ and $\Delta_A \cong \mathrm{PGL}_n(q)$ (diagonally embedded in $\mathrm{Out}_{\tilde{\mathbf{C}}}(\tilde{G})/A \cong \mathrm{PGL}_n(q)^2$). The details are worked out in [V4], Section 2.5. □

Remark 9.22

(a) The theorem yields the only known (regular) Galois realizations over $\mathbb{Q}$ for Lie type groups with arbitrarily many outer automorphisms (since q can be an arbitrarily high power of p). Simple groups covered by the theorem are the groups $\mathrm{PGL}_n(q)$ where $q > 4$ is a power of 4, and $n \geq \varphi(q-1)$ is even and prime to $q-1$; and the groups $\mathrm{PU}_n(q)$ where $q = 2^{2s}$ with s odd, and $n \geq \varphi(\sqrt{q}+1)$ is even and prime to $\sqrt{q}+1$, and $n \geq 4$. The parallel case of the symplectic braid group (see Remark 9.7(c)) yields similar results for the simple groups $\mathrm{Sp}_{2n}(4^s)$ (see [SV2]).
(b) We obtain GAL-realizations over $\mathbb{Q}_{ab}$ for the groups $\mathrm{PGL}_n(q)$ and $\mathrm{PU}_n(q)$ under weaker conditions on n and q (see [V4]). If $q = p^s$ with p a prime, and $H = \mathrm{PGL}_n(q)$ with $n > 2$, then $\mathrm{Out}(H)$ is the direct product of two cyclic groups of order s and 2, respectively. Thus if s is even and > 2 then the group $\mathrm{Out}(H)$ is noncyclic abelian of order > 4, hence does not embed into $\mathrm{PGL}_2(\bar{\mathbb{Q}})$ (see the last exercise in Section 2.2.1). Coupled with (a), this yields examples of simple groups with a GAR-realization over $\mathbb{Q}$ and $\mathbb{Q}_{ab}$ (in several variables) that cannot have such a realization in one variable.
(c) The Galois realizations in Theorem 9.21 can also be interpreted as describing the Galois action on certain subspaces of the p-division points of the Jacobians of certain curves defined over a purely transcendental field $\mathbb{Q}(x_1, \ldots, x_r)$ (see [V6]).
(d) As an application of the GAL-property, note that for the groups $\mathrm{PGL}_n(q)$ covered by the theorem we get the following: Every group H between

$\mathrm{PGL}_n(q)$ and its automorphism group occurs regularly over $\mathbb{Q}$; in particular, $H = P\Gamma L_n(q)$. Indeed, the outer automorphism group of $\mathrm{PGL}_n(q)$ is abelian, hence $H/\,\mathrm{PGL}_n(q)$ occurs regularly over $\mathbb{Q}$ by Corollary 8.13. By Corollary 8.16 and Lemma 8.17 it follows that H also occurs regularly over $\mathbb{Q}$.

10
Moduli Spaces for Covers of the Riemann Sphere

By Riemann's existence theorem, finite Galois extensions of $\mathbb{C}(x)$ correspond to Galois covers of the Riemann sphere. Those with fixed branch points and fixed Galois group G are finite in number, and can be parametrized by certain generating systems of G. (This parametrization is not canonical, however.) If we let the branch points vary we obtain a moduli space (for covers with group G and with a fixed number r of branch points) that has a finite-to-one map Ψ to the space $\mathcal{O}_r$ of branch point sets. Here $\mathcal{O}_r$ is the space of subsets of $\mathbb{C}$ of cardinality r. (Note we do not allow ∞ as a branch point – a technical hypothesis, which simplifies the exposition, but is not a real restriction.) This moduli space carries a natural topology, which makes Ψ an (unramified) covering map. Such a covering is controlled by the fundamental group of the base space $\mathcal{O}_r$. This fundamental group is isomorphic to the Artin braid group $\mathcal{B}_r$.

The natural action of the fundamental group of $\mathcal{O}_r$ on the fiber of the covering Ψ over a fixed branch point set yields an action of $\mathcal{B}_r$ on the generating systems of G parametrizing this fiber. The action of the "elementary braid" generators of $\mathcal{B}_r$ yields exactly the braid operations from Chapter 9. This braiding action – which dates back to Clebsch [Cleb] and Hurwitz [Hur] – is really the heart of the whole matter. Its significance for Galois theory was first noticed by M. Fried [Fr1]. Via covering space theory, the braiding action completely determines the topology of the moduli space. In particular, the braid orbits correspond to the components of the moduli space.

If $Z(G) = 1$ then the covers parametrized by a component $\mathcal{H}$ of the moduli space $\mathcal{H}_r^{in}(G)$ can be combined in a family $\mathcal{T} \to \mathcal{H} \times \mathbb{P}^1$: For all $u \in \mathcal{H}$, the cover corresponding to u is realized as the natural map from $\mathcal{T}_u$, the fiber over u, to $\mathbb{P}^1$. The group G acts through deck transformations of the map $\mathcal{T} \to \mathcal{H} \times \mathbb{P}^1$ so that its restriction to $\mathcal{T}_u$ yields the group $\mathrm{Deck}(\mathcal{T}_u/\mathbb{P}^1)$ for each $u \in \mathcal{H}$. Thus a point $u \in \mathcal{H}$ carries more information than just the equivalence class of the associated cover f; it also specifies a particular isomorphism between $\mathrm{Deck}(f)$ and G (modulo $\mathrm{Inn}(G)$).

By the higher-dimensional version of Riemann's existence theorem, we can recognize the above spaces as algebraic varieties defined over $\bar{\mathbb{Q}}$, such that the maps between them become algebraic morphisms defined over $\bar{\mathbb{Q}}$. Then the map $\mathcal{T} \to \mathcal{H} \times \mathbb{P}^1$ induces a function field extension $L/k(x)$, Galois with group G, where $L = \bar{\mathbb{Q}}(\mathcal{T})$, $k = \bar{\mathbb{Q}}(\mathcal{H})$, and x is the identity function on $\mathbb{P}^1$. We prove that k is isomorphic to the minimal field of definition of a generic cover in the family $\mathcal{T}$. This yields the connection to Chapter 9. The proof of Theorem 9.5 follows from that.

The moduli space $\mathcal{H}_r^{in}(G)$ even has a natural structure as a (nonsingular) affine algebraic set defined over $\mathbb{Q}$. If $Z(G) = 1$ then $\mathcal{H}_r^{in}(G)$ has a $\mathbb{Q}$-rational point if and only if there is a Galois extension of $\mathbb{Q}(x)$, regular over $\mathbb{Q}$, with Galois group isomorphic to G and with r branch points (when tensored with $\bar{\mathbb{Q}}$). A survey of this and further results is given in Section 10.3.

The moduli spaces to be described in this chapter have been constructed in [FV1] and, alternatively, in [V3]. Here we present another slightly modified (and, let us hope, improved) version of these constructions.

10.1 The Topological Construction of the Moduli Spaces

We fix an integer $r \geq 2$ and a finite group G, as well as a subgroup $\mathbf{A}$ of $\mathrm{Aut}(G)$. Let $\mathbb{P}^1 = \mathbb{C} \cup \{\infty\}$, topologized as in Chapter 4.

This section is purely topological. The notation is as in Chapter 4. In particular, paths are continuous maps from the interval $[0, 1]$ into a topological space, and the path product $\gamma_2\gamma_1$ means that first γ_1 is traversed, and then γ_2. In our notation we will not always distinguish between a path and its homotopy class.

10.1.1 A Construction of Coverings

Suppose $f : R \to S$ is a Galois covering with group G. Then with each $b \in R$ there is associated a surjective homomorphism $\Phi_b : \pi_1(S, p) \to G$, where $p = f(b)$ (see Proposition 4.19). If $Z(G) = 1$ then for $b, b' \in f^{-1}(p)$ we have $\Phi_b = \Phi_{b'}$ iff $b = b'$. This allows us to reconstruct R as the set of all pairs (p, Φ_b) (suitably topologized). A "continuous version" of this construction will later yield the family $\mathcal{T}_r(G)$ (see Proposition 10.6).

If ω is a path in S going from p to p' then we let ω^* denote the isomorphism $\pi_1(S, p) \to \pi_1(S, p')$, $\gamma \mapsto \omega\gamma\omega^{\mathrm{inv}}$. For $g \in G$ we let $\mathrm{Inn}(g)$ denote the inner automorphism $h \mapsto ghg^{-1}$ of G.

Lemma 10.1 *Assume $Z(G) = 1$. Suppose S is a connected manifold. Pick a base point $p_0 \in S$, and let $\varphi_0 : \pi_1(S, p_0) \to G$ be a surjective homomorphism. Let R be the set of pairs (p, φ), where $p \in S$ and $\varphi = \varphi_0 \circ \omega^* : \pi_1(S, p) \to G$ for some path ω in S going from p to p_0. We specify a basis of neighborhoods for each point $(p, \varphi) \in R$ as follows: For each simply connected neighborhood D of p in S, let $\mathcal{U}(D)$ consist of all (p', φ'), where $p' \in D$ and $\varphi' = \varphi\delta^*$ for δ a path in D going from p' to p. This defines a topology on R, making the map $f : R \to S$, $(p, \varphi) \mapsto p$, a Galois covering. We obtain an isomorphism $\mu : G \to Deck(f)$ by sending $g \in G$ to the deck transformation $(p, \varphi) \mapsto (p, Inn(g) \circ \varphi)$. Then for $b_0 := (p_0, \varphi_0) \in R$ we have $\Phi_{b_0} = \mu \circ \varphi_0$. Here Φ_{b_0} is the map associated with the covering f by Proposition 4.19.*

Proof. For two neighborhoods of (p, φ) of the form $\mathcal{U}(D)$ and $\mathcal{U}(D')$ we have $\mathcal{U}(D) \subset \mathcal{U}(D')$ if and only if $D \subset D'$. Hence these $\mathcal{U}(D)$ form a neighborhood basis for a topology on R.

Note that all paths δ in D going from p' to p are homotopic, hence induce the same map δ^* of fundamental groups. Hence $f^{-1}(D)$ is the disjoint union of the $\mathcal{U}(D)$ corresponding to the various possible choices of φ (for fixed p), and f maps each such $\mathcal{U}(D)$ homeomorphically onto D. Therefore f is a covering.

For each $(p, \varphi) \in R$ we have

$$f^{-1}(p) = \{(p, \mathrm{Inn}(g) \circ \varphi) : g \in G\}. \tag{10.1}$$

Indeed, we have $\varphi = \varphi_0 \circ \omega^*$ with ω a path from p to p_0. Any other path from p to p_0 is of the form $\omega' = \gamma\omega$, with γ a closed path based at p_0. Thus $f^{-1}(p) = \{(p, \varphi_0 \circ \gamma^*) \circ \omega^* : \gamma \in \pi_1(S, p_0)\} = \{(p, \mathrm{Inn}(\varphi_0(\gamma)) \circ \varphi) : \gamma \in \pi_1(S, p_0)\}$, which implies (10.1).

From (10.1) it is clear that for each $g \in G$ the map $(p, \varphi) \mapsto (p, \mathrm{Inn}(g) \circ \varphi)$ is a deck transformation of f (since it permutes the basic neighborhoods $\mathcal{U}(D)$ over D). This yields an embedding $\mu : G \to \mathrm{Deck}(f)$. By (10.1), G acts transitively on each fiber $f^{-1}(p)$. Once we have shown that R is connected, this implies that f is a Galois covering, and μ is an isomorphism (see Lemma 4.17).

Let ω be a closed path in S based at p_0. Clearly, the lift $\tilde{\omega}$ of ω via f with initial point $b_0 = (p_0, \varphi_0)$ is the path $t \mapsto (\omega(t), \varphi_t)$, where $\varphi_t = \varphi_0 \circ (\omega_t^{\mathrm{inv}})^*$. Here ω_t is the path that agrees with ω for $t' \leq t$, and remains constant for $t' \geq t$. The endpoint of $\tilde{\omega}$ is $(p_0, \varphi_0 \circ (\omega^{\mathrm{inv}})^*) = (p_0, \mathrm{Inn}(g^{-1}) \circ \varphi_0)$ where $g = \varphi_0(\omega)$. Thus each point of $f^{-1}(p_0)$ occurs as endpoint of some $\tilde{\omega}$ (whose initial point is b_0). This means that R is connected (Theorem 4.12(c)).

The deck transformation $\mu(g)$ induced by $g = \varphi_0(\omega)$ maps the endpoint $(p_0, \mathrm{Inn}(g^{-1}) \circ \varphi_0)$ of $\tilde{\omega}$ to its initial point b_0. This means that $\Phi_{b_0}(\omega) = \mu(g) = \mu\varphi_0(\omega)$. Hence $\Phi_{b_0} = \mu\varphi_0$. □

10.1.2 Distinguished Conjugacy Classes in the Fundamental Group of a Punctured Sphere

Let P be a finite subset of $\mathbb{C}$, and $p \in P$, $p_0 \in \mathbb{P}^1 \setminus P$. Choose $d > 0$ such that $|p - p'| > d$ for any $p' \neq p$ in P. Choose a path χ in $\mathbb{P}^1 \setminus P$ from p_0 to $p + d$. The path $\lambda(t) = p + d \exp(2\pi\sqrt{-1}t)$ (with $t \in [0, 1]$) is a closed path based at $p + d$, winding once around p in counterclockwise direction. Then $\gamma := \chi^{\mathrm{inv}}\lambda\chi$ represents an element $[\gamma]$ of $\pi_1(\mathbb{P}^1 \setminus P, p_0)$, and these elements $[\gamma]$ run through a conjugacy class of $\pi_1(\mathbb{P}^1 \setminus P, p_0)$ as χ and d vary. We denote this class by $\Sigma_p = \Sigma_p(\mathbb{P}^1 \setminus P, p_0)$.

If ω is a path in $\mathbb{P}^1 \setminus P$ from p_0 to p_0' then the isomorphism $\omega^* : \pi_1(\mathbb{P}^1 \setminus P, p_0) \to \pi_1(\mathbb{P}^1 \setminus P, p_0')$ maps the class $\Sigma_p(\mathbb{P}^1 \setminus P, p_0)$ to $\Sigma_p(\mathbb{P}^1 \setminus P, p_0')$.

Lemma 10.2

(i) Set $P' = P \setminus \{p\}$. Then the (normal) subgroup of $\Gamma = \pi_1(\mathbb{P}^1 \setminus P, p_0)$ generated by Σ_p consists of all $[\gamma] \in \Gamma$ for which the path γ is null-homotopic in $\mathbb{P}^1 \setminus P'$.

(ii) Suppose $f : R \to \mathbb{P}^1 \setminus P$ is a finite Galois covering. Then for each $b \in f^{-1}(p_0)$ the surjection $\Phi_b : \pi_1(\mathbb{P}^1 \setminus P, p_0) \to Deck(f)$ (from Proposition 4.19) maps Σ_p to the class C_p of $Deck(f)$ from Proposition 4.23.

Proof. (ii) follows from Proposition 4.23(5). For (i), let $v : \Gamma \to \Gamma' := \pi_1(\mathbb{P}^1 \setminus P', p_0)$ be the map induced by inclusion. Clearly, the group N generated by Σ_p lies in the kernel of v. For the converse, label the elements of P as $p_1, \dots, p_r$ with $p_1 = p$. Identify paths and their homotopy classes, for simplicity. By Corollary 4.29, the group Γ is freely generated by elements $\gamma_1, \dots, \gamma_{r-1}$ with $\gamma_i \in \Sigma_{p_i}$. Also, Γ' is freely generated by (the classes of) $\gamma_2, \dots, \gamma_{r-1}$. Since Γ/N is generated by the images of $\gamma_2, \dots, \gamma_{r-1}$, and the induced map $\bar{v} : \Gamma/N \to \Gamma'$ sends these elements to free generators of Γ', it follows that $\bar{v}$ is an isomorphism. This proves (i). □

10.1.3 The Moduli Spaces as Abstract Sets

10.1.3.1 The Spaces

Let $\mathcal{O}_r$ be the set of all subsets of $\mathbb{C}$ of cardinality r. For $P \in \mathcal{O}_r$ and $p_0 \in \mathbb{P}^1 \setminus P$, we call a map $\varphi : \pi_1(\mathbb{P}^1 \setminus P, p_0) \to G$ **admissible** if it is a surjective homomorphism, and $\varphi(\Sigma_p) \neq 1$ for each $p \in P$. Here $\Sigma_p = \Sigma_p(\mathbb{P}^1 \setminus P, p_0)$ is the conjugacy class from Section 10.1.2.

Consider pairs (P, φ), where $P \in \mathcal{O}_r$, and $\varphi : \pi_1(\mathbb{P}^1 \setminus P, \infty) \to G$ is admissible. Two such pairs (P, φ) and (P', φ') are called **A**-equivalent iff $P = P'$ and $\varphi' = A \circ \varphi$ for some $A \in \mathbf{A}$. (Here $\mathbf{A}$ is our fixed subgroup of Aut(G).) Clearly, this defines an equivalence relation on those pairs. Denote the equivalence class of the pair (P, φ) by $[P, \varphi]_{\mathbf{A}}$. We denote the set of these equivalence classes by $\mathcal{H}_r^{(\mathbf{A})}(G)$.

Further, let $\mathcal{T}_r(G)$ be the set of triples (P, q, φ), where $P \in \mathcal{O}_r, q \in \mathbb{P}^1 \setminus P$, and $\varphi : \pi_1(\mathbb{P}^1 \setminus P, q) \to G$ is admissible.

Finally, define $\mathcal{O}(r+1)$ and $\mathcal{S}_r^{(\mathbf{A})}(G)$ to be the subsets of $\mathcal{O}_r \times \mathbb{P}^1$ and $\mathcal{H}_r^{(\mathbf{A})}(G) \times \mathbb{P}^1$, respectively, consisting of all (P, q), respectively, $([P, \varphi]_{\mathbf{A}}, q)$, with $q \notin P$.

The most important case is when $\mathbf{A} = \mathrm{Inn}(G)$. In this case, we write $\mathcal{H}_r^{(\mathbf{A})}(G) = \mathcal{H}_r^{in}(G)$ and $\mathcal{S}_r^{(\mathbf{A})}(G) = \mathcal{S}_r^{in}(G)$.

10.1.3.2 Maps

Let $\Psi_{\mathbf{A}} : \mathcal{H}_r^{(\mathbf{A})}(G) \to \mathcal{O}_r$ be the map sending $[P, \varphi]_{\mathbf{A}}$ to P. For $\mathbf{A} \subset \mathbf{A}'$ we have a map $\mathcal{H}_r^{(\mathbf{A})}(G) \to \mathcal{H}_r^{(\mathbf{A}')}(G)$ sending $[P, \varphi]_{\mathbf{A}}$ to $[P, \varphi]_{\mathbf{A}'}$. This map composed with $\Psi_{\mathbf{A}'}$ equals $\Psi_{\mathbf{A}}$.

Define $\Pi : \mathcal{T}_r(G) \to \mathbb{P}^1$, $(P, q, \varphi) \mapsto q$ and $\tilde{\Pi} : \mathcal{T}_r(G) \to \mathcal{O}(r+1)$, $(P, q, \varphi) \mapsto (P, q)$. If $\mathrm{Inn}(G) \subset \mathbf{A}$ we further get a map $\Omega_{\mathbf{A}} : \mathcal{T}_r(G) \to \mathcal{H}_r^{(\mathbf{A})}(G)$, defined as follows. Given (P, q, φ), choose a path ω in $\mathbb{P}^1 \setminus P$ from ∞ to q. Let $\varphi_\omega := \varphi \circ \omega^* : \pi_1(\mathbb{P}^1 \setminus P, \infty) \to G$ (where ω^* is as in Section 10.1.1). If φ is admissible then so is φ_ω. Another choice of ω results in φ_ω being composed with an inner automorphism of G. Hence the element $[P, \varphi_\omega]_{\mathbf{A}}$ is independent of the choice of ω (since $\mathrm{Inn}(G) \subset \mathbf{A}$). Thus we can define $\Omega_{\mathbf{A}}$ by sending (P, q, φ) to $[P, \varphi_\omega]_{\mathbf{A}}$. Whenever we write $\Omega_{\mathbf{A}}$ it is understood that we assume $\mathrm{Inn}(G) \subset \mathbf{A}$.

If $\mathbf{A} = \mathrm{Inn}(G)$ we drop the subscript in the notation $\Psi_{\mathbf{A}}$, $\Omega_{\mathbf{A}}$ and $[P, \varphi]_{\mathbf{A}}$ (i.e., we just write Ψ, Ω and $[P, \varphi]$).

For each $A \in \mathrm{Aut}(G)$ we define $\delta_A : \mathcal{T}_r(G) \to \mathcal{T}_r(G)$ and $\epsilon_A : \mathcal{H}_r^{in}(G) \to \mathcal{H}_r^{in}(G)$ by sending (P, q, φ) to $(P, q, A \circ \varphi)$, and $[P, \varphi]$ to $[P, A \circ \varphi]$, respectively. This yields an action of Aut(G) on $\mathcal{T}_r(G)$ and on $\mathcal{H}_r^{in}(G)$.

10.1.3.3 Objects Classified by the Points of $\mathcal{H}_r^{in}(G)$

Let $P \in \mathcal{O}_r$. Consider pairs (f, μ), where $f : R \to \mathbb{P}^1 \setminus P$ is a Galois covering and $\mu : G \to \mathrm{Deck}(f)$ is an isomorphism. Call two such pairs (f, μ) and (f', μ') equivalent (where $f' : R' \to \mathbb{P}^1 \setminus P$) if there is a homeomorphism $\delta : R \to R'$ with $f' \circ \delta = f$ and $\mu'(g) = \delta \mu(g) \delta^{-1}$ for all $g \in G$. Clearly, this defines an equivalence relation.

Similarly, we define an equivalence relation on pairs (L, ν), where L is a finite Galois extension of $\mathbb{C}(x)$ and $\nu : G \to G(L/\mathbb{C}(x))$ an isomorphism. Call two such pairs (L, ν) and (L', ν') equivalent if there is a $\mathbb{C}(x)$-isomorphism $\rho : L \to L'$ with $\nu'(g) = \rho\nu(g)\rho^{-1}$ for all $g \in G$. Then Theorem 5.14 yields a 1–1 correspondence between the following objects:

1. The equivalence classes of pairs (L, ν), where L is a finite Galois extension of $\mathbb{C}(x)$ with branch points contained in P, and $\nu : G \to G(L/\mathbb{C}(x))$ an isomorphism.
2. The equivalence classes of pairs (f, μ), where $f : R \to \mathbb{P}^1 \setminus P$ is a Galois covering and $\mu : G \to \mathrm{Deck}(f)$ is an isomorphism.
3. The surjective homomorphisms $\varphi : \pi_1(\mathbb{P}^1 \setminus P, \infty) \to G$ modulo composition with inner automorphisms of G.

The correspondence between (2) and (3) is given by $\varphi = \mu^{-1} \circ \Phi_b$, where $\Phi_b : \pi_1(\mathbb{P}^1 \setminus P, \infty) \to \mathrm{Deck}(f)$ is the map from Proposition 4.19, for any $b \in f^{-1}(\infty)$. (Note that varying b over $f^{-1}(\infty)$ means composing Φ_b with inner automorphisms of $\mathrm{Deck}(f)$.) The correspondence between (1) and (2) is given by taking $L = \mathcal{M}(\bar{R})$ and $\nu = \iota \circ \mu$, where ι is as in Theorem 5.9.

Those φ from (3) which are admissible (as defined at the beginning of this subsection) correspond to those f for which the class C_p of $\mathrm{Deck}(f)$ associated with any $p \in P$ is nontrivial (by Lemma 10.2(ii)); and they correspond to those L whose branch point set is *exactly* P (by Theorem 5.9 Addendum). Thus if (L, ν) is as in (1), and L has branch point set equal to P, then the homomorphisms φ in (3) corresponding to (L, ν) yield a point $u = [P, \varphi]$ of $\mathcal{H}_r^{in}(G)$.

Proposition 10.3 *The above yields a* 1-1 *correspondence between the points u of $\mathcal{H}_r^{in}(G)$ and the equivalence classes of pairs (L, ν), where L is a finite Galois extension of $\mathbb{C}(x)$ with r branch points, all different from ∞, and $\nu : G \to G(L/\mathbb{C}(x))$ an isomorphism. Under this correspondence, $\Psi(u)$ is the branch point set of $L/\mathbb{C}(x)$. Further, $\epsilon_A(u)$ corresponds to $(L, \nu \circ A^{-1})$ for each $A \in Aut(G)$.*

10.1.4 The Topology of the Moduli Spaces

When we speak of a disc in $\mathbb{P}^1$ we mean an open disc (as in Section 4.2.2). Let $P = \{p_1, \ldots, p_r\} \in \mathcal{O}_r$. For each $[P, \varphi]_{\mathbf{A}}$ in $\mathcal{H}_r^{(\mathbf{A})}(G)$ we specify a basis of neighborhoods as follows: Given disjoint discs $D_1, \ldots, D_r$ in $\mathbb{C}$ around $p_1, \ldots, p_r$, let $\mathcal{N}(D_1, \ldots, D_r)$ be the set of all $[P', \varphi']_{\mathbf{A}}$, where $P' = \{p'_1, \ldots, p'_r\}$, such that $p'_i \in D_i$ for all i, and φ' is the composition of φ with the canonical isomorphisms

$$\pi_1(\mathbb{P}^1 \setminus P', \infty) \to \pi_1(\mathbb{P}^1 \setminus (D_1 \cup \cdots \cup D_r), \infty) \to \pi_1(\mathbb{P}^1 \setminus P, \infty). \quad (10.2)$$

Similarly, for (P, p_0, φ) in $\mathcal{T}_r(G)$ we specify a basis of neighborhoods as follows: Given disjoint discs $D_0, \ldots, D_r$ in $\mathbb{P}^1$ around $p_0, \ldots, p_r$, let $\mathcal{M}(D_0, \ldots, D_r)$ be the set of all (P', p_0', φ'), where $P' = \{p_1', \ldots, p_r'\}$, such that $p_i' \in D_i$ for $i = 0, \ldots, r$, and φ' is the composition of φ with the canonical isomorphisms

$$\begin{aligned} &\pi_1(\mathbb{P}^1 \setminus P', p_0') \to \pi_1(\mathbb{P}^1 \setminus (D_1 \cup \cdots \cup D_r), p_0') \\ &\quad \to \pi_1(\mathbb{P}^1 \setminus (D_1 \cup \cdots \cup D_r), p_0) \to \pi_1(\mathbb{P}^1 \setminus P, p_0). \quad (10.3) \end{aligned}$$

The second to last isomorphism is obtained by conjugating with some path from p_0' to p_0 inside D_0. This does not depend on the choice of this path since D_0 is simply connected. All other maps in (10.2) and (10.3) are induced by the inclusions between the respective spaces. Clearly, φ' is admissible if φ is.

The intersection of any two such neighborhoods of a given point is again of the same type, hence there is a unique topology on our spaces having these sets as neighborhood bases.

Similarly, we topologize the space $\mathcal{O}_r$ by using the neighborhoods $\mathcal{N}_0(D_1, \ldots, D_r)$ consisting of all $P' = \{p_1', \ldots, p_r'\}$ with $p_i' \in D_i$ for all i. Here $D_1, \ldots, D_r$ are any disjoint discs in $\mathbb{C}$. We equip the sets $\mathcal{O}(r+1)$ and $\mathcal{S}_r^{(\mathbf{A})}(G)$ with the induced topology as subspaces of $\mathcal{O}_r \times \mathbb{P}^1$ and $\mathcal{H}_r^{(\mathbf{A})}(G) \times \mathbb{P}^1$, respectively. This induced topology has neighborhood bases of the form $\mathcal{N}_0(D_1, \ldots, D_r) \times D_0$ and $\mathcal{N}(D_1, \ldots, D_r) \times D_0$, respectively, where $D_0, D_1, \ldots, D_r$ are disjoint discs in $\mathbb{P}^1$ (with $D_1, \ldots, D_r \subset \mathbb{C}$).

All maps defined in 10.1.3 are continuous with respect to those topologies. Moreover:

Lemma 10.4

(i) The maps $\Psi_{\mathbf{A}} : \mathcal{H}_r^{(\mathbf{A})}(G) \to \mathcal{O}_r, \tilde{\Pi} : \mathcal{T}_r(G) \to \mathcal{O}(r+1)$ and $\Omega_{\mathbf{A}} \times \Pi : \mathcal{T}_r(G) \to \mathcal{S}_r^{(\mathbf{A})}(G)$ are coverings. Also, the map $\mathcal{H}_r^{(\mathbf{A})}(G) \to \mathcal{H}_r^{(\mathbf{A}')}(G)$, defined for $\mathbf{A} \subset \mathbf{A}'$, is a covering.

(ii) The ϵ_A and δ_A with $A \in Aut(G)$ are deck transformations of Ψ and $\tilde{\Pi}$, respectively. Those δ_A with $A \in \mathbf{A}$ are deck transformations of the covering $\Omega_{\mathbf{A}} \times \Pi : \mathcal{T}_r(G) \to \mathcal{S}_r^{(\mathbf{A})}(G)$, and they form a group that acts transitively on each fiber of this covering.

Proof. (i) The set $\Psi_{\mathbf{A}}^{-1}(\mathcal{N}_0(D_1,\dots,D_r))$ is the disjoint union of the $\mathcal{N}(D_1,\dots,D_r)$ corresponding to the various admissible maps $\varphi : \pi_1(\mathbb{P}^1 \setminus (D_1 \cup \cdots \cup D_r), \infty) \to G$ modulo $\mathbf{A}$. Clearly, $\Psi_{\mathbf{A}}$ maps each such $\mathcal{N}(D_1,\dots,D_r)$ homeomorphically onto $\mathcal{N}_0(D_1,\dots,D_r)$ (and the latter space is homeomorphic to $D_1 \times \cdots \times D_r$). This proves that $\Psi_{\mathbf{A}}$ is a covering.

The case of $\tilde{\Pi}$ (resp., $\Omega_{\mathbf{A}} \times \Pi$) is similar, using that the inverse image of $\mathcal{N}_0(D_1,\dots,D_r) \times D_0$ (resp., $\mathcal{N}(D_1,\dots,D_r) \times D_0$) is a disjoint union of neighborhoods of the form $\mathcal{M}(D_0,\dots,D_r)$. Similar arguments apply for the map $\mathcal{H}_r^{(\mathbf{A})}(G) \to \mathcal{H}_r^{(\mathbf{A}')}(G)$.

(ii) The δ_A and ϵ_A are homeomorphisms (because they are bijections permuting the basic neighborhoods in the respective spaces). Clearly, $\Psi \circ \epsilon_A = \Psi$ and $\tilde{\Pi} \circ \delta_A = \tilde{\Pi}$. This proves the first claim in (ii). The second claim follows easily from the definitions in Section 10.1.3. □

Let $(\theta_t)_{t\in[0,1]}$ be a family of homeomorphisms of a topological space R. We say this family is continuous if the map $(t, p) \mapsto \theta_t(p)$ is a continuous map from $[0, 1] \times R \to R$.

Lemma 10.5

(a) Each homeomorphism $\theta : \mathbb{P}^1 \to \mathbb{P}^1$ fixing ∞ induces a homeomorphism $\hat{\theta} : \mathcal{H}_r^{(\mathbf{A})}(G) \to \mathcal{H}_r^{(\mathbf{A})}(G)$, mapping $[P, \varphi]_{\mathbf{A}}$ to $[\theta(P), \varphi\theta^{-1}]_{\mathbf{A}}$. Here $\varphi\theta^{-1}$ is the composition of φ with the map $\pi_1(\mathbb{P}^1 \setminus \theta(P), \infty) \to \pi_1(\mathbb{P}^1 \setminus P, \infty)$ induced by θ^{-1}.

(b) If $(\theta_t)_{t\in[0,1]}$ is a continuous family of homeomorphisms of $\mathbb{P}^1$ fixing ∞, then the family $(\hat{\theta}_t)_{t\in[0,1]}$ is also continuous.

Proof. All fundamental groups in this proof will be based at ∞, hence we drop the base point and simply write $\pi_1(\mathbb{P}^1 \setminus P)$, and so on.

(a) Clearly, $\varphi\theta^{-1}$ is a surjective homomorphism $\pi_1(\mathbb{P}^1 \setminus \theta(P)) \to G$. By Lemma 10.2(i), the map $\pi_1(\mathbb{P}^1 \setminus \theta(P)) \to \pi_1(\mathbb{P}^1 \setminus P)$ induced by θ^{-1} sends the subgroup generated by $\Sigma_{\theta(p)}$ to that generated by Σ_p. Hence $\varphi\theta^{-1}$ is admissible.

Thus $\hat{\theta}$ is well defined. (Also $\mathbf{A}$-equivalence causes no problem.) Since $\widehat{\theta^{-1}}$ is inverse to $\hat{\theta}$, it only remains to show that $\hat{\theta}$ is continuous. Clearly, the map $P \mapsto \theta(P)$ is a homeomorphism $\mathcal{O}_r \to \mathcal{O}_r$. Thus, given a neighborhood of $[\theta(P), \varphi\theta^{-1}]_{\mathbf{A}}$ of the form $\mathcal{N}(\tilde{D}_1,\dots,\tilde{D}_r)$, there are disjoint discs $D_1,\dots,D_r$ around the points of P such that $\theta(D_i) \subset \tilde{D}_i$. Then the neighborhood $\mathcal{N}(D_1,\dots,D_r)$ of $[P, \varphi]_{\mathbf{A}}$ is mapped into $\mathcal{N}(\tilde{D}_1,\dots,\tilde{D}_r)$ under $\hat{\theta}$.

Indeed, let $[P', \varphi']_{\mathbf{A}}$ in $\mathcal{N}(D_1,\dots,D_r)$, and $\gamma \in \pi_1(\mathbb{P}^1 \setminus (\tilde{D}_1 \cup \cdots \cup \tilde{D}_r))$. Then $\theta^{-1}(\gamma) \in \pi_1(\mathbb{P}^1 \setminus (D_1 \cup \cdots \cup D_r))$. By definition of the basic neighborhood

$\mathcal{N}(D_1, \ldots, D_r)$ we have $\varphi'(\theta^{-1}\gamma) = A\varphi(\theta^{-1}\gamma)$ for some $A \in \mathbf{A}$. Writing this as $\varphi'\theta^{-1}(\gamma) = A\varphi\theta^{-1}(\gamma)$, it follows that $[\theta(P'), \varphi'\theta^{-1}]_{\mathbf{A}}$ lies in $\mathcal{N}(\tilde{D}_1, \ldots, \tilde{D}_r)$. This proves that $\hat{\theta}$ maps $\mathcal{N}(D_1, \ldots, D_r)$ into $\mathcal{N}(\tilde{D}_1, \ldots, \tilde{D}_r)$. Hence $\hat{\theta}$ is continuous. This proves (a).

(b) Fix $t \in [0, 1]$ and $[P, \varphi]_{\mathbf{A}}$ in $\mathcal{H}_r^{(\mathbf{A})}(G)$. Consider a basic neighborhood $\mathcal{N}(\tilde{D}_1, \ldots, \tilde{D}_r)$ of $[\theta_t(P), \varphi\theta_t^{-1}]_{\mathbf{A}}$. Since the family $(\theta_\tau)_{\tau\in[0,1]}$ is continuous, there is $\epsilon > 0$ and there are disjoint discs $D_1, \ldots, D_r$ around the points of P such that $\theta_{t'}(D_i) \subset \tilde{D}_i$ for all $t' \in [0, 1]$ with $|t - t'| < \epsilon$. Then by the proof of (a), the $\hat{\theta}_{t'}$ with $|t - t'| < \epsilon$ map the neighborhood $\mathcal{N}(D_1, \ldots, D_r)$ of $[P, \varphi]_{\mathbf{A}}$ into $\mathcal{N}(\tilde{D}_1, \ldots, \tilde{D}_r)$. This proves (b). □

10.1.5 Families of Covers of the Riemann Sphere

The following proposition shows that if $Z(G) = 1$ then the map $\Omega\times\Pi : \mathcal{T}_r(G) \to \mathcal{S}_r^{in}(G) \subset \mathcal{H}_r^{in}(G) \times \mathbb{P}^1$ can be viewed as the family of all unramified Galois covers of $\mathbb{P}^1 \setminus P$ with group G, where P ranges over $\mathcal{O}_r$ (and there is true ramification over each point of P). The remark after the proposition shows that the process of "filling in the missing points" over the branch points (see Section 5.2.1) can also be made continuous. It yields the family $\bar{\mathcal{T}}_r(G) \to \mathcal{H}_r^{in}(G) \times \mathbb{P}^1$ of all (compact) branched Galois covers of $\mathbb{P}^1$ with group G and with r branch points $\neq \infty$. Compare with the alternative construction in [Fr8].

We do not formulate a universal property of this family, since that would lead us too far afield. We need the family (of unramified covers) for studying the algebraic structure of $\mathcal{H}_r^{in}(G)$ (see Sections 10.2.2. and 10.3.1).

Proposition 10.6 *Assume $Z(G) = 1$. Fix a point $u = [P, \varphi_0]$ of $\mathcal{H}_r^{in}(G)$. Let $R = \Omega^{-1}(u)$ (viewed as subspace of $\mathcal{T}_r(G)$). Then Π restricts to a Galois covering $f : R \to \mathbb{P}^1 \setminus P$. Its deck transformation group consists of the restrictions to R of the δ_A, $A \in Inn(G)$. Hence we get an isomorphism $\mu : G \to Deck(f)$ by mapping $g \in G$ to the restriction of $\delta_{Inn(g)}$. Set $b_0 = (P, \infty, \varphi_0) \in f^{-1}(\infty)$. Then the associated map $\Phi_{b_0} : \pi_1(\mathbb{P}^1 \setminus P, \infty) \to Deck(f)$ (see Proposition 4.19) equals $\mu \circ \varphi_0$. Thus under the correspondence from Proposition 10.3, the point u of $\mathcal{H}_r^{in}(G)$ corresponds to the pair (L, ν), where $L = \mathcal{M}(\bar{R})$ is the field of meromorphic functions on the compactification $\bar{R}$ of R, and $\nu = \iota \circ \mu$ as in Section 10.1.3.*

Proof. R consists of all (P, p, φ) with $p \in S := \mathbb{P}^1 \setminus P$, and $\varphi = \varphi_0 \circ \omega^*$ for some path ω in S from p to ∞. Dropping the first coordinate (which is constant on R), we can identify R with the space that was called R in Lemma 10.1, for $p_0 := \infty$. Also the topology on R induced from $\mathcal{T}_r(G)$ coincides with that from Lemma 10.1. Now the claim follows from that lemma. The last assertion is clear from the definitions in Section 10.1.3. □

Remark 10.7 The (unramified) coverings $\mathcal{T}_r(G) \to \mathcal{S}_r^{in}(G) \to \mathcal{O}(r+1)$ extend to ramified coverings of complex manifolds $\bar{\mathcal{T}}_r(G) \to \mathcal{H}_r^{in}(G) \times \mathbb{P}^1 \to \mathcal{O}_r \times \mathbb{P}^1$ with the following property: For any $u \in \mathcal{H}_r^{in}(G)$, the inverse image $\bar{R}$ of $\{u\} \times \mathbb{P}^1$ in $\bar{\mathcal{T}}_r(G)$ is a compact Riemann surface, and the induced map $\bar{R} \to \mathbb{P}^1$ is the standard extension (constructed in Chapter 5) of the unramified covering $R \to \mathbb{P}^1 \setminus P$ from the proposition. Further, G acts on $\bar{\mathcal{T}}_r(G)$, inducing $\mathrm{Deck}(\bar{R}/\mathbb{P}^1)$ on the fiber over each $u \in \mathcal{H}_r^{in}(G)$. Thus $\bar{\mathcal{T}}_r(G) \to \mathcal{H}_r^{in}(G) \times \mathbb{P}^1$ is the **family of all branched Galois covers of $\mathbb{P}^1$ with group G and with r branch points $\neq \infty$**. (This family is usually not connected; its connected components are described in Section 10.1.7 and Lemma 10.8.)

Since this remark will not be used in our main development, we only sketch the proof. Note first that the (natural) complex manifold structure on $\mathcal{O}(r+1)$ induces such unique structures on $\mathcal{T}_r(G)$ and $\mathcal{S}_r^{in}(G)$ such that the coverings $\mathcal{T}_r(G) \to \mathcal{S}_r^{in}(G) \to \mathcal{O}(r+1)$ are analytic maps (as in Lemma 5.6). Now fix some $P = \{p_1, \dots, p_r\} \in \mathcal{O}_r$, and let $D_1, \dots, D_r$ be open discs in $\mathbb{C}$ around $p_1, \dots, p_r$ such that the closures of these discs are mutually disjoint. Then (P, p_r) is a (general) point in $(\mathcal{O}_r \times \mathbb{P}^1) \setminus \mathcal{O}(r+1)$. A neighborhood of this point in $\mathcal{O}_r \times \mathbb{P}^1$ is the following set $\mathcal{D}$: It consists of all $(\{p_1', \dots, p_r'\}, p')$, where $p_i' \in D_i$ and $|p' - p_r'| < \epsilon$, for suitable $\epsilon > 0$. Let $D(\epsilon)$ be the disc around 0 of radius ϵ, and set $D(\epsilon)^* = D(\epsilon) \setminus \{0\}$. The map $\mathcal{D} \to D_1 \times \cdots \times D_r \times D(\epsilon)$, $(\{p_1', \dots, p_r'\}, p') \mapsto (p_1', \dots, p_r', p' - p_r')$ is a homeomorphism, and it maps $\mathcal{D} \cap \mathcal{O}(r+1)$ onto $D_1 \times \cdots \times D_r \times D(\epsilon)^*$.

Each connected covering of $D_1 \times \cdots \times D_r \times D(\epsilon)^*$ of finite degree e is equivalent to the map $D_1 \times \cdots \times D_r \times D(\epsilon^{1/e})^* \to D_1 \times \cdots \times D_r \times D(\epsilon)^*$, $(a_1, \dots, a_r, a) \mapsto (a_1, \dots, a_r, a^e)$. This extends to a ramified covering $D_1 \times \cdots \times D_r \times D(\epsilon^{1/e}) \to D_1 \times \cdots \times D_r \times D(\epsilon)$. Such patches can be used to add the "missing points" to $\mathcal{T}_r(G)$. The details are similar to the case of a single cover of $\mathbb{P}^1$ (Lemma 5.7).

Lemma 10.8 *Assume $Z(G) = 1$.*

(a) Let $\mathcal{H}$ be a connected component of $\mathcal{H}_r^{in}(G)$. Then $\mathcal{T} := \Omega^{-1}(\mathcal{H})$ is a component of $\mathcal{T}_r(G)$, and $\mathcal{S} := \{(u, q) \in \mathcal{S}_r^{in}(G) : u \in \mathcal{H}\}$ is a component of $\mathcal{S}_r^{in}(G)$. Further, the restriction of $\Omega \times \Pi$ yields a Galois covering $\mathcal{T} \to \mathcal{S}$ with deck transformation group $\{\delta_A : A \in Inn(G)\} \cong Inn(G)$.

(b) Let $\mathbf{A}$ be a subgroup of $Aut(G)$ containing $Inn(G)$, and set $\mathbf{A}_{\mathcal{H}} = \{A \in \mathbf{A} : \epsilon_A(\mathcal{H}) = \mathcal{H}\}$. Then the image $\mathcal{H}_{\mathbf{A}}$ of $\mathcal{H}$ in $\mathcal{H}_r^{(\mathbf{A})}(G)$ is a component of

the latter space. The induced map $\mathcal{H} \to \mathcal{H}_{\mathbf{A}}$ is a Galois covering with deck transformation group $\{\epsilon_A : A \in \mathbf{A}_{\mathcal{H}}\} \cong \mathbf{A}_{\mathcal{H}}/Inn(G)$.

Proof. First note that our spaces are (topological) manifolds, hence connected components are open and closed. It follows from Remark 4.8 (a) and Corollary 4.13 that a covering $\lambda : X \to Y$ maps each component X^* of X onto a component Y^* of Y, and the induced map $X^* \to Y^*$ is a covering.

(a) The space $\mathcal{T} = \Omega^{-1}(\mathcal{H})$ is open and closed in $\mathcal{T}_r(G)$. Once we have shown that it is connected, it follows that $\mathcal{T}$ is a component of $\mathcal{T}_r(G)$. Assume this for the moment. Since $\mathcal{S} = (\Omega \times \Pi)(\mathcal{T})$, it follows by the preceding paragraph and Lemma 10.4(i) that $\mathcal{S}$ is a component of $\mathcal{S}_r^{in}(G)$, and $\Omega \times \Pi$ restricts to a covering $\mathcal{T} \to \mathcal{S}$. By Lemma 10.4(ii), the δ_A, $A \in \mathrm{Inn}(G)$, form a subgroup of $\mathrm{Deck}(\mathcal{T}/\mathcal{S})$ that acts transitively on each fiber of $\mathcal{T}$ over $\mathcal{S}$. This implies (a).

It remains to show that $\mathcal{T}$ is connected. First note that Ω is an open map (composition of the covering $\Omega \times \Pi$ with projection $\mathcal{H}_r^{in}(G) \times \mathbb{P}^1 \to \mathcal{H}_r^{in}(G)$). The fibers $\Omega^{-1}(u)$ of the map $\mathcal{T} \to \mathcal{H}$ are connected by the preceding proposition. Since $\mathcal{H}$ is connected, it follows that $\mathcal{T}$ is connected. (Indeed, if $M \subset \mathcal{T}$ is open and closed in $\mathcal{T}$ then $M \cap \Omega^{-1}(u)$ is either empty or all of $\Omega^{-1}(u)$ (for all $u \in \mathcal{H}$), hence M is a union of fibers $\Omega^{-1}(u)$. This implies that $\Omega(M)$ and $\Omega(\mathcal{T} \setminus M)$ are disjoint. Since both are open, it follows that one of them is empty, that is, either M or $\mathcal{T} \setminus M$ is empty.) This completes the proof of (a).

(b) It follows as above that $\mathcal{H}_{\mathbf{A}}$ is a component of $\mathcal{H}_r^{(\mathbf{A})}(G)$, and the map $\mathcal{H} \to \mathcal{H}_{\mathbf{A}}$ is a covering. Further it is clear that the ϵ_A with $A \in \mathbf{A}_{\mathcal{H}}$ act as deck transformations of this covering. Conversely, if $u, u' \in \mathcal{H}$ have the same image in $\mathcal{H}_{\mathbf{A}}$ then $u' = \epsilon_A(u)$ for some $A \in \mathbf{A}$ (by the definitions). Since the homeomorphism ϵ_A of $\mathcal{H}_r^{in}(G)$ permutes the components of this space, and maps $u \in \mathcal{H}$ into $\mathcal{H}$, it follows that ϵ_A fixes $\mathcal{H}$. Thus the group$\{\epsilon_A : A \in \mathbf{A}_{\mathcal{H}}\}$ acts transitively on each fiber of the covering $\mathcal{H} \to \mathcal{H}_{\mathbf{A}}$. This proves (b). □

10.1.6 The Braid Group

View $\mathbb{C}^r$ as equipped with its usual **Euclidean topology** (which has a basis consisting of the sets $D_1 \times \cdots \times D_r$, where the D_i are open discs in $\mathbb{C}$). Let $\mathcal{O}^{(r)}$ be the subspace consisting of all $(p_1, \ldots, p_r) \in \mathbb{C}^r$ with $p_i \neq p_j$ for all $i \neq j$. Then the map $\mathcal{O}^{(r)} \to \mathcal{O}^{(r-1)}$, $(p_1, \ldots, p_r) \to (p_1, \ldots, p_{r-1})$, is continuous, open, and surjective. Its fibers are homeomorphic to $\mathbb{C} \setminus P'$ with $P' \in \mathcal{O}_{r-1}$, hence are connected. Thus if $\mathcal{O}^{(r-1)}$ is connected so is $\mathcal{O}^{(r)}$. By induction we get that $\mathcal{O}^{(r)}$ is connected. Then $\mathcal{O}_r$ is also connected, by the following lemma.

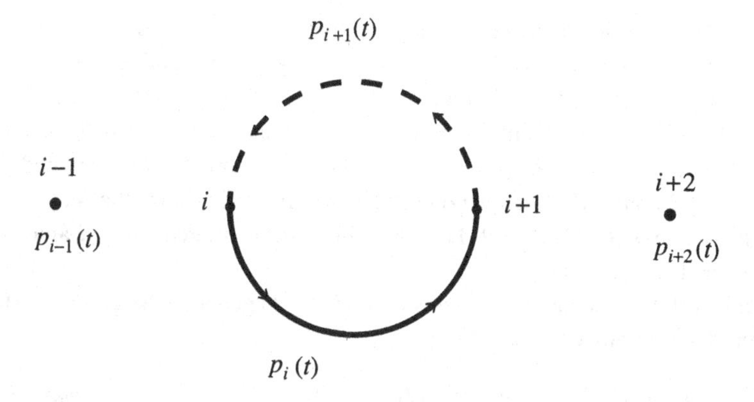

Fig. 10.1. The path Q_i.

Lemma 10.9 *The map $\vartheta : \mathcal{O}^{(r)} \to \mathcal{O}_r, (p_1, \ldots, p_r) \mapsto \{p_1, \ldots, p_r\}$, is a Galois covering. Its deck transformation group is isomorphic to the symmetric group S_r, via the map that associates with $\sigma \in S_r$ the deck transformation $(p_1, \ldots, p_r) \mapsto (p_{\sigma^{-1}(1)}, \ldots, p_{\sigma^{-1}(r)})$.*

Proof. Let $D_1, \ldots, D_r$ be disjoint discs in $\mathbb{C}$. The inverse image in $\mathcal{O}^{(r)}$ of the neighborhood $\mathcal{N}_0(D_1, \ldots, D_r)$ is the disjoint union of the $D_{\sigma(1)} \times \cdots \times D_{\sigma(r)}$, $\sigma \in S_r$. Clearly, ϑ maps each $D_{\sigma(1)} \times \cdots \times D_{\sigma(r)}$ homeomorphically onto $\mathcal{N}_0(D_1, \ldots, D_r)$. Hence ϑ is a covering.

Clearly, S_r embeds into $\mathrm{Deck}(\vartheta)$ via the map given in the lemma. Since S_r acts transitively on each fiber $\vartheta^{-1}(P)$, $P \in \mathcal{O}_r$, it follows that ϑ is a Galois covering, and S_r induces all of $\mathrm{Deck}(\vartheta)$ (Lemma 4.17(iv)). □

We fix $P_0 = \{1, 2, \ldots, r\}$ as a base point in $\mathcal{O}_r$. For $i \in \mathbb{N}$ with $1 \leq i \leq r-1$, define a closed path Q_i in $\mathcal{O}_r$ as follows: $Q_i(t) = \{p_1(t), \ldots, p_r(t)\}$, where $p_j(t)$ is the constant path based at j for $j \neq i$, and

$$p_i(t) = \frac{(2i+1) - \exp(\pi\sqrt{-1}t)}{2}, \qquad p_{i+1}(t) = \frac{(2i+1) + \exp(\pi\sqrt{-1}t)}{2}$$

for $t \in [0, 1]$. Note that Q_i is continuous by the above lemma. We let Q_i also denote the corresponding element of $\mathcal{B}_r := \pi_1(\mathcal{O}_r, P_0)$. We will prove that $\mathcal{B}_r$ is generated by $Q_1, \ldots, Q_{r-1}$ (Figure 10.1).

Set $B_0 = (1, 2, \ldots, r) \in \vartheta^{-1}(P_0)$, our base point in $\mathcal{O}^{(r)}$. From Proposition 4.19 we get the map $\Phi_{B_0} : \mathcal{B}_r \to \mathrm{Deck}(\vartheta)$. Composing with the isomorphism $\mathrm{Deck}(\vartheta) \to S_r$ we get a surjective homomorphism $\kappa : \mathcal{B}_r \to S_r$. This map can be described as follows.

Let Q be a closed path in $\mathcal{O}_r$ based at P_0, and let $\tilde{Q}$ be its lift to $\mathcal{O}^{(r)}$ with initial point B_0. We have $\tilde{Q}(t) = (q_1(t), \ldots, q_r(t))$, where q_j is a path in $\mathbb{C}$ with initial point j. Thus $Q(t)$ can be written as $Q(t) = \{q_1(t), \ldots, q_r(t)\}$. The endpoint of Q equals its initial point $P_0 = \{1, 2, \ldots, r\}$, hence there is $\sigma \in S_r$ such that q_j has endpoint $\sigma(j)$ (for $j = 1, \ldots, r$). Then the element of Deck(ϑ) induced by σ (see Lemma 10.9) maps $\tilde{Q}(1)$ to $\tilde{Q}(0) = B_0$. This means that $\kappa(Q) = \sigma$. In particular, $\kappa(Q_i) = (i, i+1) =$ the transposition interchanging i and $i+1$ (for $i = 1, \ldots, r-1$).

It is well known that the transpositions $(i, i+1)$ generate the group S_r. Here is the "continuous version" of this:

Theorem 10.10 *The elements $Q_1, \ldots, Q_{r-1}$ generate the group $\mathcal{B}_r = \pi_1(\mathcal{O}_r, P_0)$.*

Proof. Let Q be a closed path in $\mathcal{O}_r$ based at P_0, and write it in the form $Q(t) = \{q_1(t), \ldots, q_r(t)\}$ as above. We need to show that Q is homotopic to a product of certain Q_i's and their inverses. Since S_r is generated by the $\kappa(Q_i) = (i, i+1)$, we can assume $\kappa(Q) = \mathrm{id}$. Then each q_j is a closed path in $\mathbb{C}$ based at j. Define

$$q_j^{(s)}(t) = q_j(t) + s(r - q_r(t)) \tag{10.4}$$

for $j = 1, \ldots, r$ and $s, t \in [0, 1]$. Then $Q^{(s)}$, defined by $Q^{(s)}(t) = \{q_1^{(s)}(t), \ldots, q_r^{(s)}(t)\}$, is a path in $\mathcal{O}_r$ for each $s \in [0, 1]$. Thus the $Q^{(s)}$ yield a homotopy between $Q^{(0)} = Q$ and $Q^{(1)}$. This $Q^{(1)}$ has the property that $Q^{(1)}(t) = \{q_1^{(1)}(t), \ldots, q_r^{(1)}(t)\}$ with $q_r^{(1)}$ the constant path r.

Case 1 $r = 2$.

After applying the homotopy (10.4) we may assume $Q(t) = \{q_1(t), 2\}$, where $q_1(t)$ is a closed path in $\mathbb{C} \setminus \{2\}$ based at 1. Applying the same homotopy to Q_1^2 in place of Q we obtain the path $\{p_1'(t), 2\}$, where $p_1'(t) = 2 - \exp(2\pi\sqrt{-1}t)$. The group $\pi_1(\mathbb{C} \setminus \{2\}, 1)$ is generated by the class of p_1' (see Proposition 4.20). Thus q_1 is homotopic in $\mathbb{C} \setminus \{2\}$ to a power of p_1'. This yields a homotopy in $\mathcal{O}_2$ between Q and a power of Q_1^2.

Back in the general case, let T be the set of $t \in [0, 1]$ for which there exist distinct $i, j \in \{1, \ldots, r\}$ with $\mathrm{Re}\, q_i(t) = \mathrm{Re}\, q_j(t)$. Here Re means "real part". Replacing Q by a homotopic path we may assume that T is finite, and that for each $t \in T$ there is exactly one pair (i, j) with $i \neq j$ and $\mathrm{Re}\, q_i(t) = \mathrm{Re}\, q_j(t)$. Indeed, by Lemma 4.26 we can write Q as the product of finitely many paths A_j each of which runs entirely inside a neighborhood of the form $\mathcal{N}_0(D_1, \ldots, D_r)$. Since those neighborhoods are simply connected, we can replace A_j by any other path in that neighborhood with the same initial and endpoint. Clearly, this allows us to achieve the desired normalization. From now on we only consider paths that are normalized in this way. We use induction on the cardinality N of T, as well as on r.

Case 2 $N = 0$.

If $N = 0$ then $\operatorname{Re} q_j(t) < \operatorname{Re} q_{j+1}(t)$ for each $j = 1, \ldots, r-1$, and for each t. Setting

$$q_j(s,t) = q_j(t) + s(j - q_j(t))$$

we obtain a homotopy $Q(s,t) = \{q_1(s,t), \ldots, q_r(s,t)\}$ between Q and the constant path P_0. We need only check that $Q(s,t) \in \mathcal{O}_r$ for each s, t. This holds because

$$\operatorname{Re} q_{j+1}(s,t) - \operatorname{Re} q_j(s,t) = (1-s)(\operatorname{Re} q_{j+1}(t) - \operatorname{Re} q_j(t)) + s > 0.$$

Case 3 $r > 2$, $N > 0$.

Let t_0 be the smallest element of T. Then clearly $\operatorname{Re} q_i(t_0) = \operatorname{Re} q_{i+1}(t_0)$ for a (unique) i with $1 \le i \le r-1$. If $i < r-1$ we proceed as follows; if $i = r-1$ we need to replace r by 1 in the following. Choose $t_1 \in [0,1]$ with $t_1 > t_0$, but $t_1 < t'$ for each $t' \in T \setminus \{t_0\}$. (Such t_1 exists since $1 \notin T$.) Then

$$\operatorname{Re} q_1(t) < \cdots < \operatorname{Re} q_{i-1}(t) < \begin{cases} \operatorname{Re} q_i(t) \\ \operatorname{Re} q_{i+1}(t) \end{cases} <$$

$$\operatorname{Re} q_{i+2}(t) < \cdots < \operatorname{Re} q_r(t)$$

for all $t \le t_1$. This condition (as well as the above normalization) is preserved under the homotopy (10.4), hence we may additionally assume that q_r is the constant path r. Then $\operatorname{Re} q_j(t) < r$ for all $j < r$, $t \le t_1$. (Note that we are in the case $i < r-1$.) Hence those $q_j(t)$ lie in the half plane $H := \{z \in \mathbb{C} : \operatorname{Re} z \le r'\}$ for some real number r' with $r - 1 < r' < r$.

Define paths U, V in $\mathcal{O}_r$ as follows: $V(t) = Q(t)$ and $U(t) = Q(t_1)$ for $t \le t_1$, while $V(t) = Q(t_1)$ and $U(t) = Q(t)$ for $t \ge t_1$. Thus Q is homotopic to UV. Let $W(t) = \{w_1(t), \ldots, w_r(t)\}$ be a path in $\mathcal{O}_r$ going from $P_0 = Q(0)$ to $Q(t_1)$ such that $w_j(t) \in H$ for $j < r$ and $w_r(t) \equiv r$; further, $\operatorname{Re} w_j(t) < \operatorname{Re} w_{j+1}(t)$ for $j = 1, \ldots, r-1$. Let $Q' = W^{\mathrm{inv}} V$. Then Q' is a closed path in $\mathcal{O}_r$ based at P_0. We have $Q'(t) = \{q_1'(t), \ldots, q_r'(t)\}$, where $q_j'(t) \in H$ for $j < r$, and $q_r'(t) \equiv r$. By induction we may assume there is a homotopy in $\mathcal{O}_{r-1}$ between the path $\{q_1'(t), \ldots, q_{r-1}'(t)\}$ and a product of certain $Q_j^{\pm 1}$ with $j \le r-2$. (We can naturally view those Q_j as elements of $\mathcal{B}_{r-1}$.) This homotopy can be written as $\{q_1'(s,t), \ldots, q_{r-1}'(s,t)\}$ for continuous $q_j'(s,t)$ (by Theorem 4.12 and Lemma 10.9). Applying a retraction $\mathbb{C} \to H$ (a continuous map that is the identity on H) we may assume that all

$q'_j(s,t)$ lie in H. (An example of such a retraction is $z \mapsto 2r' + z - 2\mathrm{Re}\, z$, for $z \in \mathbb{C} \setminus H$.) Then $\{q'_1(s,t), \ldots, q'_{r-1}(s,t), r\}$ yields a homotopy in $\mathcal{O}_r$ between Q' and a product of certain $Q_j^{\pm 1}$.

Clearly, the N-invariant of $Q'' := UW$ is less than that of Q. Thus we may assume by induction that Q'' is homotopic to a product of certain $Q_j^{\pm 1}$. Then the same holds for Q, since Q is homotopic to $Q''Q'$. □

Remark 10.11 There is a well-known presentation of $\mathcal{B}_r$ in terms of the generators Q_i, by the relations

$$Q_i Q_{i+1} Q_i = Q_{i+1} Q_i Q_{i+1} \quad \text{for } i = 1, \ldots, r-2 \qquad \text{and}$$
$$Q_i Q_j = Q_j Q_i \quad \text{for } |i-j| > 1.$$

This yields the isomorphism with the (abstract) Artin braid group introduced in Chapter 9. The interpretation as braids can be explained as follows. Consider a closed path $Q(t) = \{q_1(t), \ldots, q_r(t)\}$ in $\mathcal{O}_r$ based at $\{1, \ldots, r\}$. Identify the complex plane $\mathbb{C}$ with $\mathbb{R}^2$ in the standard way. Then the $(q_i(t), t)$ form r intertwining paths in $\mathbb{R}^3$ – a "braid" – going from the plane at level $t = 0$ to that at level $t = 1$. Homotopy classes of braids, equipped with the multiplication given by concatenation, form the braid group on r strings as originally defined by Artin. This braid group is isomorphic to the fundamental group of $\mathcal{O}_r$ via the above correspondence. For this and much more on the braid group, see [Bir] and [Han]. For the purpose of this book, all we need to know from this is Theorem 10.10.

10.1.7 The Braiding Action on Generating Systems

Set $\Gamma_0 = \pi_1(\mathbb{P}^1 \setminus P_0, \infty)$, where $P_0 = \{1, 2, \ldots, r\}$ is our base point in $\mathcal{O}_r$. First we specify generators of Γ_0. For $j = 1, \ldots, r$ let C_j be the circle around j of radius $1/4$, and R_j the ray $\{j - b\sqrt{-1} : b \in \mathbb{R}, b \geq 1/4\}$. Let γ_j be a closed path in $\mathbb{P}^1$ that runs from ∞ towards j on the ray R_j until it reaches C_j, then travels once around C_j in counterclockwise direction, and returns to ∞ along R_j. We let γ_j also denote the corresponding element of Γ_0. Clearly, γ_j lies in the conjugacy class Σ_j of Γ_0 associated with the point $j \in P_0$ (see Section 10.1.2).

The product $\gamma_1 \cdots \gamma_r$ is homotopic to a path going once around all of the points from P_0, hence is null-homotopic in $\mathbb{P}^1 \setminus P_0$. Thus $\gamma_1 \cdots \gamma_r = 1$ in Γ_0. Applying Corollary 4.29 after a coordinate transformation mapping r to ∞, it follows that $\gamma_1, \ldots, \gamma_{r-1}$ are free generators of Γ_0.

Now we return to studying the covering $\Psi_{\mathbf{A}} : \mathcal{H}_r^{(\mathbf{A})}(G) \to \mathcal{O}_r$. Recall that there is a natural action of the fundamental group $\mathcal{B}_r = \pi_1(\mathcal{O}_r, P_0)$ on the fiber

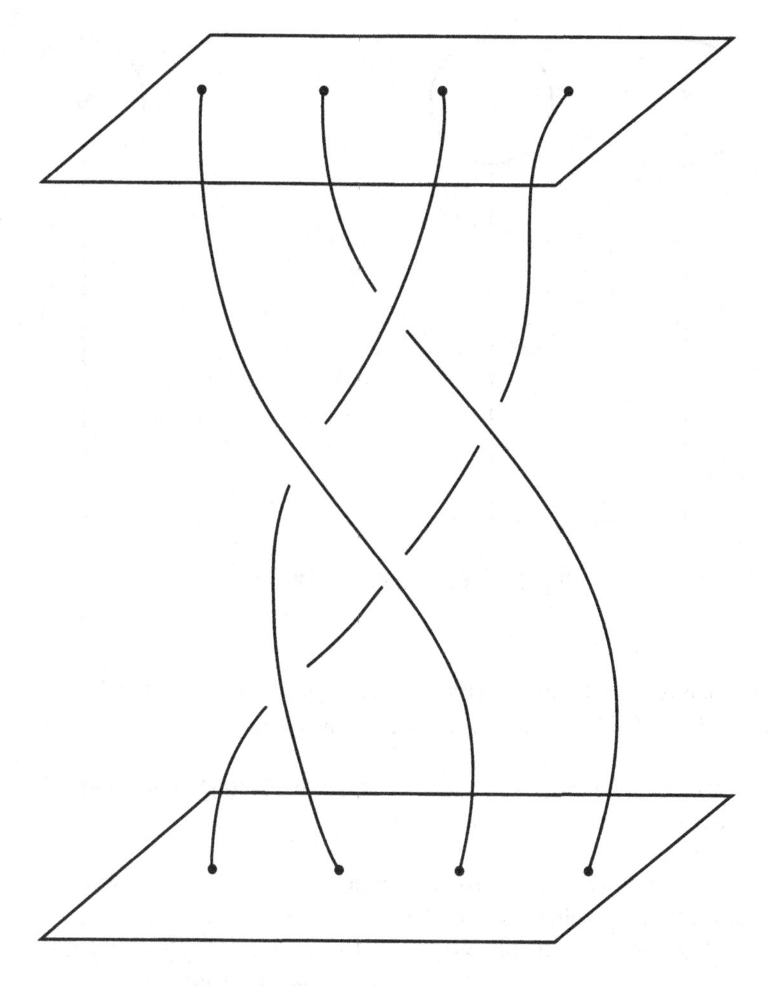

Fig. 10.2. A braid on four strings.

over the base point P_0 (see Theorem 4.12). This action determines $\mathcal{H}_r^{(\mathbf{A})}(G)$ topologically as a covering of $\mathcal{O}_r$ (by general covering space theory). To describe this action explicitly, we need a parametrization of the fiber $\Psi_{\mathbf{A}}^{-1}(P_0)$. This fiber consists of all $[P_0, \varphi]_{\mathbf{A}}$, where $\varphi : \Gamma_0 \to G$ is admissible. Such φ is determined by its values $g_1, \dots, g_r$ on the generators $\gamma_1, \dots, \gamma_r$. These tuples $(g_1, \dots, g_r)$ yield the desired parametrization. Set

$$\mathcal{E}_r(G) = \{(g_1, \dots, g_r) \in G^r : G = \langle g_1, \dots, g_r \rangle, \\ g_1 \cdots g_r = 1, g_i \neq 1 \qquad \text{for all } i\}.$$

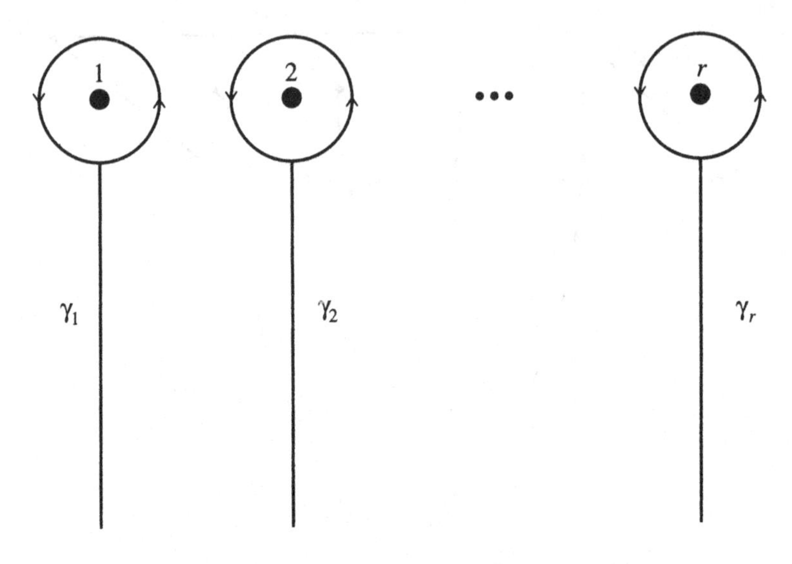

Fig. 10.3. Generators of $\pi_1(\mathbb{P}^1 \setminus P_0, \infty)$

The group $\mathbf{A}$ acts on this set with $A \in \mathbf{A}$ mapping $(g_1, \ldots, g_r)$ to $(A(g_1), \ldots, A(g_r))$. Let $\mathcal{E}_r^{(\mathbf{A})}(G)$ denote the set of $\mathbf{A}$-orbits on $\mathcal{E}_r(G)$.

Lemma 10.12 *We obtain a bijection $\Psi_{\mathbf{A}}^{-1}(P_0) \to \mathcal{E}_r^{(\mathbf{A})}(G)$ by mapping $[P_0, \varphi]_{\mathbf{A}}$ to the class of $(g_1, \ldots, g_r)$, where $g_i = \varphi(\gamma_i)$.*

Proof. By definition, a surjective homomorphism $\varphi : \Gamma_0 \to G$ is admissible iff φ takes nontrivial values on the classes Σ_j of Γ_0 associated with the points of P_0. Since $\gamma_j \in \Sigma_j$, it follows that φ is admissible iff $\varphi(\gamma_j) \neq 1$ for all j. Further, φ and φ' correspond to the same element of $\Psi_{\mathbf{A}}^{-1}(P_0)$ iff $\varphi' = A\varphi$ for some $A \in \mathbf{A}$. For the corresponding tuples, this means that $(g'_1, \ldots, g'_r) = (A(g_1), \ldots, A(g_r))$. Hence the map in the lemma is well defined and injective. It is surjective because Γ_0 is free on $\gamma_1, \ldots, \gamma_{r-1}$, hence each $(g_1, \ldots, g_r) \in \mathcal{E}_r(G)$ gives rise to some admissible φ. □

Via the bijection in the lemma, the action of $\mathcal{B}_r$ on $\Psi_{\mathbf{A}}^{-1}(P_0)$ yields an action on $\mathcal{E}_r^{(\mathbf{A})}(G)$. Here the Q_i's act through the braid operations from Chapter 9:

Theorem 10.13 *In the above action of $\mathcal{B}_r$ on $\mathcal{E}_r^{(\mathbf{A})}(G)$, the element Q_i $(1 \leq i \leq r-1)$ maps the class of $(g_1, \ldots, g_r)$ to that of*

$$(g_1, \cdots, g_{i-1}, g_i g_{i+1} g_i^{-1}, g_i, \cdots, g_r). \tag{10.5}$$

Proof. Fix i with $1 \leq i \leq r-1$. Let $h : \mathbb{R}_+ \to \mathbb{R}_+$ be a continuous function with $h(u) = 1$ for $u \leq 3/4$ and $h(u) = 0$ for $u \geq 1$. Set

$$\theta_t(z) = \frac{2i+1}{2} + \left(z - \frac{2i+1}{2}\right) \exp\left(h\left(\left|z - \frac{2i+1}{2}\right|\right)\pi\sqrt{-1}t\right),$$

for $z \in \mathbb{C}$, $t \in [0,1]$.

Setting $\theta_t(\infty) = \infty$, we obtain a continuous family $(\theta_t)_{t\in[0,1]}$ of homeomorphisms of $\mathbb{P}^1$ (as in Section 10.1.4). On the disc around $\frac{2i+1}{2}$ of radius $3/4$, the map θ_t is rotation around $\frac{2i+1}{2}$ by the angle $t\pi$ in counterclockwise direction. Outside the disc around $\frac{2i+1}{2}$ of radius 1, θ_t is the identity.

The path $t \mapsto \theta_t(P_0)$ in $\mathcal{O}_r$ is exactly the path Q_i. By Lemma 10.5, the $\hat{\theta}_t$ form a continuous family of homeomorphisms of $\mathcal{H}_r^{(\mathbf{A})}(G)$. Thus $t \mapsto \hat{\theta}_t(B_0)$ is the lift of Q_i via $\Psi_{\mathbf{A}}$ with initial point $B_0 \in \Psi_{\mathbf{A}}^{-1}(P_0)$. Its endpoint is $\hat{\theta}_1(B_0)$. Thus Q_i, in its natural action on $\Psi_{\mathbf{A}}^{-1}(P_0)$, coincides with $\hat{\theta}_1$. (By definition of this action, see Theorem 4.12.)

Let $\varphi : \Gamma_0 \to G$ be admissible, and let $(g_1, \ldots, g_r) = (\varphi(\gamma_1), \ldots, \varphi(\gamma_r))$ be the associated tuple. By definition, $\hat{\theta}_1$ maps $[P_0, \varphi]_{\mathbf{A}}$ to $[\theta_1(P_0), \varphi\theta_1^{-1}]_{\mathbf{A}} = [P_0, \varphi\theta_1^{-1}]_{\mathbf{A}}$. Thus in its action on $\mathcal{E}_r^{(\mathbf{A})}(G)$, Q_i maps the class of $(g_1, \ldots, g_r)$ to that of $(g'_1, \ldots, g'_r)$, where $g'_j = \varphi\theta_1^{-1}(\gamma_j)$. Since θ_1 is the identity outside the disc around $\frac{2i+1}{2}$ of radius 1, we have $\theta_1^{-1}(\gamma_j) = \gamma_j$ for $j \neq i, i+1$, hence $g'_j = g_j$ for those j. Further it is clear that $\theta_1^{-1}(\gamma_{i+1})$ is homotopic to γ_i, hence $g'_{i+1} = g_i$. Since $g'_1 \cdots g'_r = 1$, we must have $g'_i = g_i g_{i+1} g_i^{-1}$. This completes the proof. □

10.1.8 Components of $\mathcal{H}_r^{(\mathbf{A})}(G)$, and the Example of Simple Covers

If $\mathcal{H}$ is a (connected) component of $\mathcal{H}_r^{(\mathbf{A})}(G)$, then $\Psi_{\mathbf{A}}$ restricts to a covering $\mathcal{H} \to \mathcal{O}_r$ (Corollary 4.13). Thus the fiber in $\mathcal{H}$ over P_0 corresponds to a $\mathcal{B}_r$-orbit on $\mathcal{E}_r^{(\mathbf{A})}(G)$, under the bijection from Lemma 10.12 (by Theorem 4.12 (c)). This defines a 1-1 correspondence between the components of $\mathcal{H}_r^{(\mathbf{A})}(G)$ and the $\mathcal{B}_r$-orbits on $\mathcal{E}_r^{(\mathbf{A})}(G)$.

For any r-tuple $\mathbf{C} = (C_1, \ldots, C_r)$ of nontrivial conjugacy classes of G, the set $\mathrm{Ni}(\mathbf{C})$ (from Section 9.2) is invariant under the action of $\mathcal{B}_r$. Thus the image $\mathrm{Ni}^{(\mathbf{A})}(\mathbf{C})$ of $\mathrm{Ni}(\mathbf{C})$ in $\mathcal{E}_r^{(\mathbf{A})}(G)$ is a union of $\mathcal{B}_r$-orbits. The components of $\mathcal{H}_r^{(\mathbf{A})}(G)$ corresponding to these orbits can be described as follows.

Proposition 10.14 *Let $\mathbf{C} = (C_1, \ldots, C_r)$ be an r-tuple of nontrivial conjugacy classes of G. Let $\mathcal{H}^{(\mathbf{A})}(\mathbf{C})$ be the subset of $\mathcal{H}_r^{(\mathbf{A})}(G)$ consisting of all $[P, \varphi]_{\mathbf{A}}$*

such that the elements of P can be labeled as $p_1, \dots, p_r$ with $\varphi(\Sigma_{p_i}) \in C_i$ for $i = 1, \dots, r$. (Here Σ_{p_i} is the class of $\pi_1(\mathbb{P}^1 \setminus P, \infty)$ defined in Section 10.1.2.)

(a) Then $\mathcal{H}^{(\mathbf{A})}(\mathbf{C})$ is a union of connected components of $\mathcal{H}_r^{(\mathbf{A})}(G)$. Under the bijection from Lemma 10.12, the fiber in $\mathcal{H}^{(\mathbf{A})}(\mathbf{C})$ over P_0 corresponds to the set $Ni^{(\mathbf{A})}(\mathbf{C})$. This yields a 1-1 correspondence between the components of $\mathcal{H}^{(\mathbf{A})}(\mathbf{C})$ and the braid orbits on $Ni^{(\mathbf{A})}(\mathbf{C})$. In particular, $\mathcal{H}^{(\mathbf{A})}(\mathbf{C})$ is connected if and only if $\mathcal{B}_r$ acts transitively on $Ni^{(\mathbf{A})}(\mathbf{C})$.

(b) For $\mathbf{A} = Inn(G)$ we write $\mathcal{H}^{(\mathbf{A})}(\mathbf{C}) = \mathcal{H}^{in}(\mathbf{C})$. The above induces a 1-1 correspondence between the components $\mathcal{H}$ of $\mathcal{H}^{in}(\mathbf{C})$ and the braid orbits on Ni($\mathbf{C}$). For $A \in Aut(G)$ we have $\epsilon_A(\mathcal{H}) = \mathcal{H}$ if and only if A leaves the corresponding braid orbit invariant (acting as $(g_1, \dots, g_r) \mapsto (A(g_1), \dots, A(g_r))$).

(c) Suppose under the correspondence in Proposition 10.3, the pair (L, ν) is associated with $u \in \mathcal{H}_r^{in}(G)$. Then u lies in $\mathcal{H}^{in}(\mathbf{C})$ if and only if the branch points of $L/\mathbb{C}(x)$ can be labeled as $p_1, \dots, p_r$ such that $\nu(C_i)$ is the class of $G(L/\mathbb{C}(x))$ associated with the branch point p_i, for $i = 1, \dots, r$.

Proof.

(a) The set $\mathcal{H}^{(\mathbf{A})}(\mathbf{C})$ is open in $\mathcal{H}_r^{(\mathbf{A})}(G)$ since all points in a basic neighborhood $\mathcal{N}(D_1, \dots, D_r)$ correspond to the same map $\pi_1(\mathbb{P}^1 \setminus (D_1 \cup \dots \cup D_r), \infty) \to G$. By the same argument, the complement of $\mathcal{H}^{(\mathbf{A})}(\mathbf{C})$ is open. Thus $\mathcal{H}^{(\mathbf{A})}(\mathbf{C})$ is open and closed, hence a union of connected components of $\mathcal{H}_r^{(\mathbf{A})}(G)$. The rest of (a) is clear (again using Theorem 4.12).

(b) By Lemma 9.4, the natural map Ni($\mathbf{C}$) $\to$ Ni^{in}($\mathbf{C}$) induces a bijection on braid orbits. This yields the first part of (b). By the proof of Lemma 10.12, the action of ϵ_A on $\Psi^{-1}(P_0)$ corresponds to the action of A on $\mathcal{E}_r(G)$ via $(g_1, \dots, g_r) \mapsto (A(g_1), \dots, A(g_r))$. This proves (b).

(c) Follows from the definitions, Lemma 10.2 (ii) and Theorem 5.9 Addendum.

□

Exercise In the situation of the proposition, suppose the tuple $\mathbf{C}$ is weakly rigid, and $\mathbf{A} = \mathrm{Aut}_C(G)$ (=the subgroup of Aut(G) fixing each C_i).

(a) Prove that the covering $\Psi_{\mathbf{A}} : \mathcal{H}^{(\mathbf{A})}(\mathbf{C}) \to \mathcal{O}_r$ is equivalent to the covering $\mathcal{O}^{(r)} \to \mathcal{O}_r$ from Lemma 10.9 (resp., is an isomorphism) if the C_i are mutually distinct (resp., if $C_1 = \dots = C_r$).

(b) More generally, show that we have a sequence of coverings $\mathcal{O}^{(r)} \to \mathcal{H}^{(\mathbf{A})}(\mathbf{C}) \to \mathcal{O}_r$, and Deck($\mathcal{O}^{(r)}/\mathcal{H}^{(\mathbf{A})}(\mathbf{C})$) embeds into S_n as the subgroup $S_{r_1} \times$

$S_{r_2} \times \cdots$, where we have labeled the C_i such that $C_1 = \cdots = C_{r_1}, C_{r_1+1} = \cdots = C_{r_2}$, etc. (and there are no further equalities between the C_i).

It is a basic problem to decide when $\mathcal{H}^{(\mathbf{A})}(\mathbf{C})$ is connected, that is, when $\mathcal{B}_r$ acts transitively on $\mathrm{Ni}^{(\mathbf{A})}(\mathbf{C})$. In Section 9.4 we saw that this is certainly not always true. It is true, however, in a sufficiently general situation, by a theorem of Conway and Parker (see [FV1], Appendix). This theorem requires the technical hypothesis that the Schur multiplier of G be generated by commutators. (This hypothesis can always be achieved by replacing G by a group with quotient G.) Under this hypothesis, $\mathcal{B}_r$ acts transitively on $\mathrm{Ni}(\mathbf{C})$ provided each conjugacy class of G occurs sufficiently often among the classes in $\mathbf{C}$.

In the remainder of this subsection, we consider the case of simple covers of $\mathbb{P}^1$, and explain its classical application to the connectedness of the space of curves of genus g. This will not be used elsewhere in the book. Simple covers are associated with the class of 2-cycles (transpositions) in S_n. The following lemma on this is due to Clebsch [Cleb] and Hurwitz [Hur]; we essentially follow the proof in [BF2], Appendix. See Fried [Fr6] for the parallel case of 3-cycles in the alternating group.

Lemma 10.15 (Clebsch 1872) *Let $G = S_n$ $(n \geq 2)$, and let $\mathbf{C} = (C, \ldots, C)$ be the r-tuple consisting of r times the class C of transpositions. Then $Ni(\mathbf{C})$ is nonempty if and only if r is even and $r \geq 2(n-1)$. If this holds then $\mathcal{B}_r$ acts transitively on $Ni(\mathbf{C})$.*

Proof.

Claim 1 If T is any set of transpositions generating S_n then $|T| \geq n-1$, and there are $n-1$ elements in T that generate S_n. Also, there are $n-2$ elements in T that generate a subgroup of S_n conjugate to S_{n-1} (canonically embedded into S_n).

Proof. This uses standard arguments from [Hup], Kap. II, 4. We prove inductively that for each $j = 1, \ldots, n-1$ there are $t_1, \ldots, t_j \in T$, and there is a subset B_j of $\{1, \ldots, n\}$ of cardinality $j+1$, such that the group $\langle t_1, \ldots, t_j \rangle$ induces the full symmetric group on B_j and fixes all elements not in B_j. The case $j = 1$ is clear. Given such data for $j < n-1$, there must be $t_{j+1} \in T$ with $t_{j+1}(B_j) \neq B_j$. Then the induction step is accomplished by setting $B_{j+1} = B_j \cup t_{j+1}(B_j)$. For $j = n-1$ and $j = n-2$, respectively, we obtain Claim 1.

Claim 2 Consider r-tuples $\tau = (\tau_1, \ldots, \tau_r)$ of transpositions in S_n with $\tau_1 \cdots \tau_r = \mathrm{id}$. (Clearly, such τ exist only for even r.) Each braid orbit on those tuples contains some τ with $\tau_{2i} = \tau_{2i-1}$ for $i = 1, \ldots, r/2$.

Proof. Write transpositions as (c, d) with $c < d$. Accordingly, order them lexicographically (i.e., $(1, 2) < (1, 3) < \cdots < (2, 3) < \cdots$). With that order

on the transpositions, also order the tuples $(\tau_1, \ldots, \tau_r)$ lexicographically. If $\tau_i < \tau_{i+1}$ for some i with $1 \le i \le r-1$ then Q_i^{-1} maps τ to a tuple that is strictly bigger than τ (in the lexicographic order). Thus if we take τ maximal in its braid orbit then $\tau_1 \ge \tau_2 \ge \cdots \ge \tau_r$. The latter condition implies that if $\tau_r = (c, d) \ne \tau_{r-1}$ then the product $\tau_1 \cdots \tau_r$ maps the number c to some strictly bigger number. But we have $\tau_1 \cdots \tau_r = \text{id}$, hence $\tau_r = \tau_{r-1}$. Applying induction to the tuple $(\tau_1, \ldots, \tau_{r-2})$ yields Claim 2.
Claim 3 Ni($\mathbf{C}$) is nonempty if and only if r is even and $r \ge 2(n-1)$.
Proof. If r is as specified then the r transpositions

$$(1,2), (1,2), (2,3), (2,3), \ldots, (n-1,n), (n-1,n), \ldots, (n-1,n) \quad (10.6)$$

form an element of Ni($\mathbf{C}$). Thus the condition is sufficient. Claims 1 and 2 imply it is necessary. (This also follows from the Riemann–Hurwitz formula.)
Conclusion. Suppose $\tau = (\tau_1, \ldots, \tau_r)$ is as in Claim 2, and $\tau_1, \ldots, \tau_r$ generate S_n. Applying Claim 1 and a sequence of braid operations that effect a cyclic permutation of $\tau_1, \ldots, \tau_r$ (see the proof of Lemma 9.4) we may assume the following: $\tau_1, \ldots, \tau_{r-2}$ generate S_n or a subgroup conjugate to S_{n-1}, according to whether $r > 2(n-1)$ or $r = 2(n-1)$. By Lemma 9.4 we may further assume that $\tau_r = (n-1, n)$. If $r > 2(n-1)$ then by induction on r, the tuple $(\tau_1, \ldots, \tau_{r-2})$ is in the braid orbit of the tuple (10.6) of length $r-2$. Hence τ is in the braid orbit of the tuple (10.6) of length r.

If $r = 2(n-1)$ then we can again use Lemma 9.4 to assume that $\tau_1, \ldots, \tau_{r-2}$ generate the canonically embedded S_{n-1} (keeping the condition that $\tau_r = (n-1, n)$). Then by induction on n, the tuple $(\tau_1, \ldots, \tau_{r-2})$ is in the braid orbit of $((1,2), (1,2), (2,3), (2,3), \ldots, (n-2, n-1), (n-2, n-1))$. Hence again τ is in the braid orbit of the tuple (10.6) of length r. This completes the proof. □

Remark 10.16 (Simple Covers) Let $n \ge 3$. An analytic map $Y \to \mathbb{P}^1$, where Y is a compact Riemann surface, is called a simple (branched) cover of $\mathbb{P}^1$ of degree n if the number of pre-images of any point of $\mathbb{P}^1$ is either n or $n-1$. Then the case $n-1$ occurs only for a finite set P of points of $\mathbb{P}^1$. (These are the branch points.) It is an exercise to show the following: Let $f : R \to \mathbb{P}^1 \setminus P$ be a Galois covering of ramification type $[S_n, P, \mathbf{C}]$, where $\mathbf{C}$ is the r-tuple consisting of r times the class of transpositions. Identify S_n with Deck(f) accordingly, and also with the extension of Deck(f) to a group of isomorphisms of the compactification $\bar{R}$. Let Y be the quotient of $\bar{R}$ by the action of S_{n-1} (naturally embedded in S_n). Then the induced map $Y \to \mathbb{P}^1$ is a simple cover, and each simple cover of $\mathbb{P}^1$ arises in this way.

Set $\mathcal{H}_{r,n} := \mathcal{H}^{in}(\mathbf{C})$, where $\mathbf{C}$ (as above) is r times the class of transpositions in $G := S_n$. By the above lemma and theorem, $\mathcal{H}_{r,n}$ is connected. Let $\mathcal{T}_{r,n}$ be the component of $\mathcal{T}_r(G)$ over $\mathcal{H}_{r,n}$, and let $\mathcal{S}_{r,n}$ be the image of $\mathcal{T}_{r,n}$ in $\mathcal{S}_r^{in}(G)$ (see Lemma 10.8). Then we get a Galois covering $\mathcal{T}_{r,n} \to \mathcal{S}_{r,n}$ whose group can be identified with S_n. Let $\bar{\mathcal{T}}_{r,n}$ be the closure of $\mathcal{T}_{r,n}$ in $\bar{\mathcal{T}}_r(G)$ (see Remark 10.7). The Galois covering $\mathcal{T}_{r,n} \to \mathcal{S}_{r,n}$ extends to a branched covering $\bar{\mathcal{T}}_{r,n} \to \mathcal{H}_{r,n} \times \mathbb{P}^1$, whose deck transformation group is still S_n.

For each $u \in \mathcal{H}_{r,n}$, its pre-image in $\mathcal{T}_{r,n}$ yields a Galois covering f_u of $\mathbb{P}^1 \setminus P$, $P = \Psi(u)$, whose group is the restriction of $\mathrm{Deck}(\mathcal{T}_{r,n}/\mathcal{S}_{r,n}) \cong S_n$ (see Lemma 10.8). By Lemma 10.6 and Section 10.1.3, each Galois covering of $\mathbb{P}^1 \setminus P$ of ramification type $[S_n, P, \mathbf{C}]$ is equivalent to f_u for one and only one $u \in \mathcal{H}_{r,n}$ (with $\Psi(u) = P$). (Here one needs that $\mathrm{Inn}(G) = \mathrm{Aut}(G)$ for $n \neq 6$, while for $n = 6$ it is still true that each automorphism of S_n fixing the class of transpositions is inner.) By the paragraph before the previous one, we obtain a 1-1 correspondence between the points of $\mathcal{H}_{r,n}$ and the equivalence classes of simple covers of $\mathbb{P}^1$ of degree n with r branch points. All those covers occur in the continuous (connected) family $\bar{\mathcal{T}}_{r,n}/S_{n-1} \to \mathcal{H}_{r,n} \times \mathbb{P}^1$.

The importance of simple covers comes from the fact that the "general" branched cover of $\mathbb{P}^1$ is simple; that is, picking a meromorphic function on a compact Riemann surface "at random," it will "almost certainly" yield a simple cover. In particular, every compact Riemann surface Y admits a simple cover $Y \to \mathbb{P}^1$. (Choose a birational map from Y to a plane curve Y' having only nodes as singularities, and follow this by projecting to a line in the plane from a suitable point outside Y'; see [FJ2].) Further, it can be shown that any two compact Riemann surfaces Y_1 and Y_2 of the same genus admit simple covers to $\mathbb{P}^1$ with the same invariants n and r. Thus Y_1 and Y_2 occur in the same connected family $\bar{\mathcal{T}}_{r,n}/S_{n-1} \to \mathcal{H}_{r,n} \times \mathbb{P}^1$. This yields the classical proof due to Hurwitz [Hur] for the connectedness of the moduli space of curves of genus g. (See [BF2] and [Fu] for this and generalizations.)

10.2 The Algebraic Structure of the Moduli Spaces

We keep the notation from Section 10.1. In particular, G is a finite group and r an integer ≥ 2.

We have studied the topological coverings $\Psi_{\mathbf{A}} : \mathcal{H}_r^{(\mathbf{A})}(G) \to \mathcal{O}_r$ and $\tilde{\Pi} : \mathcal{T}_r(G) \to \mathcal{O}(r+1)$. However, the spaces $\mathcal{O}_r$ and $\mathcal{O}(r+1)$ carry much more structure than just the topology; these spaces embed into affine space as affine varieties. By the generalization of Riemann's existence theorem to higher dimensions, this structure carries over to the covering spaces $\mathcal{H}_r^{(\mathbf{A})}(G)$ and $\mathcal{T}_r(G)$.

Recall that for each $r \geq 2$ there is a polynomial $\Delta_r \in \mathbb{Z}[X_1, \ldots, X_r]$ such that the discriminant of any monic polynomial $f(Z) = Z^r + a_1 Z^{r-1} + \cdots + a_r$ of degree r (over any field) is given as $\Delta_r(a_1, \ldots, a_r)$. The polynomial $f(Z)$ has r distinct roots if and only if $\Delta_r(a_1, \ldots, a_r) \neq 0$.

Lemma 10.17 *We obtain a homeomorphism*

$$\mathcal{O}_r \to \{(a_1, \ldots, a_r) \in \mathbb{C}^r : \Delta_r(a_1, \ldots, a_r) \neq 0\}$$

by sending $\{p_1, \ldots, p_r\}$ *to* $(a_1, \ldots, a_r)$, *where* $(Z - p_1) \cdots (Z - p_r) = Z^r + a_1 Z^{r-1} + \cdots + a_r$.

Proof. Bijectivity follows from the remarks before the lemma. Continuity follows because the a_i's are continuous functions of the p_i's (namely, the elementary symmetric functions). For the continuity of the inverse, consider a compact ball B in $\{\mathbf{a} \in \mathbb{C}^r : \Delta_r(\mathbf{a}) \neq 0\}$. One sees easily that the inverse image C of B in $\mathcal{O}_r$ has bounded p_i's, and also has the $|p_i - p_j|$ $(i \neq j)$ bounded away from zero (using that $\Delta_r(a_1, \ldots, a_r) = \prod_{i \neq j}(p_i - p_j)$). Thus C is compact (by using the map from Lemma 10.9). Hence the map $C \to B$ is homeomorphic (being continuous and bijective). It follows that the map in the lemma is open, hence homeomorphic. □

10.2.1 Coverings of Affine Varieties

To continue we need some notions from algebraic geometry. A subset A of $\mathbb{C}^n$ is called an **algebraic set** if it is defined by a system of equations $f_1(X_1, \ldots, X_n) = 0, \ldots, f_s(X_1, \ldots, X_n) = 0$, with $f_i \in \mathbb{C}[X_1, \ldots, X_n]$. Here $X_1, \ldots, X_n$ are the coordinate functions on $\mathbb{C}^n$. We say A is **defined over some subfield** K of $\mathbb{C}$ if the f_i lie in $K[X_1, \ldots, X_n]$. Then the **coordinate ring** $K[A]$ of A over K is defined as the ring of all functions $A \to K$ that are induced by the polynomial functions in $K[X_1, \ldots, X_n]$.

We say A is **irreducible** if the ring $\mathbb{C}[A]$ is a domain. If this holds, we call A an **affine variety**. Such A is called **nonsingular** if the matrix $(\partial f_i / \partial x_j)_{i=1,\ldots,s, j=1,\ldots,n}$ has the same rank at all points of A. Now suppose A is an affine variety defined over K. Then $K[A]$ is a domain, and we define the **function field** $K(A)$ of A over K to be the field of fractions of $K[A]$.

A **polynomial map** $\mathbb{C}^m \to \mathbb{C}^n$ is a map of the form $(a_1, \ldots, a_m) \mapsto (g_1(a_1, \ldots, a_m), \ldots, g_n(a_1, \ldots, a_m))$, with $g_j \in \mathbb{C}[X_1, \ldots, X_m]$. A polynomial map $g : B \to A$, where $A \subset \mathbb{C}^n$ and $B \subset \mathbb{C}^m$ are algebraic sets, is the restriction of a polynomial map $\mathbb{C}^m \to \mathbb{C}^n$ (that maps B into A). We say such a map is **defined**

over K if all $g_j \in K[X_1, \ldots, X_m]$. If this holds, then $f \mapsto f \circ g$ defines a K-algebra homomorphism $g^* : K[A] \to K[B]$. If g is surjective then g^* is injective, hence induces an embedding of $K(A)$ into $K(B)$ for irreducible A, B.

An algebraic set $A \subset \mathbb{C}^n$ carries the **complex topology** – induced from the Euclidean topology on $\mathbb{C}^n$ – as well as the **Zariski-topology** – whose closed sets are the algebraic sets contained in A. We let $A(\mathbb{C})$ denote the topological space consisting of A together with its complex topology.

Now we can state the version of the generalized Riemann existence theorem that we shall need (due essentially to Grauert and Remmert, see [Hsh], App. B, Th. 3.2, or [Se1], Th. 6.1.4; one further needs the descent from $\mathbb{C}$ to K, see [Se1], Th. 6.3.3.). For $A = \mathbb{P}^1 \setminus P$ it follows essentially from the analytic version of RET (Chapter 6) plus the constructions of Chapter 5. The higher-dimensional case is much deeper, and we will not attempt to give a proof here.

Theorem 10.18 *Let A be a nonsingular affine variety defined over the algebraically closed subfield K of $\mathbb{C}$. Then we have:*

(a) Each finite (topological) covering $R \to A(\mathbb{C})$, with R connected, is equivalent to a covering $g : B(\mathbb{C}) \to A(\mathbb{C})$, where B is a nonsingular affine variety defined over K, and g is a polynomial map defined over K.

(b) Suppose B, B' are nonsingular affine varieties defined over K, and $g : B \to A$, $g' : B' \to A$ are polynomial maps defined over K that are coverings in the complex topology. Then each homeomorphism $\alpha : B(\mathbb{C}) \to B'(\mathbb{C})$ with $g' \circ \alpha = g$ is a polynomial map defined over K (with polynomial inverse).

(c) Each g as above induces an embedding of $K[A]$ into $K[B]$, hence of $K(A)$ into $K(B)$. Here $[K(B) : K(A)]$ equals the degree of the covering $g : B \to A$. Further, $K[B]$ is the integral closure of $K[A]$ in $K(B)$. The extension $K(B)/K(A)$ is Galois if and only if g is a Galois covering, and in this case the Galois group of $K(B)/K(A)$ is isomorphic to Deck$(B(\mathbb{C})/A(\mathbb{C}))$ via pulling back functions (as in Theorem 5.9).

(The theorem is true without the restriction to the affine case.) Actually, the generalized Riemann existence theorem can be avoided in our context (see [V3]), but one still needs some nontrivial results from the theory of several complex variables. The exposition becomes shorter and clearer if we proceed as follows.

10.2.2 The Action of Field Automorphisms on the Points of $\mathcal{H}_r^{in}(G)$

First a lemma that leads us back into the situation of Chapter 5. We use the notation from Chapter 5.

Lemma 10.19 *Let $f : R \to \mathbb{C} \setminus P$ be a Galois covering of finite degree n, where $P \in \mathcal{O}_r$. Let $\mathcal{M}$ be the field of meromorphic functions on R that extend to a meromorphic function on the compact Riemann surface $\bar{R}$.*

(i) If $h : R \to \mathbb{C}$ is a continuous function that satisfies an algebraic equation $F(f, h) = 0$, where $F(X, Y) \in \mathbb{C}[X, Y]$ is of degree > 0 in Y, then $h \in \mathcal{M}$.
(ii) If $h \in \mathcal{M}$ takes n distinct values on $f^{-1}(p)$ for some $p \in \mathbb{C} \setminus P$ then h generates $\mathcal{M}$ over $\mathbb{C}(f)$.

Proof. (i) We may assume F is irreducible and monic in Y (Lemma 1.3). Then its Y-discriminant $D(X)$ is a nonzero element of $\mathbb{C}[X]$. Adding the zeroes of $D(X)$ to the set P we may assume that $D(X)$ has no zeroes on $\mathbb{C} \setminus P$. By analyticity of simple roots (Theorem 1.18) it follows that h is locally an analytic function of f. Hence h is meromorphic on R. By Lemma 5.5, it extends to a meromorphic function on $\bar{R}$. This proves (i). Claim (ii) follows as in the proof of Theorem 5.9. □

If A is an affine variety in $\mathbb{C}^n$ defined over K, and $f \neq 0$ in $K[A]$ then the set $A_f = \{a \in A : f(a) \neq 0\}$ has a natural structure as affine variety defined over K, with coordinate ring $K[A][f^{-1}]$. This comes from the embedding of A_f in $\mathbb{C}^{n+1}$ as the set of all (a, b) with $a \in A$, $b \in \mathbb{C}$ and $f(a)b = 1$. If A is nonsingular then so is A_f.

Thus the map from Lemma 10.17 embeds $\mathcal{O}_r$ into affine space as a nonsingular affine variety defined over $\bar{\mathbb{Q}}$. By generalized RET, it follows that each component $\mathcal{H}$ of $\mathcal{H}_r^{in}(G)$ embeds in the same way, such that the covering $\Psi : \mathcal{H} \to \mathcal{O}_r$ becomes a polynomial map defined over $\bar{\mathbb{Q}}$ in the coordinates of these affine spaces.

Consider the group $\text{Aut}(\mathbb{C}/\bar{\mathbb{Q}})$, consisting of all automorphisms of the field $\mathbb{C}$ that are the identity on $\bar{\mathbb{Q}}$. (In order to avoid the use of Zorn's Lemma later on, we could replace $\mathbb{C}$ here by a suitable algebraically closed subfield of finite transcendence degree; it is more convenient, however, to work with $\mathbb{C}$.) If B is an affine variety in $\mathbb{C}^m$ defined over $\bar{\mathbb{Q}}$, then each $\alpha \in \text{Aut}(\mathbb{C}/\bar{\mathbb{Q}})$ preserves the defining equations for B. Hence α permutes the points of B, by sending $(b_1, \ldots, b_m)$ to $(\alpha(b_1), \ldots, \alpha(b_m))$. This yields an action of $\text{Aut}(\mathbb{C}/\bar{\mathbb{Q}})$ on the points of $\mathcal{H}$, via the above embedding of $\mathcal{H}$ into affine space. The following theorem shows how this action (which depends on the variety structure of $\mathcal{H}$) relates to the moduli space structure of $\mathcal{H}$.

Theorem 10.20 *Suppose $Z(G) = 1$. Let $\mathcal{H}$ be a component of $\mathcal{H}_r^{in}(G)$, and consider the above action of $\text{Aut}(\mathbb{C}/\bar{\mathbb{Q}})$ on $\mathcal{H}$. Let $\alpha \in \text{Aut}(\mathbb{C}/\bar{\mathbb{Q}})$ and $u \in \mathcal{H}$. If, under the correspondence from Proposition 10.3, u corresponds to the (class* *of the) pair (L, ν), and $\alpha(u)$ corresponds to the pair (L', ν'), then there is an α-isomorphism $\lambda : L \to L'$ with $\nu' = \lambda^* \circ \nu$.*

Here, as usual, λ^* denotes the isomorphism $G(L/\mathbb{C}(x)) \to G(L'/\mathbb{C}(x))$ given by conjugating with λ.

Proof. Let $\mathcal{S} = \{(u, q) \in \mathcal{H} \times \mathbb{C} : q \notin \Psi(u)\}$ and $\mathcal{T} = (\Omega \times \Pi)^{-1}(\mathcal{S})$. These spaces $\mathcal{S}$ and $\mathcal{T}$ are the inverse images of $\{(P, q) \in \mathcal{O}(r+1) : q \neq \infty\}$ in the spaces that were called $\mathcal{S}$ and $\mathcal{T}$ in Lemma 10.8. (The reason for this change in notation is merely technical, to ensure we remain in the realm of affine varieties.) The same proof shows that $\mathcal{T}$ is connected, and $\Omega \times \Pi : \mathcal{T} \to \mathcal{S}$ is a Galois covering with deck transformation group $\{\delta_A : A \in \mathrm{Inn}(G)\}$. Here and in the following, we let δ_A, Ω and Π denote the restriction to $\mathcal{T}$ of the original maps. Further, for $g \in G$ we denote $\delta_{\mathrm{Inn}(g)}$ simply by δ_g.

Let $B \subset \mathbb{C}^m$ be the image of $\mathcal{H}$ under the given embedding into affine space. Consider the polynomial map $\Psi_{\mathrm{alg}} : B \to \{\mathbf{a} \in \mathbb{C}^r : \Delta_r(\mathbf{a}) \neq 0\}$ induced by $\Psi : \mathcal{H} \to \mathcal{O}_r$. Nonvanishing of the function $B \times \mathbb{C} \to \mathbb{C}$, $(b, q) \mapsto q^r + a_1 q^{r-1} + \cdots + a_r$, where $(a_1, \ldots, a_r) = \Psi_{\mathrm{alg}}(b)$, defines an open subvariety S of $B \times \mathbb{C}$. This is again a nonsingular affine variety defined over $\bar{\mathbb{Q}}$. Clearly, $S(\mathbb{C})$ is naturally homeomorphic to $\mathcal{S}$. By generalized RET, it also follows that $\mathcal{T}$ embeds into affine space as nonsingular affine variety defined over $\bar{\mathbb{Q}}$, such that the map $\Omega \times \Pi : \mathcal{T} \to \mathcal{S}$ becomes a polynomial map defined over $\bar{\mathbb{Q}}$. Let $\mathbb{C}[\mathcal{T}]$ (resp., $\bar{\mathbb{Q}}[\mathcal{T}]$) be the ring of functions $\mathcal{T} \to \mathbb{C}$ that become polynomial functions defined over $\mathbb{C}$ (resp., $\bar{\mathbb{Q}}$) under this embedding. Choose analogous notation for $\mathcal{S}$ (in place of $\mathcal{T}$). Then by Theorem 10.18(c), $\mathbb{C}[\mathcal{T}]$ is integral over $\mathbb{C}[\mathcal{S}]$.

Now fix some $u \in \mathcal{H}$, and let $P = \Psi(u)$, $R = \Omega^{-1}(u)$. By Proposition 10.6, the map Π restricts to a Galois covering $f : R \to \mathbb{C} \setminus P$, and the map $\mu : G \to \mathrm{Deck}(f)$, $g \mapsto \delta_g|_R$ is an isomorphism. Let $\mathcal{M}$ be as in Lemma 10.19 (the field of functions $R \to \mathbb{P}^1$ that extend to meromorphic functions on the compactification of R).

Claim 1 Restriction to R gives a homomorphism from $\mathbb{C}[\mathcal{T}]$ onto a subring of $\mathcal{M}$ having $\mathcal{M}$ as field of fractions.

Proof. Since $\mathbb{C}[\mathcal{T}]$ is integral over $\mathbb{C}[\mathcal{S}]$, each $h \in \mathbb{C}[\mathcal{T}]$ satisfies an equation $\tilde{F}(h) = 0$ where $\tilde{F}$ is a monic polynomial with coefficients in $\mathbb{C}[\mathcal{S}]$. The restrictions to $\{u\} \times (\mathbb{C} \setminus P)$ of the functions in $\mathbb{C}[\mathcal{S}]$ are contained in the field $\mathbb{C}(\pi)$, where $\pi : \{u\} \times (\mathbb{C} \setminus P) \to \mathbb{C}$ is projection $(u, q) \mapsto q$. Under the embedding $\mathbb{C}[\mathcal{S}] \to \mathbb{C}[\mathcal{T}]$, this π corresponds to the above Π. Since $f = \Pi|_R$, it follows that the element $h_R := h|_R$ satisfies an equation $F(f, h_R) = 0$, where $F(X, Y) \in \mathbb{C}(X)[Y]$ is of degree > 0 in Y. Hence $h_R \in \mathcal{M}$ by Lemma 10.19(i). Clearly, we can choose h such that h_R satisfies the hypothesis of Lemma 10.19(ii). Then $\mathcal{M} = \mathbb{C}(f, h_R)$ by that Lemma. This proves Claim 1.

The embeddings of $\mathcal{T}$ and $\mathcal{S}$ into affine space yield actions of $\mathrm{Aut}(\mathbb{C}/\bar{\mathbb{Q}})$ on $\mathcal{T}$ and $\mathcal{S}$ (as for $\mathcal{H}$). The maps $\Omega : \mathcal{T} \to \mathcal{H}$, $\Pi : \mathcal{T} \to \mathbb{C}$ and $\Omega \times \Pi : \mathcal{T} \to \mathcal{S}$ all commute with the action of $\mathrm{Aut}(\mathbb{C}/\bar{\mathbb{Q}})$ (since the corresponding polynomial maps are defined over $\bar{\mathbb{Q}}$). Further, $\mathrm{Aut}(\mathbb{C}/\bar{\mathbb{Q}})$ acts on $\mathbb{C}[\mathcal{T}]$, with $\alpha \in \mathrm{Aut}(\mathbb{C}/\bar{\mathbb{Q}})$ sending h to the function $\alpha(h) : t \mapsto \alpha(h(\alpha^{-1}(t)))$. (For the corresponding polynomial functions, this just means applying α to their coefficients.)

Now fix some $\alpha \in \mathrm{Aut}(\mathbb{C}/\bar{\mathbb{Q}})$, and let $u' = \alpha(u)$, $R' = \Omega^{-1}(u')$, $P' = \Psi(u')$, $f' = \Pi|_{R'} : R' \to \mathbb{C} \setminus P'$ and $\mathcal{M}'$ be defined accordingly.

Claim 2 The map $h_R \mapsto \alpha(h)_{R'}$, where $h \in \mathbb{C}[\mathcal{T}]$, extends to a field isomorphism $\lambda : \mathcal{M} \to \mathcal{M}'$ with $\lambda(f) = f'$ and $\lambda|_{\mathbb{C}} = \alpha$.

Proof. We have $\alpha(R) = \alpha(\Omega^{-1}(u)) = \Omega^{-1}(\alpha(u)) = \Omega^{-1}(u') = R'$. Thus $h_R = 0$ if and only if $\alpha(h)_{R'} = 0$. Hence the map in Claim 2 is well defined and injective. It extends to a field isomorphism $\mathcal{M} \to \mathcal{M}'$ by Claim 1. We have $\lambda(f) = \lambda(\Pi|_R) = \alpha(\Pi)|_{R'} = \Pi|_{R'} = f'$, and clearly $\lambda|_{\mathbb{C}} = \alpha$.

Claim 3 Let $\nu : G \to G(\mathcal{M}/\mathbb{C}(f))$ be the isomorphism mapping $g \in G$ to that element of $G(\mathcal{M}/\mathbb{C}(f))$ that sends $d \in \mathcal{M}$ to $d \circ \delta^{-1}$, where δ is the restriction of δ_g to R. Let $\nu' : G \to G(\mathcal{M}'/\mathbb{C}(f'))$ be defined analogously. Then $\nu' = \lambda^* \circ \nu$.

Proof. It suffices to check this on elements $d \in \mathcal{M}$ of the form $d = h_R$ with $h \in \mathbb{C}[\mathcal{T}]$. Let $g \in G$. We get $\lambda^*(\nu(g))(\lambda(d)) = (\lambda\nu(g)\lambda^{-1})(\lambda(d)) = \lambda(\nu(g)(d)) = \lambda((h \circ \delta_g^{-1})|_R) = \alpha(h \circ \delta_g^{-1})_{R'} = (\alpha(h) \circ \delta_g^{-1})_{R'} = (\alpha(h))_{R'} \circ (\delta')^{-1} = \lambda(d) \circ (\delta')^{-1} = \nu'(g)(\lambda(d))$, hence $\lambda^* \circ \nu = \nu'$. Here δ' is the restriction of δ_g to R'. We have used the fact that $\alpha(h \circ \delta_g^{-1}) = \alpha(h) \circ \delta_g^{-1}$, which follows from the fact that the deck transformations δ_g are polynomial maps defined over $\bar{\mathbb{Q}}$ (Theorem 10.18(b)).

Conclusion. Now take $L := \mathcal{M}$, and view it as an extension of $\mathbb{C}(x)$ by identifying x with f. Then the pair (L, ν), where ν is as in Claim 3, is in the class associated with u (by Proposition 10.6). Analogously for $L' := \mathcal{M}'$, and ν'. Now the theorem follows from Claim 2 and Claim 3. (Note that the assertion in the theorem remains true if we replace (L, ν) and (L', ν') by equivalent pairs.) □

As mentioned before, we have to use Zorn's Lemma because we are working with the huge group $\mathrm{Aut}(\mathbb{C}/\bar{\mathbb{Q}})$. The particular consequence of Zorn's Lemma we need is that any automorphism of a subfield of $\mathbb{C}$ can be extended to an automorphism of $\mathbb{C}$. This implies that for each subfield K of $\mathbb{C}$, the fixed field of $\mathrm{Aut}(\mathbb{C}/K)$ equals K.

For $u \in \mathcal{H}$ define $\bar{\mathbb{Q}}(u)$ to be the subfield of $\mathbb{C}$ generated by $\bar{\mathbb{Q}}$ and the coordinates $u_1, \ldots, u_m$ of the point $(u_1, \ldots, u_m) \in \mathbb{C}^m$ corresponding to u (under the given embedding of $\mathcal{H}$ into affine space).

Corollary 10.21 *In the situation of the theorem, again let $u \in \mathcal{H}$ correspond to the pair (L, ν). Then $\bar{\mathbb{Q}}(u)$ is the minimal field k between $\bar{\mathbb{Q}}$ and $\mathbb{C}$ such that L is defined over k.*

Proof. Let $\alpha \in \mathrm{Aut}(\mathbb{C}/\bar{\mathbb{Q}})$. By the theorem and Proposition 10.3, we have $\alpha(u) = u$ if and only if there exists an α-automorphism λ of L with $\lambda^* = \mathrm{id}$. The latter condition means that α lies in $\mathrm{Aut}(\mathbb{C}/k)$, where k is the minimal field occurring in the corollary (by the proof of Proposition 7.12). The former condition $\alpha(u) = u$ means that α lies in $\mathrm{Aut}(\mathbb{C}/\bar{\mathbb{Q}}(u))$. Thus $\mathrm{Aut}(\mathbb{C}/\bar{\mathbb{Q}}(u)) = \mathrm{Aut}(\mathbb{C}/k)$. By the remarks before the corollary, it follows that $\bar{\mathbb{Q}}(u) = k$. □

Remark 10.22

(a) The hypothesis $Z(G) = 1$ in the theorem (not in the corollary, however) can be dropped, see [FV1] and [V3].
(b) Note $\mathcal{O}_r$ is actually a variety defined over $\mathbb{Q}$ (not only $\bar{\mathbb{Q}}$). Also $\mathcal{H}_r^{in}(G)$ can be shown to have a natural structure as an algebraic set defined over $\mathbb{Q}$, with $G(\bar{\mathbb{Q}}/\mathbb{Q})$ permuting its components. Then the theorem and corollary remain true for $\mathbb{Q}$ in place of $\bar{\mathbb{Q}}$, and for $\mathcal{H}$ a $\mathbb{Q}$-component of $\mathcal{H}_r^{in}(G)$. See Section 10.3.

10.2.3 The Algebraic Structure of $\mathcal{H}_r^{(\mathbf{A})}(G)$, and the Proof of Theorem 9.5

In this subsection we assume $Z(G) = 1$. Again fix a component $\mathcal{H}$ of $\mathcal{H}_r^{in}(G)$. Let $\mathbf{A}$ be a subgroup of $\mathrm{Aut}(G)$ containing $\mathrm{Inn}(G)$, and $\mathcal{H}_{\mathbf{A}}$ the image of $\mathcal{H}$ in $\mathcal{H}_r^{(\mathbf{A})}(G)$. Then $\mathcal{H}_{\mathbf{A}}$ is a component of $\mathcal{H}_r^{(\mathbf{A})}(G)$, and the map $\mathcal{H} \to \mathcal{H}_{\mathbf{A}}$ is a Galois covering with deck transformation group $\{\epsilon_A : A \in \mathbf{A}, \epsilon_A(\mathcal{H}) = \mathcal{H}\}$ (see Lemma 10.8). We first apply generalized RET to the covering $\Psi_{\mathbf{A}} : \mathcal{H}_{\mathbf{A}} \to \mathcal{O}_r$, obtaining an embedding of $\mathcal{H}_{\mathbf{A}}$ into affine space as nonsingular variety defined over $\bar{\mathbb{Q}}$, and then again apply generalized RET to the covering $\mathcal{H} \to \mathcal{H}_{\mathbf{A}}$, to get a compatible embedding of $\mathcal{H}$. This allows us to define the rings $\bar{\mathbb{Q}}[\mathcal{H}]$ and $\bar{\mathbb{Q}}[\mathcal{H}_{\mathbf{A}}]$ of polynomial functions defined over $\bar{\mathbb{Q}}$ on the respective space. For $u \in \mathcal{H}$, the field $\bar{\mathbb{Q}}(u)$ equals the subfield of $\mathbb{C}$ generated by the values $h(u)$, $h \in \bar{\mathbb{Q}}[\mathcal{H}]$. True also for $u' \in \mathcal{H}_{\mathbf{A}}$, and so on.

The coverings $\mathcal{H} \to \mathcal{H}_{\mathbf{A}} \to \mathcal{O}_r$ yield embeddings $\bar{\mathbb{Q}}[\mathcal{O}_r] \subset \bar{\mathbb{Q}}[\mathcal{H}_{\mathbf{A}}] \subset \bar{\mathbb{Q}}[\mathcal{H}]$. Thus if $u \in \mathcal{H}$, and $u_{\mathbf{A}}$ (resp., P) is its image in $\mathcal{H}_{\mathbf{A}}$ (resp., $\mathcal{O}_r$), then $\bar{\mathbb{Q}}(P) \subset$

$\bar{\mathbb{Q}}(u_{\mathbf{A}}) \subset \bar{\mathbb{Q}}(u)$. Here $\bar{\mathbb{Q}}(P) = \bar{\mathbb{Q}}(a_1, \ldots, a_r)$, where the tuple $(a_1, \ldots, a_r)$ corresponds to P under the embedding from Lemma 10.17; that is, $Y^r + a_1 Y^{r-1} + \cdots + a_r = (Y - p_1) \cdots (Y - p_r)$, where $P = \{p_1, \ldots, p_r\}$.

By the above corollary, $k := \bar{\mathbb{Q}}(u)$ is the minimal field of definition containing $\bar{\mathbb{Q}}$ of the extension $L/\mathbb{C}(x)$ associated with u. Thus there is a unique subfield L_k of L such that L_k is Galois over $k(x)$, and restriction gives an isomorphism $\mathrm{res}_{L/L_k} : G(L/\mathbb{C}(x)) \to G(L_k/k(x))$ (see Lemma 3.1).

Lemma 10.23 *Again let $u \in \mathcal{H}$ correspond to the pair (L, ν). Let $k = \bar{\mathbb{Q}}(u)$ and $k_{\mathbf{A}} = \bar{\mathbb{Q}}(u_{\mathbf{A}})$, where $u_{\mathbf{A}}$ is the image of u in $\mathcal{H}_{\mathbf{A}}$.*

(i) Identify G with $G(L_k/k(x))$ via the isomorphism $\mathrm{res}_{L/L_k} \circ \nu$. Then L_k is Galois over $k_{\mathbf{A}}(x)$, and $H := G(L_k/k_{\mathbf{A}}(x))$ embeds into Aut(G) via conjugation action on $G = G(L_k/k(x))$. The image of H in Aut(G) lies in $\mathbf{A}_{\mathcal{H}} := \{A \in \mathbf{A} : \epsilon_A(\mathcal{H}) = \mathcal{H}\}$, and it equals $\mathbf{A}_{\mathcal{H}}$ if the branch points $p_1, \ldots, p_r$ of $L/\mathbb{C}(x)$ are algebraically independent over $\mathbb{Q}$.

(ii) Identify G with $G(L/\mathbb{C}(x))$ via the isomorphism ν. Let $\Psi(u) = P = \{p_1, \ldots, p_r\}$ (=the set of branch points of $L/\mathbb{C}(x)$). For $p \in P$, let C_p be the associated conjugacy class of $G = G(L/\mathbb{C}(x))$. Let $\mathbf{C} = (C_{p_1}, \ldots, C_{p_r})$ and assume $\mathbf{A} = \mathrm{Aut}_{\mathbf{C}}(G)$ (=the subgroup of Aut(G) fixing each C_p). If $\mathbf{C}$ is weakly rigid then $\bar{\mathbb{Q}}(u_{\mathbf{A}})$ equals the minimal field k' between $\bar{\mathbb{Q}}$ and $\mathbb{C}$ such that the ramification type of $L/\mathbb{C}(x)$ is k'-rational (cf. Proposition 9.1).

Proof.

Claim 1 $\mathrm{Aut}(\mathbb{C}/k_{\mathbf{A}})$ is the group of all $\alpha \in \mathrm{Aut}(\mathbb{C}/\bar{\mathbb{Q}})$ for which there is an α-automorphism λ of L with $\nu^{-1}\lambda^*\nu \in \mathbf{A}$. Then automatically $\nu^{-1}\lambda^*\nu \in \mathbf{A}_{\mathcal{H}}$.

Proof. Let $\alpha \in \mathrm{Aut}(\mathbb{C}/\bar{\mathbb{Q}})$. We have $\alpha \in \mathrm{Aut}(\mathbb{C}/k_{\mathbf{A}})$ if and only if $\alpha(u_{\mathbf{A}}) = u_{\mathbf{A}}$, which holds if and only if $\alpha(u) = \epsilon_A(u)$ for some $A \in \mathbf{A}_{\mathcal{H}}$ (since Deck $(\mathcal{H}/\mathcal{H}_{\mathbf{A}}) = \{\epsilon_A : A \in \mathbf{A}_{\mathcal{H}}\}$). If $\alpha(u) = \epsilon_A(u)$ for some $A \in \mathbf{A}$ then we must have $A \in \mathbf{A}_{\mathcal{H}}$ because $\alpha(u) \in \mathcal{H}$. Recall that the point $\epsilon_A(u)$ is associated with the pair $(L, \nu \circ A^{-1})$ (Proposition 10.3). Thus by Theorem 10.20 we have $\alpha(u) = \epsilon_A(u)$ if and only if there is an α-automorphism λ of L with $\nu \circ A^{-1} = \lambda^* \circ \nu$. This proves Claim 1.

Claim 2 If $p_1, \ldots, p_r$ are algebraically independent then

$$[k : k_{\mathbf{A}}] = \deg(\mathcal{H}/\mathcal{H}_{\mathbf{A}}) = [\mathbf{A}_{\mathcal{H}} : \mathrm{Inn}(G)].$$

Proof. Suppose $p_1, \ldots, p_r$ are algebraically independent. Then so are the $a_1, \ldots, a_r$ (the coefficients of the polynomial with roots $p_1, \ldots, p_r$, see the opening paragraph of this subsection). Thus the evaluation map $\bar{\mathbb{Q}}[\mathcal{O}_r] \to \mathbb{C}$,

$h \mapsto h(P) = h(a_1, \ldots, a_r)$, is injective. This yields an isomorphism between $\bar{\mathbb{Q}}(\mathcal{O}_r)$ (the field of fractions of $\bar{\mathbb{Q}}[\mathcal{O}_r]$) and $\bar{\mathbb{Q}}(P)$.

Let R be the subring of $\bar{\mathbb{Q}}(\mathcal{H})$ consisting of all h_1/h_2 with $h_i \in \bar{\mathbb{Q}}[\mathcal{H}]$ and $h_2(u) \neq 0$. Then by the preceding paragraph, R contains $\bar{\mathbb{Q}}(\mathcal{O}_r)$ and $\bar{\mathbb{Q}}[\mathcal{H}]$, hence equals $\bar{\mathbb{Q}}(\mathcal{H})$ (since $\bar{\mathbb{Q}}[\mathcal{H}]$ contains a primitive element for $\bar{\mathbb{Q}}(\mathcal{H})/\bar{\mathbb{Q}}(\mathcal{O}_r)$). This implies that the evaluation map $\bar{\mathbb{Q}}[\mathcal{H}] \to \mathbb{C}$, $h \mapsto h(u)$ is also injective. Hence the isomorphism between $\bar{\mathbb{Q}}(\mathcal{O}_r)$ and $\bar{\mathbb{Q}}(P)$ extends to an isomorphism between $\bar{\mathbb{Q}}(\mathcal{H})$ and $k = \bar{\mathbb{Q}}(u)$. The latter restricts to an isomorphism between $\bar{\mathbb{Q}}(\mathcal{H}_{\mathbf{A}})$ and $k_{\mathbf{A}} = \bar{\mathbb{Q}}(u_{\mathbf{A}})$. Thus $G(k/k_{\mathbf{A}}) \cong G(\bar{\mathbb{Q}}(\mathcal{H})/\bar{\mathbb{Q}}(\mathcal{H}_{\mathbf{A}})) \cong \mathrm{Deck}(\mathcal{H}/\mathcal{H}_{\mathbf{A}})$ by Theorem 10.18(c). The latter group equals $\{\epsilon_A : A \in \mathbf{A}_{\mathcal{H}}\}$, hence is isomorphic to $\mathbf{A}_{\mathcal{H}}/\mathrm{Inn}(G)$ (Lemma 10.8). This proves Claim 2.

(i). Each $\beta \in G(\overline{k_{\mathbf{A}}}/k_{\mathbf{A}})$ extends to some $\alpha \in \mathrm{Aut}(\mathbb{C}/k_{\mathbf{A}})$. By Claim 1, this α extends to an automorphism λ of L fixing x and satisfying $\nu^{-1}\lambda^*\nu \in \mathbf{A}_{\mathcal{H}}$. By uniqueness of the minimal field of definition and the associated L_k, the automorphism λ leaves k and L_k invariant. In particular, $G(\overline{k_{\mathbf{A}}}/k_{\mathbf{A}})$ leaves k invariant, that is, k is Galois over $k_{\mathbf{A}}$. Furthermore, each element of $G(k/k_{\mathbf{A}})$ extends to an automorphism of L_k fixing x, hence L_k is Galois over $k_{\mathbf{A}}(x)$.

Using the minimality of k, we see, as in Proposition 9.2(b), that the centralizer of $G = G(L_k/k(x))$ in $H = G(L_k/k_{\mathbf{A}}(x))$ is trivial. Since k is Galois over $k_{\mathbf{A}}$, the group G is normal in H, hence H embeds into $\mathrm{Aut}(G)$ via conjugation action. The group H is generated modulo G by the $\lambda|_{L_k}$, where λ is as in the preceding paragraph. Recall that λ^* is the automorphism of $G(L/\mathbb{C}(x))$ given by conjugating with λ. Thus $\nu^{-1}\lambda^*\nu$ is the automorphism of $G = G(L_k/k(x))$ given by conjugating with $\lambda|_{L_k}$ (under the identifications in (i)). Since $\nu^{-1}\lambda^*\nu \in \mathbf{A}_{\mathcal{H}}$ (see the preceding paragraph), it follows that the subgroup H' of $\mathrm{Aut}(G)$ induced by conjugation action of H lies in $\mathbf{A}_{\mathcal{H}}$.

Now assume $p_1, \ldots, p_r$ are algebraically independent. Then $[H' : \mathrm{Inn}(G)] = [H : G] = [k : k_{\mathbf{A}}] = [\mathbf{A}_{\mathcal{H}} : \mathrm{Inn}(G)]$ by Claim 2. Thus $H' = \mathbf{A}_{\mathcal{H}}$. This proves (i).

(ii). Both $k_{\mathbf{A}}$ and k' contain $k_0 := \bar{\mathbb{Q}}(P) = \bar{\mathbb{Q}}(a_1, \ldots, a_r)$ (see the start of this subsection and Proposition 9.1). From the proof of Proposition 9.1 we see that $\mathrm{Aut}(\mathbb{C}/k')$ is the group of all $\alpha \in \mathrm{Aut}(\mathbb{C}/k_0)$ with $C_{\alpha(p)} = C_p$ for all $p \in P$. By Claim 1 and the identifications in (ii), $\mathrm{Aut}(\mathbb{C}/k_{\mathbf{A}})$ is the group of all $\alpha \in \mathrm{Aut}(\mathbb{C}/k_0)$ for which there is an α-automorphism λ of L with $\lambda^* \in \mathbf{A}$. We use this to show that $\mathrm{Aut}(\mathbb{C}/k_{\mathbf{A}})$ equals $\mathrm{Aut}(\mathbb{C}/k')$. Then (ii) follows.

If λ is an α-automorphism of L with $\lambda^* \in \mathbf{A} = \mathrm{Aut}_C(G)$ then $C_{\alpha(p)} = \lambda^*(C_p) = C_p$ by Lemma 2.8. Conversely, let $\alpha \in \mathrm{Aut}(\mathbb{C}/k_0)$ with $C_{\alpha(p)} = C_p$ for all $p \in P$. By Lemma 3.5 there is an α-isomorphism λ from L to some

$L'/\mathbb{C}(x)$. Then $C'_{\alpha(p)} = \lambda^*(C_p) = \lambda^*(C_{\alpha(p)})$ by Lemma 2.8. Hence L and L' are of the same ramification type (via the isomorphism λ^*). By weak rigidity, we may assume $L = L'$ (Theorem 2.17). Then λ is an α-automorphism of L with $\lambda^* \in \mathrm{Aut}_{\mathbf{C}}(G)$. □

Now we can give the

Proof of Theorem 9.5. We give the proof for $\bar{\ell} = \bar{\mathbb{Q}}$. In the general case, just replace $\bar{\mathbb{Q}}$ by $\bar{\ell}$ in all of Section 10.2.

In the situation of Theorem 9.5, set $\mathbf{A} = \mathrm{Aut}_{\mathbf{C}}(G)$, and let $\nu : G \to G(L/\mathbb{C}(x))$ be the identity map. Then the pair (L, ν) corresponds to a point u of $\mathcal{H}^{in}(\mathbf{C})$ (by Proposition 10.14(c)). Let $\mathcal{H}$ be the component of $\mathcal{H}^{in}(\mathbf{C})$ containing u, and let $\mathcal{H}_{\mathbf{A}}$ (resp., $u_{\mathbf{A}}$) be the image of $\mathcal{H}$ (resp., u) in $\mathcal{H}_r^{(\mathbf{A})}(G)$.

Let k and $k_{\mathcal{T}}$ be as in the theorem. By Corollary 10.21 and Lemma 10.23(ii), we have $k = \bar{\mathbb{Q}}(u)$ and $k_{\mathcal{T}} = \bar{\mathbb{Q}}(u_{\mathbf{A}})$. Furthermore, the embedding of $G(k/k_{\mathcal{T}})$ into $\mathrm{Out}_{\mathbf{C}}(G)$ in the theorem coincides with the embedding of H/G into $\mathrm{Out}(G)$ from Lemma 10.23(i). Thus the group Δ in the theorem equals $\mathbf{A}_{\mathcal{H}}/\ \mathrm{Inn}(G)$ (by that lemma). The latter group is the stabilizer in $\mathrm{Out}_{\mathbf{C}}(G)$ of a braid orbit on $\mathrm{Ni}(\mathbf{C})$ by Proposition 10.14(b). □

10.3 Digression: The Inverse Galois Problem and Rational Points on Moduli Spaces

In this section we summarize some extensions of the above results (with proofs only sketched or omitted). Most of them have appeared in [FV1] and [FV2].

10.3.1 The $\mathbb{Q}$-Structure on $\mathcal{H}_r^{in}(G)$

Each algebraic set A in $\mathbb{C}^m$ is the union of its (finitely many) **irreducible components** (= maximal irreducible algebraic subsets). We sometimes call them absolutely irreducible components (to distinguish them from the $\mathbb{Q}$-components of an algebraic set defined over $\mathbb{Q}$, which we do not need here). We call A nonsingular iff its irreducible components are mutually disjoint, and each of them is a nonsingular variety.

We want to introduce an algebraic structure defined over $\mathbb{Q}$ (not only $\bar{\mathbb{Q}}$) on $\mathcal{H}_r^{in}(G)$, compatible with the natural $\mathbb{Q}$-structure on $\mathcal{O}_r$ (induced by the embedding from Lemma 10.17). First, we note that the group $\mathrm{Aut}(\mathbb{C})$ acts naturally on the set $\mathcal{H}_r^{in}(G)$, extending the action of $\mathrm{Aut}(\mathbb{C}/\bar{\mathbb{Q}})$ from Theorem 10.20. Indeed, one easily checks that there is an (abstract) action of $\mathrm{Aut}(\mathbb{C})$ on the set of equivalence classes of the pairs (L, ν) from Proposition 10.3 such that the following holds: $\alpha \in \mathrm{Aut}(\mathbb{C})$ maps the class of (L, ν) to that of (L', ν') iff there is an α-isomorphism $\lambda : L \to L'$ with $\nu' = \lambda^* \circ \nu$. Via the 1-1 correspondence in Proposition 10.3, this yields an action of Aut($\mathbb{C}$) on $\mathcal{H}_r^{in}(G)$. It is proved in [FV1] (and alternatively in [V3]) that this action arises from an algebraic structure on $\mathcal{H}_r^{in}(G)$ defined over $\mathbb{Q}$:

Theorem 10.24 *There is a homeomorphism $\phi : \mathcal{H}_r^{in}(G) \to B(\mathbb{C})$, for some nonsingular algebraic set B defined over $\mathbb{Q}$, such that the following holds: First, the composition of ϕ^{-1} with $\Psi : \mathcal{H} \to \mathcal{O}_r$, followed by the embedding $\mathcal{O}_r \to \{\mathbf{a} \in \mathbb{C}^r : \Delta_r(\mathbf{a}) \neq 0\}$, is a polynomial map defined over $\mathbb{Q}$. Second, the natural action of Aut($\mathbb{C}$) on B corresponds via ϕ to the above action on $\mathcal{H}_r^{in}(G)$.*

Idea of Proof. We outline a proof in the case $Z(G) = 1$. (The general case reduces to that.) By generalized RET, each connected component $\mathcal{H}$ of $\mathcal{H}_r^{in}(G)$ has a structure as affine variety defined over $\bar{\mathbb{Q}}$. Thus $\mathcal{H}_r^{in}(G)$ has a structure as affine algebraic set defined over $\bar{\mathbb{Q}}$, whose irreducible components coincide with the above components $\mathcal{H}$ (connected components in the complex topology). We just call them components from now on.

The coordinate ring $R_{\bar{\mathbb{Q}}} := \bar{\mathbb{Q}}[\mathcal{H}_r^{in}(G)]$ is the direct sum of the rings $\bar{\mathbb{Q}}[\mathcal{H}]$, $\mathcal{H}$ a component of $\mathcal{H}_r^{in}(G)$. To introduce a $\mathbb{Q}$-structure on $\mathcal{H}_r^{in}(G)$, it suffices to exhibit a $\mathbb{Q}$-subalgebra $R_{\mathbb{Q}}$ of $R_{\bar{\mathbb{Q}}}$ such that the inclusion map $R_{\mathbb{Q}} \to R_{\bar{\mathbb{Q}}}$ extends to an isomorphism $R_{\mathbb{Q}} \otimes_{\mathbb{Q}} \bar{\mathbb{Q}} \to R_{\bar{\mathbb{Q}}}$. By Theorem 10.18(c), $\bar{\mathbb{Q}}[\mathcal{H}]$ is the integral closure of $\bar{\mathbb{Q}}[\mathcal{O}_r]$ in the function field $\bar{\mathbb{Q}}(\mathcal{H})$. Thus $R_{\bar{\mathbb{Q}}}$ is the integral closure of $\bar{\mathbb{Q}}[\mathcal{O}_r]$ in the total ring of fractions $K_{\bar{\mathbb{Q}}}$ of $R_{\bar{\mathbb{Q}}}$. This $K_{\bar{\mathbb{Q}}}$ is the direct sum of the fields $\bar{\mathbb{Q}}(\mathcal{H})$. Hence we can define $R_{\mathbb{Q}}$ as the integral closure of $\mathbb{Q}[\mathcal{O}_r]$ in $K_{\mathbb{Q}}$, once we have found a suitable subring $K_{\mathbb{Q}}$ of $K_{\bar{\mathbb{Q}}}$ with $K_{\mathbb{Q}} \otimes_{\mathbb{Q}} \bar{\mathbb{Q}} \cong K_{\bar{\mathbb{Q}}}$.

The key to finding $K_{\mathbb{Q}}$ is Corollary 10.21. Let u be a generic point of $\mathcal{H}$ over $\bar{\mathbb{Q}}$ (as in the proof of Lemma 10.23, Claim 2). Then $\bar{\mathbb{Q}}(u) \cong \bar{\mathbb{Q}}(\mathcal{H})$ is the minimal field of definition over $\bar{\mathbb{Q}}$ of the extension $L/\mathbb{C}(x)$ corresponding to u. Now let κ be the minimal field of definition over $\mathbb{Q}$ of this extension. If κ is regular over $\mathbb{Q}$ then $\kappa \otimes_{\mathbb{Q}} \bar{\mathbb{Q}} \cong \bar{\mathbb{Q}}\kappa = \bar{\mathbb{Q}}(u) \cong \bar{\mathbb{Q}}(\mathcal{H})$, and we get a structure as (absolutely irreducible) affine variety defined over $\mathbb{Q}$ on $\mathcal{H}$ with $\mathbb{Q}(\mathcal{H}) \cong \kappa$. In general, we can identify $\kappa \otimes_{\mathbb{Q}} \bar{\mathbb{Q}}$ with the direct sum of all $\bar{\mathbb{Q}}(\mathcal{H}')$, where $\mathcal{H}'$ ranges over the $G(\bar{\mathbb{Q}}/\mathbb{Q})$-conjugates of $\mathcal{H}$. (The latter action of $G(\bar{\mathbb{Q}}/\mathbb{Q})$ on the components of $\mathcal{H}_r^{in}(G)$ comes from the given action of Aut($\mathbb{C}$) on the points of $\mathcal{H}_r^{in}(G)$.) Finally, the desired $K_{\mathbb{Q}}$ is obtained as the direct sum of such fields κ, one for each $G(\bar{\mathbb{Q}}/\mathbb{Q})$-orbit of components of $\mathcal{H}_r^{in}(G)$. □

Actually, the algebraic set B is unique up to polynomial isomorphism defined over $\mathbb{Q}$. From now on we identify $\mathcal{H}_r^{in}(G)$ with B (via ϕ). For $u \in \mathcal{H}_r^{in}(G)$ define $\mathbb{Q}(u)$ to be the subfield of $\mathbb{C}$ generated by the coordinates $u_1, \ldots, u_m$ of the point $\phi(u) = (u_1, \ldots, u_m)$. The same proof as for Corollary 10.21 (using Theorem 10.24 instead of Theorem 10.20) yields part (a) of

Corollary 10.25 *Assume $Z(G) = 1$.*

(a) Let $u \in \mathcal{H}_r^{in}(G)$ correspond to the pair (L, ν). Then $\mathbb{Q}(u)$ is the minimal subfield k of $\mathbb{C}$ such that L is defined over k.
(b) The group G occurs regularly over $\mathbb{Q}$ if and only if for some $r \geq 2$ the algebraic set $\mathcal{H}_r^{in}(G)$ has a $\mathbb{Q}$-rational point.

Note that if there is a Galois extension $L/\mathbb{C}(x)$ with group isomorphic to G, defined over $\mathbb{Q}$, then there is such L whose branch points are different from ∞. (Use a suitable coordinate transformation with rational coefficients.) Then part (b) of the corollary is an immediate consequence of (a), because of the 1-1 correspondence in Proposition 10.3.

The hypothesis $Z(G) = 1$ is not so serious as it looks, since every finite group H is the quotient of a finite group G with $Z(G) = 1$. Clearly, if G occurs regularly over $\mathbb{Q}$ then so does H. Therefore, the above corollary reduces the regular version of the Inverse Galois Problem to finding $\mathbb{Q}$-rational points on certain algebraic sets defined over $\mathbb{Q}$. (A $\mathbb{Q}$-rational point u is defined by the condition that $\mathbb{Q}(u) = \mathbb{Q}$.) Unfortunately, the algebraic structure of the spaces $\mathcal{H}_r^{in}(G)$ is hard to describe explicitly, which makes it difficult to decide whether there are $\mathbb{Q}$-rational points. The only thing we can say right away is that each $\mathbb{Q}$-rational point of $\mathcal{H}_r^{in}(G)$ lies on an (absolutely) irreducible component defined over $\mathbb{Q}$. Thus the first task is to find such components. This is done in the next subsection.

Example 10.26 Let us consider the algebraic structure on a subspace of $\mathcal{H}_r^{in}(G)$ of the form $\mathcal{H}_r^{in}(\mathbf{C})$ (see Section 10.1.8). This subspace is a union of irreducible components of $\mathcal{H}_r^{in}(G)$, hence is a Zariski closed subset. It is defined over $\mathbb{Q}$ if and only if the tuple $\mathbf{C}$ is rational (an easy consequence of Lemma 2.8; see [FV1], Th. 1).

Now assume that $\mathbf{C}$ is rigid. If additionally the classes in $\mathbf{C}$ are all rational, then the sequence $\mathcal{O}^{(r)} \to \mathcal{H}^{in}(\mathbf{C}) \to \mathcal{O}_r$ from the exercise in Section 10.1.8 consists of coverings defined over $\mathbb{Q}$. Thus $\mathcal{H}^{in}(\mathbf{C})$ has a lot of $\mathbb{Q}$-rational points, as predicted by the Rational Rigidity Criterion. If the tuple $\mathbf{C}$ is rational, but does not consist of rational classes, then the map $\mathcal{O}^{(r)} \to \mathcal{H}^{in}(\mathbf{C})$ is not defined over $\mathbb{Q}$; but still $\mathcal{H}^{in}(\mathbf{C})$ is a rational variety defined over $\mathbb{Q}$. (Follows from Lemma 8.7.) Thus $\mathcal{H}^{in}(\mathbf{C})$ still has many rational points, which we know by the Rigidity Criterion over $\mathbb{Q}$ (Corollary 3.18).

10.3.2 Absolutely Irreducible Components of $\mathcal{H}_r^{in}(G)$ Defined over $\mathbb{Q}$

For each group G there is some r (actually infinitely many r) such that $\mathcal{H}_r^{in}(G)$ has an irreducible component defined over $\mathbb{Q}$. There are two ways of seeing this. First, by a theorem of Harbater (see Theorem 11.31 in Chapter 11), G occurs regularly over the field $k = \mathbb{Q}((t))$. Choose some embedding of k into $\mathbb{C}$. Then there exists a Galois extension $L/\mathbb{C}(x)$ defined over k, and an isomorphism $\nu : G \to G(L/\mathbb{C}(x))$. Again we can arrange that L has all branch points different from ∞. Let r be the number of those branch points. Then the point $u \in \mathcal{H}_r^{in}(G)$ corresponding to the pair (L, ν) has $\mathbb{Q}(u) \subset k$ (by Corollary 10.25; the hypothesis $Z(G) = 1$ is not necessary for this direction). Hence $\mathbb{Q}(u)$ is regular over $\mathbb{Q}$. By standard facts of algebraic geometry, this implies that u lies on an irreducible component of $\mathcal{H}_r^{in}(G)$ that is defined over $\mathbb{Q}$. The number r can be bounded effectively in terms of G (see Remark 11.24).

The second method relies on a result of Conway and Parker (see [FV1], Appendix) about the braid orbits on generating systems of a finite group. Combined with [FV1], Lemma 2, this implies that each finite group is the quotient of a finite group G with $Z(G) = 1$ and with the following property: If $\mathbf{C} = (C_1, \ldots, C_r)$ is a tuple of conjugacy classes of G containing each class of G sufficiently many times, then the set $\mathrm{Ni}(\mathbf{C})$ is a single braid orbit. Thus $\mathcal{H}^{in}(\mathbf{C})$ is a connected component, hence an irreducible component, of $\mathcal{H}_r^{in}(G)$ (Proposition 10.14). This component is defined over $\mathbb{Q}$ if and only if $\mathbf{C}$ is rational (by Lemma 2.8). The latter holds if each class of G occurs the same (sufficiently large) number of times in $\mathbf{C}$.

Having an absolutely irreducible component defined over $\mathbb{Q}$ of the form $\mathcal{H}^{in}(\mathbf{C})$ makes things more explicit (although r may have to be taken very large). Using this, the above-mentioned theorem of Harbater (for Galois realizations over $\mathbb{Q}_p(x)$) and results of Dèbes and Fried on Galois groups over $\mathbb{R}(x)$ ([DF1]) one gets (see [FV1] and [Desch])

Theorem 10.27 *For each finite group G there are infinitely many $r \in \mathbb{N}$ with the following property: The algebraic set $\mathcal{H}_r^{in}(G)$ has an absolutely irreducible component $\mathcal{H}$ defined over $\mathbb{Q}$, and $\mathcal{H}$ has an $\mathbb{R}$-rational point and a $\mathbb{Q}_p$-rational point for each prime p.*

Here $\mathbb{Q}_p$ is the p-adic field. Recall that if $Z(G) = 1$ and $\mathcal{H}$ has a $\mathbb{Q}$-rational point then G occurs regularly over $\mathbb{Q}$. The existence of $\mathbb{Q}$-rational points would follow from the theorem if one could prove a "local-global principle" for $\mathcal{H}$. In general, however, one cannot expect this to hold as shown by examples of M. Fried (see [DF2]).

Contrary to the case of $\mathbb{Q}$, lower bounds are known for the number of $\mathbb{F}_p$-rational points on an absolutely irreducible variety over the finite field $\mathbb{F}_p$. This and reduction mod p yields the following (see [FV1], Cor. 2):

Theorem 10.28 *Let G be a finite group. Then G occurs as a Galois group over the rational function field $\mathbb{F}_p(x)$ for all but finitely many primes p.*

10.3.3 The Application to PAC-Fields

A field P is called PAC (pseudo algebraically closed) if every absolutely irreducible (nonempty) variety defined over P has infinitely many P-rational points. Corollary 10.25 and Theorem 10.27, together with the trick of replacing a given group by a group with trivial center that maps onto the given group, yield the following: If $P \subset \mathbb{C}$ is a PAC-field then every finite group G occurs regularly over P. Hence if P is also hilbertian then every finite group is a Galois group over P. Using the interplay between $\mathcal{H}_r^{in}(G)$ and $\mathcal{H}_r^{(\mathrm{A})}(G)$ for $\mathrm{A} = \mathrm{Aut}(G)$, this has been extended in [FV2] as follows:

Theorem 10.29 *Suppose P is PAC, hilbertian, and of characteristic 0. Then all finite embedding problems over P are solvable. Hence if P is countable then its absolute Galois group is free profinite of countable rank.*

Recently this has been generalized to fields of positive characteristic by Pop [P2].

A construction of Fried and Jarden (see [FV2]) yields a hilbertian PAC-field $P \subset \bar{\mathbb{Q}}$ that is Galois over $\mathbb{Q}$ with Galois group isomorphic to $\prod_{n=2}^{\infty} S_n$ (the direct product over all symmetric groups S_n). By the theorem, $G(\bar{\mathbb{Q}}/P)$ is isomorphic to the free profinite group $\hat{F}_\omega$ of countable rank. Hence we get an exact sequence for $G(\bar{\mathbb{Q}}/\mathbb{Q})$:

Corollary 10.30 *There is an exact sequence*

$$1 \to \hat{F}_\omega \to G(\bar{\mathbb{Q}}/\mathbb{Q}) \to \prod_{n=2}^{\infty} S_n \to 1.$$

11

Patching over Complete Valued Fields

This final chapter presents a "patching method" for constructing Galois extensions of $k(x)$, where k is a field that is complete with respect to an ultrametric absolute value. Using this method, we prove that each finite group is a Galois group over $k(x)$ (regularly). Examples of such k are the p-adic fields $\mathbb{Q}_p$ as well as the fields $k_0((t))$ of formal Laurent series over any field k_0.

This approach is due to Harbater [Ha1], who phrased things in the language of formal geometry (i.e., formal schemes). Serre [Se1], Th. 8.4.6 and Liu [Liu] translated it into the language of rigid analytic geometry, which is more intuitive, but still requires quite a substantial machinery (see [FP]). The elementary approach presented here has been worked out by D. Haran and the author [HV]. It uses some basic ultrametric analysis which we develop in Sections 11.1 and 11.2.

To describe the underlying idea, consider an algebraic equation $f(x, y) = 0$, separable in y, with coefficients in some field k. By Hensel's Lemma, this equation has a formal Laurent series solution of the form $y = \sum_{i=n}^{\infty} a_i (x - p)^i$ for almost all $p \in k$, with $a_i \in \bar{k}$. If $k = \mathbb{C}$, these series are convergent and this leads to the analytic theory presented in Chapters 4 – 6. Now there is a notion of convergence over each field k that has an **absolute value**. If k is complete then the corresponding analytic theory resembles that over $\mathbb{C}$ to some extent. In particular, the series $y = \sum_{i=n}^{\infty} a_i (x - p)^i$ is again convergent.

The main difference from the classical case is that there is no analogue for the topological theory of the fundamental group. Thus we only get a weak substitute of RET which, however, applies to more general fields (including fields of positive characteristic). It allows us to generalize Theorem 7.14 to algebraically closed fields ℓ of arbitrary characteristic: Each embedding problem over $\ell(x)$ is solvable (Theorem 11.32). For $\ell = \bar{\mathbb{F}}_p$ this is the geometric case of the conjecture of Shafarevich. It was first proved by Harbater [Ha3] and Pop [P3] in a more general setting, using formal and rigid geometry, respectively.

The above development has culminated in the proof of Abhyankar's conjecture, due to Raynaud [Ra] and Harbater [Ha2] (see the concluding remarks in Section 11.3). This requires deeper methods which go beyond the framework of this book.

From Section 11.2 on, we fix a field k that is complete with respect to a nontrivial ultrametric absolute value $|\cdot|$.

11.1 Power Series over Complete Rings

11.1.1 Absolute Values

Let R be an integral domain equipped with a nontrivial ultrametric absolute value $|\cdot|$. This means that $a \mapsto |a|$ is a map from R to the nonnegative real numbers, satisfying:

- $|a| = 0$ if and only if $a = 0$.
- There is $a \in R$ with $0 < |a| < 1$.
- $|ab| = |a| \cdot |b|$.
- $|a + b| \leq \max(|a|, |b|)$.

The second axiom excludes the trivial case that $|a| = 1$ for all $a \neq 0$ in R. The ultrametric inequality $|a + b| \leq \max(|a|, |b|)$ implies the usual triangle inequality $|a + b| \leq |a| + |b|$.

The notion of convergence of sequences and series in R is defined just as in the classical case, with $|\cdot|$ replacing the usual absolute value on $\mathbb{C}$. The usual properties hold, with the same proofs. The stronger ultrametric inequality allows for some simplifications. As in the classical case, we have the notion of Cauchy sequence, and each convergent sequence is Cauchy. (When we say a sequence or series in R converges we mean it converges to an element of R.) We further assume:

- R is complete, this is, every Cauchy sequence in R converges.

Here are some useful consequences of the above axioms. It is especially the first property that makes ultrametric analysis quite pleasant:

(1) A series $\sum_{i=0}^{\infty} a_i$ with $a_i \in R$ converges if and only if $\lim a_i = 0$.

(2) As usual, we say a series of the form $\sum_{i=-\infty}^{\infty} a_i$ converges if both $\sum_{i=0}^{\infty} a_i$ and $\sum_{i=1}^{\infty} a_{-i}$ converge, and its value is the sum of the two other series. If the series $\sum_{i=-\infty}^{\infty} a_i$ and $\sum_{j=-\infty}^{\infty} b_j$ converge, so does the series $\sum_{i=-\infty}^{\infty} a_i b_{n-i}$ for each $n \in \mathbb{Z}$, and denoting the value of the latter series by c_n we have $\sum_{n=-\infty}^{\infty} c_n$ convergent with

$$\left(\sum_{i=-\infty}^{\infty} a_i\right) \cdot \left(\sum_{j=-\infty}^{\infty} b_j\right) = \sum_{n=-\infty}^{\infty} c_n.$$

(3) *If* $\lim a_i = a$ *then* $\lim |a_i| = |a|$.
(4) *If* $\sum_{i=0}^{\infty} a_i$ *converges then*

$$\left| \sum_{i=0}^{\infty} a_i \right| \leq \max |a_i|$$

and if this maximum is attained for only one value of i, *say* $|a_i| < |a_1|$ *for* $i = 2, 3, \ldots$ *then we get equality*

$$\left| \sum_{i=0}^{\infty} a_i \right| = |a_1|.$$

(5) *If* $a \in R$ *and* $|a| < 1$, *then* $1 - a$ *is a unit in* R.

Proof. (1) is an almost immediate consequence of the ultrametric inequality $|a_i + \cdots + a_{i+n}| \leq \max(|a_i|, \ldots, |a_{i+n}|)$; (2) and (3) are standard; (4) reduces to finite sums by (3), and then to the case of two summands by induction. Thus consider $a, b \in R$ with $|a| > |b|$. Then $|a| = |(a+b)-b| \leq \max(|a+b|, |b|) \leq \max(|a|, |b|) = |a|$. Hence $|a| = \max(|a+b|, |b|) = |a+b|$.

For (5), note that the series $1 + a + a^2 + a^3 + \cdots$ converges since $\lim a^i = 0$. Its sum is the inverse of $1 - a$. □

11.1.2 Power Series

Consider the ring $R[[x]]$ of formal power series over R. (It is defined as in 2.1.1, with the ring R replacing the field k.) Fix some $f(x) = \sum_{i=0}^{\infty} a_i x^i$ in $R[[x]]$.

Lemma 11.1

(i) *If* $f(c)$ *converges for some* $c \in R$, *then it converges for all* b *with* $|b| \leq |c|$. *Hence we have the notion of radius of convergence, just as in the classical case. This radius is either a nonnegative real number or* $+\infty$.
(ii) *We have* $\lim a_i = 0$ *if and only if* $f(c)$ *converges for all* $c \in R$ *of absolute value* ≤ 1. *These* f *form a subring of* $R[[x]]$ *that we denote by* $R\{\{x\}\}$. *For each* $c \in R$ *with* $|c| \leq 1$ *we have the evaluation map* $R\{\{x\}\} \to R$, $f \mapsto f(c)$. *This map is a ring homomorphism.*
(iii) *For* $f \in R\{\{x\}\}$ *define* $\|f\| = \max(|a_i|)$. *This defines an ultrametric absolute value on* $R\{\{x\}\}$, *extending that on* R, *and* $R\{\{x\}\}$ *is complete with respect to this absolute value.*

Proof. (i) follows from (1) above. The first claim in (ii) follows from (1) and (i). The rest of (ii) follows from (2) and the definition of multiplication in $R[[x]]$.

For (iii), we first check that $\|fg\| = \|f\| \cdot \|g\|$ for $f, g \in R\{\{x\}\}$. Let $f = \sum_{i=0}^{\infty} a_i x^i$ and $g = \sum_{i=0}^{\infty} b_i x^i$. We may assume $f \neq 0$ and $g \neq 0$. Clearly $\|fg\| \leq \|f\| \cdot \|g\|$. Conversely, let n, m be the largest indices such that $|a_n| = \|f\|$ and $|b_m| = \|g\|$, and consider the coefficient c_ℓ of x^ℓ in fg, where $\ell = n + m$. If $i + j = \ell$ and $(i, j) \neq (n, m)$ then $|a_i| < \|f\|$ or $|b_j| < \|g\|$, hence $|a_i| \cdot |b_j| < \|f\| \cdot \|g\|$. Thus $\max(|a_i b_j| : i + j = \ell) = |a_n| \cdot |b_m| = \|f\| \cdot \|g\|$, and this maximum it attained only when $(i, j) = (n, m)$. Hence $|c_\ell| = |\sum_{i+j=\ell} a_i b_j| = \|f\| \cdot \|g\|$ (by (4) above), and so $\|fg\| \geq \|f\| \cdot \|g\|$. □

The other axioms for an absolute value hold trivially. For the completeness, consider a Cauchy sequence (f_n) in $R\{\{x\}\}$. This yields a Cauchy sequence in each coefficient, hence (f_n) converges coefficientwise to some $f \in R[[x]]$. The usual $\epsilon/2$-argument shows that actually $f \in R\{\{x\}\}$, and $\|f - f_n\| \to 0$.

11.1.3 Algebraic Power Series Are Convergent

In this subsection we assume $R = k$ is a (complete) field. We say a formal Laurent series $f \in k((x))$ is convergent if $x^n f$ is a power series with a positive radius of convergence for some (hence all suitably large) $n \in \mathbb{N}$. Clearly, sums and products of convergent Laurent series are convergent.

Also, if $f \neq 0$ is convergent then $\frac{1}{f}$ is convergent. To prove this, we may assume $f = \sum_{i=0}^{\infty} a_i x^i \in k[[x]]$, with $a_0 = 1$. Since the absolute value on k is nontrivial there are $c \in k$ of arbitrarily small positive absolute value. For suitably small $|c|$ the power series $f(cx) = \sum_{i=0}^{\infty} (a_i c^i) x^i$ lies in $k\{\{x\}\}$, and even satisfies $|a_i c^i| < 1$ for all $i > 0$. Then $\|f(cx) - 1\| < 1$, hence $f(cx)$ is invertible in $k\{\{x\}\}$ by property (5) from Section 11.1.1. Thus $\frac{1}{f(cx)}$ lies in $k\{\{x\}\}$, hence $\frac{1}{f}$ converges for $x = c$.

It follows that if $f \in k((x))$ is convergent then the same holds for all elements of the subfield of $k((x))$ generated by f and $k(x)$. Now we are ready to prove the "implicit function theorem in ultrametric analysis."

Theorem 11.2 *Suppose the power series $\psi(x) = \sum_{i=0}^{\infty} a_i x^i$ in $k[[x]]$ is algebraic over $k(x)$ (i.e., satisfies a relation $F(x, \psi(x)) = 0$ with $F(x, y) \neq 0$ in $k[x, y]$). Then ψ has a positive radius of convergence.*

Proof. If ψ has a positive radius of convergence, then the same holds for all power series in the subfield of $k((x))$ generated by $k(x)$ and ψ. Hence we can assume by Lemma 7.8 that $F(0, y)$ is separable, of y-degree equal to that of $F(x, y)$. We may further assume $\psi(0) = 0$. Then the coefficients a_i of ψ satisfy the recursion formula from the proof of Theorem 1.18. From this recursion formula we see as in that proof that $|a_i| \le \tilde{a}_i$, where the $\tilde{a}_i \in \mathbb{R}_+$ are the coefficients of a convergent power series in $\mathbb{C}[[x]]$. Thus $\lim \tilde{a}_i s^i = 0$ for some positive real number s. Pick $c \in k$ with $0 < |c| < s$. (Here we use that the absolute value on k is nontrivial.) Then $|a_i c^i| \le \tilde{a}_i s^i$, hence $\lim a_i c^i = 0$ in k and so the series $\psi(c)$ converges. □

11.1.4 Weierstraß Division

For $f = \sum_{i=0}^{\infty} a_i x^i \neq 0$ in $R\{\{x\}\}$ define the **pseudo-degree** of f to be the integer $n = \max(i : |a_i| = \|f\|)$. We say f is **regular** if a_n is invertible in R.

Theorem 11.3 *Let $f, g \in R\{\{x\}\}$, with f regular of pseudodegree n. Then there are unique $q \in R\{\{x\}\}$ and $r \in R[x]$ such that $g = qf + r$ and $\deg r < n$. Moreover, $\|qf\| \le \|g\|$ and $\|r\| \le \|g\|$.*

Proof. Let $f = \sum_{i=0}^{\infty} a_i x^i$ and $g = \sum_{i=0}^{\infty} b_i x^i$.

Estimates. Assume $g = qf + r$, where $\deg r < n$. Let m be the pseudo-degree of q (if $q = 0$, set $m = 0$). Then $\|qf\|$ equals the absolute value of the coefficient c of z^{n+m} in qf (see the proof of Lemma 11.1(iii)). The coefficient of z^{n+m} in $g = qf + r$ equals c, since $\deg r < n + m$. Thus $\|qf\| \le \|g\|$. It follows that $\|r\| = \|g - qf\| \le \max(\|g\|, \|qf\|) \le \|g\|$.

Uniqueness. Assume $g = qf + r = q'f + r'$, where $\deg r, \deg r' < n$. Then $0 = (q - q')f + (r - r')$. By the estimates just proved, $\|(q - q')f\| \le \|0\| = 0$ and $\|r - r'\| = 0$. Hence $q = q'$ and $r = r'$.

Existence if f is a polynomial of degree n. For each integer $\mu \ge 0$ let $g_\mu = \sum_{i=0}^{\mu} b_i x^i \in R[x]$. Then $\lim g_\mu = g$ (with respect to the absolute value on $R\{\{x\}\}$ from Section 11.1.2). Since the highest coefficient a_n of f is a unit in R, ordinary division of polynomials yields $q_\mu, r_\mu \in R[x]$ with $g_\mu = q_\mu f + r_\mu$ and $\deg r_\mu < n$. Thus for all μ, μ' we have $g_\mu - g_{\mu'} = (q_\mu - q_{\mu'})f + (r_\mu - r_{\mu'})$. Thus $\|q_\mu - q_{\mu'}\| \cdot \|f\|, \|r_\mu - r_{\mu'}\| \le \|g_\mu - g_{\mu'}\|$. Thus $\{q_\mu\}_{\mu=0}^{\infty}$ and $\{r_\mu\}_{\mu=0}^{\infty}$ are Cauchy sequences in $R\{\{x\}\}$, hence converge to certain $q \in R\{\{x\}\}$ and $r \in R[x]$, respectively. Clearly $g = qf + r$, and $\deg r < n$.

Existence in the general case. Set $f' = \sum_{i=0}^{n} a_i x^i$. Then $\|f - f'\| < \|f'\|$, and f' is a polynomial with $\deg(f') = n =$ pseudo-degree(f'). Thus the above special case applies to f'. Set $g^{(0)} = g$, and given $g^{(\nu)}$ define $g^{(\nu+1)}$, q_ν and r_ν by

$$g^{(\nu)} = q_\nu f' + r_\nu = q_\nu f + r_\nu + g^{(\nu+1)}$$

where $\deg r_\nu < n$. Thus

$$\begin{aligned}\|g^{(\nu+1)}\| &= \|q_\nu\| \cdot \|f - f'\| = \|q_\nu\| \cdot \|f'\| \cdot \frac{\|f - f'\|}{\|f'\|} \\ &\le \|g^{(\nu)}\| \cdot \frac{\|f - f'\|}{\|f'\|}.\end{aligned}$$

Since $\frac{\|f-f'\|}{\|f'\|} < 1$ it follows that $\lim g^{(\nu)} = 0$. Then also $\lim q_\nu = 0 = \lim r_\nu$. Hence $q = \sum_{\nu=0}^{\infty} q_\nu$ and $r = \sum_{\nu=0}^{\infty} r_\nu$ are well-defined elements of $R\{\{x\}\}$ with $\deg r < n$. Clearly $g = qf + r$. □

Corollary 11.4 *Let $f \in R\{\{x\}\}$ be regular of pseudo-degree n. Then $f = p \cdot u$ where u is a unit of $R\{\{x\}\}$ and $p \in R[x]$ a monic polynomial of degree n with $\|p\| = 1$.*

Proof. By Theorem 11.3 we have $x^n = qf + r$ for (unique) $q, r \in R\{\{x\}\}$ with $\deg(r) < n$ and $\|r\| \le \|x^n\| = 1$. Then $p := x^n - r$ is monic of degree n and $\|p\| = 1$. Further $p = qf$. It remains to be shown that q is a unit of $R\{\{x\}\}$.

Note that p is regular of pseudo-degree n. Thus $f = q'p + r'$ for certain $q', r' \in R\{\{x\}\}$ with $\deg(r') < n$. Then $f = q'qf + r'$. But $f = 1f + 0$ as well. By the uniqueness of the decomposition $f = q'p + r'$ it follows that $q'q = 1$, hence q is a unit of $R\{\{x\}\}$. □

11.2 Rings of Converging Power Series

For the rest of this chapter, we fix a field k that is complete with respect to a nontrivial ultrametric absolute value $|\cdot|$. Here are our two main examples.

Example 11.5 Let $k = k_0((t))$, the field of formal Laurent series over any field k_0, or $k = \mathbb{Q}_p$, the field of p-adic numbers for some prime number p. We can represent the elements λ of k by expressions of the form $\sum_{i=n}^{\infty} a_i t^i$ (resp., $\sum_{i=n}^{\infty} a_i p^i$) with $a_i \in k_0$ (resp., $a_i \in \{0, 1, \ldots, p-1\}$). Addition and multiplication of these elements is defined so as to extend naturally those operations on the corresponding finite sums $\sum_{i=0}^{m} a_i t^i$ (resp., $\sum_{i=0}^{m} a_i p^i$), which represent polynomials in t (resp., integers in p-adic notation). Choose real s with $0 < s < 1$, and define

$$|\lambda| = s^{v(\lambda)}$$

for $\lambda \neq 0$, where $v(\lambda)$ is the smallest integer i with $a_i \neq 0$. It is easy to check that this satisfies the axioms from Section 11.1.1. (The dependence on s is not essential.)

11.2.1 The Basic Set-Up

The following notation will be fixed throughout this chapter.

$$A_1 = k\{\{x\}\} = \left\{ \sum_{i=0}^{\infty} a_i x^i : a_i \in k, \quad a_i \to 0 \right\};$$

$$Q_1 = \text{ field of fractions of } A_1$$

$$A_2 = k\{\{x^{-1}\}\} = \left\{ \sum_{i=0}^{\infty} a_i x^{-i} : a_i \in k, \quad a_i \to 0 \right\};$$

$$Q_2 = \text{ field of fractions of } A_2$$

$$A = \left\{ \sum_{i=-\infty}^{\infty} a_i x^i : a_i \in k, \quad a_i \to 0 \right\}; \qquad Q = \text{ field of fractions of } A$$

In the latter case, $a_i \to 0$ means that $\lim_{i\to\infty} a_i = 0 = \lim_{i\to-\infty} a_i$. The expression $\sum_{i=-\infty}^{\infty} a_i x^i$ is a formal notation. We will shortly see that this expression can actually be interpreted as a converging sum. But for the moment, we just identify it with the sequence $(a_i)_{i=-\infty}^{\infty}$ of coefficients. This makes A into a k-vector space (via componentwise operations). Now we define the multiplication on A.

Proposition 11.6 *Let $f = \sum_{i=-\infty}^{\infty} a_i x^i$, $g = \sum_{j=-\infty}^{\infty} b_j x^j$ in A.*

(a) Define multiplication on A by $f \cdot g = \sum_{n=-\infty}^{\infty} c_n x^n$, where

$$c_n = \sum_{i+j=n} a_i b_j = \sum_{i=-\infty}^{\infty} a_i b_{n-i}.$$

This is well defined and makes A into an integral domain (and k-algebra).

(b) Set

$$\|f\| = \max_{i\in\mathbb{Z}} |a_i|.$$

This extends the absolute value from k to A, and A is complete with respect to this absolute value.

(c) *View x naturally as an element of A (with $a_i = 0$ for $i \neq 1$, and $a_1 = 1$). This embeds the ring $k[x, x^{-1}]$ into A, as the set of finite sums $\sum_{i=n}^{m} a_i x^i$. Now each element $\sum_{i=-\infty}^{\infty} a_i x^i$ of A can be interpreted as a converging sum, that is, limit of the partial sums $\sum_{i=-n}^{n} a_i x^i$. In particular, $k[x, x^{-1}]$ is dense in A (i.e., for each $f \in A$ and $\epsilon > 0$ there is $g \in k[x, x^{-1}]$ with $\|f - g\| < \epsilon$).*

Proof. By (2) of Section 11.1.1 the series defining c_n converges, and $c_n \to 0$ as $n \to \pm\infty$. Thus $f \cdot g$ is a well-defined element of A. One checks easily that this makes A into an (associative) k-algebra.

One proves as in Section 11.1.2 that $\|fg\| = \|f\| \cdot \|g\|$. First, this implies that A is a domain: If $fg = 0$ then $0 = \|fg\| = \|f\| \cdot \|g\|$, hence $f = 0$ or $g = 0$. Second, it follows that $\|\cdot\|$ is an ultrametric absolute value on A. The completeness is a routine check (as in Section 11.1.2). This proves (a) and (b). (c) is easy. □

Remark 11.7 For each $f = \sum_{i=-\infty}^{\infty} a_i x^i \in A$ and $c \in k$ with $|c| = 1$ the sum $f(c)$ converges in k. (Indeed, both $\sum_{i=0}^{\infty} a_i c^i$ and $\sum_{i=1}^{\infty} a_{-i} c^{-i}$ converge; see Lemma 11.1.) Hence we can view A as a ring of functions on $\{c \in k : |c| = 1\}$, the "ring of holomorphic functions" on this set, in the sense of rigid geometry, see [FP]. Similarly, A_1 (resp., A_2) is the "ring of holomorphic functions" on $\{c \in k : |c| \leq 1\}$ (resp., on $\{c \in k : |c| \geq 1\} \cup \{\infty\}$). We will not make use of this in our formal development, but it is the underlying geometric motivation.

We view A_1 and A_2 as subsets of A in the natural way. This is compatible with the ring structure and absolute value on $A_1 = k\{\{x\}\}$ and $A_2 = k\{\{x^{-1}\}\}$ from Section 11.1.2. Thus A_1 and A_2 are complete subrings of A.

11.2.2 Structure of the Rings A, A_1, and A_2

Lemma 11.8 *Each $f \in A$ (resp., A_1) can be written as $f = pu$ with $p \in k[x]$ and u a unit of A (resp., A_1).*

Proof. For $A_1 = k\{\{x\}\}$ the claim follows from Corollary 11.4 (applied for $R = k$). Now let $f = \sum_{i=-\infty}^{\infty} a_i x^i \in A$. We may assume $f \neq 0$, and $-1 = \min(i : |a_i| = \|f\|)$ (after multiplying f by a power of x, which is a unit of A).

Now we set $R = A_1$, and introduce a new variable w over R. Consider the ring $R\{\{w\}\}$ of power series $\sum_{j=0}^{\infty} \alpha_j w^j$ with $\alpha_j \in R$ and $\|\alpha_j\| \to 0$. Setting $\alpha_0 = \sum_{i=0}^{\infty} a_i x^i$ and $\alpha_j = a_{-j}$ for $j > 0$ we obtain an element $\hat{f}$ of $R\{\{w\}\}$

that is regular of pseudo-degree 1 (see Section 11.1.4). By Corollary 11.4 we have $\hat{f} = \hat{p}\hat{u}$, where $\hat{u}$ is a unit of $R\{\{w\}\}$ and $\hat{p} = w + \beta$ for some $\beta \in R$.

Since $\|x^{-1}\| = 1$ the series $F(x^{-1}) = \sum_{j=0}^{\infty} \beta_j x^{-j}$ converges in A for each $F = \sum_{j=0}^{\infty} \beta_j w^j \in R\{\{w\}\}$. Thus we have the evaluation map $\theta : R\{\{w\}\} \to A$, $F \mapsto F(x^{-1})$. This map is a ring homomorphism (by Lemma 11.1(ii)), hence maps $\hat{u}$ to a unit u' of A. Thus $f = \theta(\hat{f}) = \theta(\hat{p})\theta(\hat{u}) = (x^{-1} + \beta)u' = (1 + x\beta)x^{-1}u'$. Replacing f by $f' = 1 + x\beta \in R = A_1$ reduces us to the case that $f \in A_1$. But this case has already been dealt with. □

Theorem 11.9 *The rings A, A_1, and A_2 are principal ideal domains. Each ideal is generated by an element of $k[x, x^{-1}]$.*

Proof. By the lemma, each ideal I of A is generated by $I' = I \cap k[x]$. This I' is an ideal of $k[x]$, hence $I' = k[x]p$ for some $p \in k[x]$ (since $k[x]$ is a principal ideal domain). Thus $I = Ap$ is a principal ideal. This proves the claim for A.

The same argument works for A_1. The case of A_2 follows by using the isomorphism $A_1 \to A_2$ that sends x to x^{-1}. □

Remark 11.10 The classes of prime elements of A_1 (modulo multiplication by units) are represented by the monic irreducible polynomials $p(x) \in k[x]$ with $\|p\| = 1$ (i.e., all coefficients of $p(x)$ having absolute value ≤ 1). Similarly, for A_2 take the $p(x^{-1})$, with p as before. We leave this sharpening of Lemma 11.8 as an exercise, as well as the corresponding statement for A. (Follows essentially from Corollary 11.4.) In the language of valuation theory, this means that there is a 1-1 correspondence between the places of Q_1 (resp., Q_2) integral on A_1 (resp., A_2), and those places of $k(x)$ of "absolute value" ≤ 1 (resp., ≥ 1).

Recall that Q_i is the field of fractions of A_i (for $i = 1, 2$). We view it as embedded into Q, the field of fractions of A.

Corollary 11.11 *The intersection of Q_1 and Q_2 inside Q equals $k(x)$.*

Proof. We have $k[x] \subset A_1$ and $k[x^{-1}] \subset A_2$, hence $k(x) \subset Q_1 \cap Q_2$. For the converse, let $f \in Q_1 \cap Q_2$. Then $f = f_1/p_1 = f_2/p_2$ with $f_i \in A_i$ and $0 \neq p_1 \in k[x]$ and $0 \neq p_2 \in k[x^{-1}]$ (by Lemma 11.8 and the above isomorphism $A_1 \to A_2$). There are $n, m \in \mathbb{N}$ such that $x^n p_2 \in k[x]$ and $x^{n-m} p_1 \in k[x^{-1}]$. Then the element $g = (x^n p_2) f_1 = x^m (x^{n-m} p_1) f_2$ lies in A_1, and $x^{-m} g$ lies in A_2. Clearly this implies that $g \in k[x]$ (of degree $\leq m$). Thus $f = f_1/p_1 = g/(x^n p_2 p_1) \in k(x)$. □

11.2.3 The Embedding of A into $k[[x - c]]$

Lemma 11.12 *Let $c \in k$ with $|c| = 1$. Then there is a k-algebra embedding $A \to k[[t]]$ mapping $x - c$ to t.*

Proof. The idea is to develop each $f \in A$ into its Taylor series around c. In order to work in the realm of convergent series, we use a little detour.

Let t be a new variable over k, and consider the complete ring $k\{\{t\}\}$ as subring of $k[[t]]$ in the usual way. Choose $b \in k$ with $0 < |b| < 1$. Then $1 + c^{-1}bt$ is a unit of $k\{\{t\}\}$ since $\|c^{-1}bt\| = |b| < 1$ (see (5) of Section 11.1.1). Then also $c(1 + c^{-1}bt) = c + bt$ is a unit of $k\{\{t\}\}$, hence $(c + bt)^i$ lies in $k\{\{t\}\}$ for each integer i. Further, $\|c + bt\| = \max(|c|, |b|) = 1$.

Now consider $f = \sum_{i=-\infty}^{\infty} a_i x^i \in A$. Then the series

$$f_{b,c} = \sum_{i=-\infty}^{\infty} a_i \, (bt + c)^i$$

converges in $k\{\{t\}\}$, since $\|a_i(bt + c)^i\| = |a_i| \cdot \|bt + c\|^i = |a_i| \to 0$. This defines a map $A \to k\{\{t\}\}$, $f \mapsto f_{b,c}$. This map is clearly k-linear, and using (2) of Section 11.1.1 we see it is a k-algebra homomorphism. Further, it maps x to $bt + c$.

The map

$$\sum_{i=0}^{\infty} \alpha_i t^i \mapsto \sum_{i=0}^{\infty} (\alpha_i b^{-i}) t^i$$

is a k-algebra automorphism of $k[[t]]$. It sends $bt + c$ to $t + c$. Composing this map with the map $f \mapsto f_{b,c}$ we obtain a k-algebra homomorphism $\Phi : A \to k[[t]]$ that sends x to $t + c$.

It remains to check that Φ is injective. Its kernel is an ideal of A, hence of the form Ap for some $p \in k[x]$ (Theorem 11.9). But for $p(x) \in k[x]$ we have $\Phi(p) = p(t + c)$, which is zero if and only if $p(x) = 0$. □

11.3 GAGA

The term GAGA has been coined by Serre to mean the interplay between analytic geometry and algebraic geometry. It can be seen as a far-reaching generalization of (the analytic version of) Riemann's existence theorem. Harbater used "formal GAGA" for his Galois realizations, and the proof of Serre and Liu uses "rigid GAGA." The goal of this section is to prove an elementary version of rigid GAGA (Theorem 11.17).

11.3.1 Cartan's Lemma

Lemma 11.13 *Let M be an associative ring with 1. Suppose $\|\cdot\| : M \to \mathbb{R}$ is a function (**norm**) such that for all $\Pi, \Omega \in M$ the following hold:*

(i) $\|\Pi\| \geq 0$, *and* $\|\Pi\| = 0$ *if and only if* $\Pi = 0$. *Further,* $\|1\| = \|-1\| = 1$.
(ii) $\|\Pi\Omega\| \leq \|\Pi\| \cdot \|\Omega\|$.
(iii) $\|\Pi + \Omega\| \leq \max(\|\Pi\|, \|\Omega\|)$.
(iv) *M is complete, that is, every Cauchy sequence (with respect to this norm) converges.*

Here convergence of sequences (with respect to the norm $\|\cdot\|$) is defined in the usual way. If $(\Pi_j)_{j=1}^{\infty}$ is a sequence in M with $\|\Pi_j\| \to 0$ and $c :=$ max $\|\Pi_j\| < 1$ then the partial products $\Omega_n := (1 + \Pi_1) \cdot \ldots \cdot (1 + \Pi_n)$ converge to some unit Ω of M.

Proof. We have $\|\Omega_n\| \leq \|1 + \Pi_1\| \cdot \ldots \cdot \|1 + \Pi_n\| \leq 1$. Thus

$$\|\Omega_n - \Omega_{n-1}\| = \|\Omega_{n-1}\Pi_n\| \leq \|\Omega_{n-1}\| \cdot \|\Pi_n\| \leq \|\Pi_n\|$$

which goes to zero. Hence the Ω_n converge to some $\Omega \in M$ (by completeness and the ultrametric inequality (iii)). Further

$$\begin{aligned} \|1 - \Omega_n\| &= \left\| \sum_{i=1}^{n} (\Omega_{i-1} - \Omega_i) \right\| \leq \max \|\Omega_{i-1} - \Omega_i\| \\ &= \max \|\Omega_{i-1}\Pi_i\| \leq \max \|\Pi_i\| \\ &= c. \end{aligned}$$

(Here $\Omega_0 = 1$.) By continuity, $\|1 - \Omega\| \leq c < 1$ as well. This implies that Ω is a unit (by the usual argument using the geometric series; see (5) of Section 11.1.1). □

Lemma 11.14 (Cartan's Lemma) *Let M be as in the preceding lemma. Let M_1 and M_2 be complete subrings of M. Suppose that for each $\Delta \in M$ there are $\Delta^+ \in M_1$ and $\Delta^- \in M_2$ with $\Delta = \Delta^+ + \Delta^-$ and $\|\Delta^+\|, \|\Delta^-\| \leq \|\Delta\|$. Then each $\Lambda \in M$ with $\|\Lambda - 1\| < 1$ can be written as $\Lambda = \Lambda_1\Lambda_2$ where Λ_i is a unit of M_i (for $i = 1, 2$).*

Proof. Set $\Delta_1 = \Lambda - 1$ and $c = \|\Delta_1\|$. Then $0 \le c < 1$. The condition

$$1 + \Delta_{j+1} = (1 - \Delta_j^+)(1 + \Delta_j)(1 - \Delta_j^-)$$

defines recursively a sequence $(\Delta_j)_{j=1}^{\infty}$ in M. From

$$\Delta_{j+1} = \Delta_j^+ \Delta_j^- - \Delta_j^+ \Delta_j - \Delta_j \Delta_j^- + \Delta_j^+ \Delta_j \Delta_j^-$$

it follows that $\|\Delta_{j+1}\| \le \|\Delta_j\|^2$. It follows that all $\|\Delta_j\| \le c < 1$, and $\|\Delta_j\| \to 0$. Further,

$$1 + \Delta_{j+1} = \left(1 - \Delta_j^+\right) \cdots \left(1 - \Delta_1^+\right) \Lambda \left(1 - \Delta_1^-\right) \cdots \left(1 - \Delta_j^-\right). \qquad (11.1)$$

We have $\|\Delta_j^-\| \le \|\Delta_j\| \le c < 1$ and $\|\Delta_j^-\| \to 0$. Hence by the preceding lemma, the partial products $(1 - \Delta_1^-) \cdots (1 - \Delta_j^-)$ converge to some unit Λ_2' of M_2. Similarly, the $(1 - \Delta_j^+) \cdots (1 - \Delta_1^+)$ converge to some unit Λ_1' of M_1. Passing to the limit in (11.1) we get $1 = \Lambda_1' \Lambda \Lambda_2'$. Hence the claim. □

Now we return to the study of the rings A, A_1, and A_2 from Section 11.2.1. Fix $n \in \mathbb{N}$, and let $M = M_n(A)$ be the ring of $n \times n$-matrices over A. For $\Lambda = (\lambda_{ij}) \in M$ define

$$\|\Lambda\| = \max \|\lambda_{ij}\|.$$

This satisfies the axioms from Lemma 11.13. Further, $M_i := M_n(A_i)$ is a complete subring of M for $i = 1, 2$. Each $\Delta \in M$ admits a decomposition $\Delta = \Delta^+ + \Delta^-$ as in Cartan's Lemma: Indeed, it suffices to prove this for $n = 1$, that is, $M = A$. (Apply it then to each matrix entry.) For $f = \sum_{i=-\infty}^{\infty} a_i x^i \in A$, set $f^+ = \sum_{i=0}^{\infty} a_i x^i \in A_1$, and $f^- = \sum_{i=1}^{\infty} a_{-i} x^{-i} \in A_2$. This has the desired properties.

Recall that Q_i is the field of fractions of A_i. Further, $\mathrm{GL}_n(A)$ is the group of units of the ring $M = M_n(A)$.

Corollary 11.15 *Let $\Lambda \in GL_n(A)$. Then there are $\Lambda_i \in GL_n(Q_i) \cap GL_n(A)$ (for $i = 1, 2$) such that $\Lambda = \Lambda_1 \Lambda_2$.*

Proof. As $k[x, x^{-1}]$ is dense in A (see Proposition 11.6), there is a matrix $\Pi \in M$ with coefficients in $k[x, x^{-1}]$ such that $\|\Pi - \Lambda^{-1}\| < \frac{1}{\|\Lambda\|}$. Then $\|\Lambda\Pi - 1\| = \|\Lambda(\Pi - \Lambda^{-1})\| \le \|\Lambda\| \cdot \|\Pi - \Lambda^{-1}\| < 1$. Hence by Cartan's Lemma there are $\Lambda_1 \in \mathrm{GL}_n(A_1)$ and $\tilde{\Lambda}_2 \in \mathrm{GL}_n(A_2)$ with $\Lambda\Pi = \Lambda_1 \tilde{\Lambda}_2$. Thus

$\Pi \in \mathrm{GL}_n(A)$, in particular $\det(\Pi) \neq 0$; since Π has coefficients in $k[x, x^{-1}] \subset Q_2$, we get $\Pi \in \mathrm{GL}_n(Q_2) \cap \mathrm{GL}_n(A)$.

Finally, $\Lambda = \Lambda_1\Lambda_2$, where $\Lambda_2 := \tilde{\Lambda}_2\Pi^{-1} \in \mathrm{GL}_n(Q_2) \cap \mathrm{GL}_n(A)$. Also, $\Lambda_1 \in \mathrm{GL}_n(A_1) \subset \mathrm{GL}_n(Q_1) \cap \mathrm{GL}_n(A)$. Hence the claim. □

11.3.2 Induced Algebras

In this subsection we collect some basic facts about algebras with group action. The proofs are very easy and are usually omitted.

11.3.2.1 Semisimple Algebras

Let M be a commutative (and associative) finite-dimensional algebra with 1 over a field K. We say M is **semi-simple** if it is the direct sum of finite field extensions of K. It is easy to see that M is semi-simple if and only if the ideal generated by all the minimal ideals of M equals M. An **idempotent** of M is a nonzero element ϵ with $\epsilon^2 = \epsilon$. We call ϵ **minimal** if it is not the sum of two idempotents ϵ', ϵ'' with $\epsilon'\epsilon'' = 0$.

Suppose $M = L_1 \oplus \cdots \oplus L_m$ is the direct sum of the finite field extensions L_j of K. Write the elements of M as tuples $u = (u_1, \ldots, u_m)$ with $u_j \in L_j$. (Those tuples add and multiply componentwise.) Such u is an idempotent if and only if all $u_j \in \{0, 1\}$, but not all $= 0$. Thus the minimal idempotents of M are exactly the "standard basis vectors" having one entry $= 1$ and all other entries $= 0$. Hence any two distinct minimal idempotents of M have product zero. More generally, if ϵ is a minimal idempotent and ϵ' any idempotent, then either $\epsilon\epsilon' = 0$ or $\epsilon\epsilon' = \epsilon$. Finally, each nonzero ideal of M is generated by a unique idempotent; the ideal is minimal if and only if the idempotent is.

11.3.2.2 Base Change

Let K'/K be a field extension. If M is a finite-dimensional K-algebra with 1 then $M' = M \otimes_K K'$ is a K'-algebra in a natural way (where K' acts on M' through its isomorphic copy $1 \otimes K'$). M embeds into M' as $M \otimes 1$, and each K-algebra automorphism of M extends uniquely to a K'-algebra automorphism of M'. Hence if M is a K-algebra with G-action for some group G, then M' is a K'-algebra with G-action in a natural way. Here $(M')^G$ (the space of G-fixed points) equals $K'M^G$; in particular, if $M^G = K$ (canonically embedded into M as $K \cdot 1$) then $(M')^G = K'$.

If M is semi-simple then so is M', provided M is a direct sum of *separable* field extensions of K. (Indeed, if M is a separable field extension of K then $M \cong K[y]/(f)$ for a separable polynomial $f \in K[y]$, and then f is also separable over K', hence is a product of distinct irreducibles f_j and we have $M' \cong K'[y]/(f) \cong \bigoplus_j K'[y]/(f_j)$.)

11.3.2.3 Induced Algebras

Let G be a finite group with subgroup H. Let L be a field with subfield K. Suppose H acts on L via K-algebra automorphisms.

Let M_0 be an L-vector space with basis ϵ_g $(g \in G)$. Make it into an L-algebra (isomorphic to the direct sum of $n = |G|$ copies of L) by requiring that the ϵ_g are the minimal idempotents. The group G acts on M_0 through L-algebra automorphisms, by the rule that $g' \in G$ sends ϵ_g to $\epsilon_{g'g}$.

The map $\Phi : L \to M_0$ mapping $u \in L$ to $\sum_{h \in H} h^{-1}(u)\epsilon_h$ is an H-equivariant K-linear multiplicative injection (but in general $\Phi(1) \neq 1$, hence we don't call it a K-algebra map). Here H-equivariant means that $\Phi(h(u)) = h(\Phi(u))$ for all $u \in L$ and $h \in H$. Thus $L_1 := \Phi(L)$ is a K-subspace of M_0 invariant under H, and is a field with multiplicative identity $\Phi(1)$. Clearly, there are exactly $m = [G : H]$ conjugates $L_1, \ldots, L_m$ of L_1 under G, and their sum M is a semi-simple K-algebra that is the direct sum of the fields $L_1, \ldots, L_m$. The group G acts on M through K-algebra automorphisms, permuting $L_1, \ldots, L_m$ transitively such that H is the stabilizer of L_1. Further, L_1 can be identified with L via the map Φ, and under this identification, the action of H on L_1 yields the original action of H on L.

Definition 11.16 *We call $M = \mathrm{Ind}_H^G(L)$ the K-algebra with G-action induced from the K-algebra L with H-action.*

Note that G permutes the minimal idempotents of M transitively (since they correspond to the minimal ideals $L_1, \ldots, L_m$ of M). The subgroup H is the stabilizer of the minimal idempotent $\Phi(1)$.

11.3.2.4

We are going to apply this in the case where L is Galois over K, and the action of H on L yields an isomorphism $H \to G(L/K)$. Then $M^G = K$ and $\dim_K(M) = |G|$. Further, if I is a nonzero G-invariant ideal of M then $I = M$. (Indeed, the unique idempotent generating I lies in $M^G = K$, hence equals 1.)

11.3.3 An Elementary Version of 1-Dimensional Rigid GAGA

Theorem 11.17 *Let G be a finite group generated by subgroups G_1 and G_2. For $i = 1, 2$, let P_i be a Galois extension of the field Q_i (from Section 11.2.1)* *with group isomorphic to G_i. Fix such an isomorphism, and form accordingly the induced Q_i-algebra with G-action $N_i = Ind_{G_i}^G(P_i)$. Let e_i be a minimal idempotent of N_i whose stabilizer in G is G_i. Fix a minimal idempotent $e^{(i)}$ of $N_i \otimes_{Q_i} Q$ with $e^{(i)}e_i = e^{(i)}$. (Here and in the following we view N_i as naturally embedded into $N_i \otimes_{Q_i} Q$, and we view the latter as Q-algebra with G-action.) Consider the following conditions:*

1. *For $i = 1, 2$, there is a basis of P_i over Q_i whose A-span in $P^{(i)} := P_i \otimes_{Q_i} Q$ equals the A-span of the idempotents of $P^{(i)}$.*
2. *There is an isomorphism $\theta : N_1 \otimes_{Q_1} Q \to N_2 \otimes_{Q_2} Q$ of Q-algebras with G-action, mapping $e^{(1)}$ to $e^{(2)}$. Further, there is a basis $\alpha_1, \dots, \alpha_n$ of N_1 over Q_1, and a basis $\beta_1, \dots, \beta_n$ of N_2 over Q_2, such that θ maps the A-span of $\alpha_1, \dots, \alpha_n$ onto the A-span of $\beta_1, \dots, \beta_n$.*

The following holds:

(a) Condition (1) implies (2).
(b) Given (2), identify $N := N_1 \otimes_{Q_1} Q$ with $N_2 \otimes_{Q_2} Q$ via the given isomorphism, and form the intersection $F = N_1 \cap N_2$ (inside N). Then F is a Galois field extension of $k(x)$ with group G. Each basis of F over $k(x)$ is a basis of N_1 over Q_1 as well as a basis of N_2 over Q_2.

Proof. We first prove (b). Clearly, G acts on F with fixed points $F^G = N_1^G \cap N_2^G = Q_1 \cap Q_2 = k(x)$ (see Section 11.3.2.4 and Corollary 11.11). First we show that F is "large enough," more precisely, G acts faithfully on F. Set $\alpha = (\alpha_1, \dots, \alpha_n)$ and $\beta = (\beta_1, \dots, \beta_n)$.

Claim 1 We may assume $\alpha = \beta$.

Condition (2) yields that $A\alpha_1 \oplus \cdots \oplus A\alpha_n = A\beta_1 \oplus \cdots \oplus A\beta_n$ (under the identification via θ). Hence there is a matrix $\Lambda \in \mathrm{GL}_n(A)$ with $\alpha\Lambda = \beta$. Write $\Lambda = \Lambda_1\Lambda_2$ as in Corollary 11.15. Then $\alpha\Lambda_1 = \beta\Lambda_2^{-1}$ is a basis for N_1 over Q_1 as well as a basis for N_2 over Q_2, satisfying condition (2). This proves Claim 1.

Assume from now on that $\alpha = \beta$. Then $\alpha_1, \dots, \alpha_n \in F$, hence G acts faithfully on F. Further, $\alpha_1, \dots, \alpha_n$ form a basis for F over $k(x)$. Indeed, any $\gamma \in N$ can uniquely be written as $\gamma = h_1\alpha_1 + \cdots + h_n\alpha_n$ with $h_j \in Q$. Then $\gamma \in N_i$ if and only if all $h_j \in Q_i$. Hence $\gamma \in F$ if and only if all $h_j \in Q_1 \cap Q_2 = k(x)$.

Assertion (b) is proved once we have shown:

Claim 2 F is a field.

F is a finite-dimensional algebra over the field $k(x)$, hence has a minimal (nonzero) ideal. The sum of all minimal ideals of F is a G-invariant ideal

$I \neq 0$ of F. Then $Q_1 I$ is a G-invariant ideal $\neq 0$ of N_1, hence equals N_1 (see Section 11.3.2.4). This implies $I = F$ (because $N_1 \cong F \otimes_{k(x)} Q_1$). It follows that F is semi-simple (see Section 11.3.2.1).

Let e be our fixed minimal idempotent $e^{(1)}$ of N (identified with $e^{(2)}$ via the isomorphism θ). There is a minimal idempotent e_F of F with $e_F e \neq 0$ (since the sum of all minimal idempotents of F is 1). Then $e_F e = e$. Since $e_i e = e$ (by construction), it follows that $e_F e_i = e_i$. If e'_F is another minimal idempotent of F then $e'_F e_i = e'_F e_F e_i = 0 e_i = 0$. Thus e_F is the unique minimal idempotent of F with $e_F e_i = e_i$. It follows that $G_i = \mathrm{Stab}_G(e_i) \subset \mathrm{Stab}_G(e_F)$. Hence e_F is fixed by $G = \langle G_1, G_2 \rangle$.

Now e_F is an idempotent in $N_1^G = Q_1$, hence $e_F = 1$. Thus 1 is a minimal idempotent of F, which means that F is a field. This proves Claim 2, and thereby assertion (b).

(a). Let $i = 1, 2$. The Q_i-linear G_i-equivariant multiplicative injection $P_i \to N_i$ (see Section 11.3.2.3) extends to a Q-linear G_i-equivariant multiplicative injection $P^{(i)} \to N^{(i)} := N_i \otimes_{Q_i} Q$. Since N_i is the direct sum of the G-conjugates of P_i, $N^{(i)}$ is also the direct sum of the G-conjugates of $P^{(i)}$.

Condition (1) implies that $P^{(i)}$ is the Q-span of its idempotents. By the preceding paragraph, it follows that $N^{(i)}$ is also the Q-span of its idempotents. Thus the minimal idempotents form a Q-basis of $N^{(i)}$. In particular, their number equals $\dim N^{(i)} = |G|$.

We have $N_i^G = Q_i$ by Section 11.3.2.4, hence $(N^{(i)})^G = Q$ by Section 11.3.2.2. The sum over (the distinct elements in) a G-orbit of minimal idempotents of $N^{(i)}$ is a G-invariant idempotent, hence lies in $(N^{(i)})^G = Q$ and thus equals 1. Hence G acts transitively on the set of minimal idempotents of $N^{(i)}$. Since their number equals $|G|$, it follows that G acts regularly on this set (i.e., transitively with trivial stabilizers). It follows that we get a G-equivariant bijection between the minimal idempotents of $N^{(1)}$ and those of $N^{(2)}$ by mapping $g(e^{(1)})$ to $g(e^{(2)})$ for each $g \in G$. Here $e^{(i)}$ is the fixed minimal idempotent of $N^{(i)}$ from the theorem. This bijection on minimal idempotents extends uniquely to a G-equivariant Q-algebra isomorphism $\theta : N^{(1)} \to N^{(2)}$.

By condition (1), each $g(P_i)$ ($g \in G$) has a Q_i-basis whose A-span in $g(P^{(i)})$ is the A-span of the idempotents in $g(P^{(i)})$. Putting those bases together as $g(P_i)$ runs over the distinct G-conjugates of P_i (i.e., as g runs over a system of representatives for the left cosets of G_i in G) yields a Q_i-basis of N_i whose A-span in $N^{(i)}$ is the A-span of the idempotents in $N^{(i)}$. Now clearly the A-span of the idempotents in $N^{(1)}$ and $N^{(2)}$, respectively, correspond under θ. This proves (2). $\square$

Remark 11.18 Suppose P_i/Q_i ($i = 1, 2$) is a finite Galois extension with group G_i, satisfying condition (1). Then $P^{(i)}$ is a direct sum of copies of Q, hence has a Q-algebra surjection to Q. This map embeds P_i into Q, so that we may view P_i as subfield of Q (containing Q_i).

If G is any finite group generated by isomorphic copies of G_1 and G_2 then the extension $F/k(x)$ with group G constructed in the theorem embeds into $P_1 \cap P_2$. Indeed, N_i is a direct sum of copies of P_i (as Q_i-algebra), and projecting to such a copy yields a $k(x)$-embedding of F into P_i. Since $F/k(x)$ is Galois, there is at most one extension of $k(x)$ inside Q that is $k(x)$-isomorphic to F. Hence F embeds into $P_1 \cap P_2$.

Conversely, let $\tilde{F}$ be a finite Galois extension of $k(x)$ inside $P_1 \cap P_2$, with group $\tilde{G}$. Then G_1 and G_2 leave $\tilde{F}$ invariant, hence induce subgroups $\tilde{G}_1$ and $\tilde{G}_2$ of $\tilde{G}$. These subgroups generate $\tilde{G}$ (since the fixed field of $\langle \tilde{G}_1, \tilde{G}_2 \rangle$ in $\tilde{F}$ is contained in $P_1^{G_1} \cap P_2^{G_2} = Q_1 \cap Q_2 = k(x)$).

To formulate the result suggested by these remarks, we need

Lemma 11.19 *Let $n \in \mathbb{N}$. Suppose the finite group G is generated by subgroups G_1 and G_2. We say G is n-**freely generated** by G_1 and G_2 if the following holds:*

(i) If H is any group of order $\leq n$ then each pair of homomorphisms $G_1 \to H$ and $G_2 \to H$ extends to a homomorphism $G \to H$.
(ii) The intersection of all normal subgroups of G of index $\leq n$ is trivial.

For any given finite groups G_1 and G_2, and $n \geq |G_1| \cdot |G_2|$, there is a finite group G that is n-freely generated by (isomorphic copies of) G_1 and G_2. If G' is a group satisfying (ii), and generated by homomorphic images G'_1 and G'_2 of G_1 and G_2, respectively, then there is a surjective homomorphism $G \to G'$ extending the given maps $G_i \to G'_i$. In particular, if $G = \langle G'_1, G'_2 \rangle$, where G'_i is a homomorphic image of G_i, then there is an automorphism of G mapping G_i to G'_i.

Proof. The proof is analogous to that of Lemma 7.1, replacing the free group $\mathcal{F}_s$ by the free product $\mathcal{F}$ of G_1 and G_2. Note that if $n \geq |G_1| \cdot |G_2|$ then the quotient G of $\mathcal{F}$ by the intersection of all normal subgroups of index $\leq n$ still maps onto $G_1 \times G_2$, hence the images of G_1 and G_2 in G are isomorphic images. □

Corollary 11.20 *Let P_i/Q_i ($i = 1, 2$) be a finite Galois extension inside Q with group G_i, satisfying condition (1) of Theorem 11.17. Let K_n be the composite of all Galois extensions of $k(x)$ inside $P_1 \cap P_2$ having degree $\leq n$ (where $n \in \mathbb{N}$*

with $n \geq |G_1| \cdot |G_2|$). Then restriction embeds G_i into $H = G(K_n/k(x))$, and H is n-freely generated by G_1 and G_2.

Proof. The proof is similar to that of Corollary 7.3. Let G_0 be n-freely generated by isomorphic copies of G_1 and G_2. By the theorem and Remark 11.18, there is a Galois extension $K/k(x)$ inside $P_1 \cap P_2$ with group isomorphic to G_0. Thus $G := G(K/k(x))$ is n-freely generated by certain isomorphic copies of G_1 and G_2. Further, G is generated by the restriction images of G_1 and G_2 (see Remark 11.18). By the lemma, it follows that these restriction images are actually isomorphic images, and G is n-freely generated by them. From now on we identify G_i with its restriction image in G. Then G is n-freely generated by G_1 and G_2.

It remains to show that $K = K_n$. Condition (ii) from the lemma means that K is generated by its subextensions of degree $\leq n$. Hence $K \subset K_n$.

Let L be any Galois extension of $k(x)$ of degree $\leq n$, inside $P_1 \cap P_2$. Let K' be the composite of K and L, and $G' = G(K'/k(x))$. Then G' is generated by the restriction images G'_1 and G'_2 of G_1 and G_2, respectively (Remark 11.18). Clearly, G' satisfies condition (ii). Hence it follows from the Lemma that $|G'| \leq |G|$. Thus $K' = K$, that is, $L \subset K$. This proves that $K = K_n$. □

Remark 11.21 The union of the above fields K_n (as $n \to \infty$) is the composite K_∞ of all finite Galois extensions of $k(x)$ inside $P_1 \cap P_2$. It follows that the Galois group of K_∞ over $k(x)$ is the free product of G_1 and G_2 (in the category of profinite groups).

We have noticed a parallel to the classical case (from Section 7.1). One can push this a little further, by also enlarging the groups G_1 and G_2. Then one inductively obtains free Galois groups over $k(x)$, as in the classical case (see Pop's "one-half Riemann existence theorem" [P4]). The difference from the classical case is that here it does not suffice to prescribe the branch points, but we must prescribe the full behavior over the "discs" $\{c \in k : |c| \leq 1\}$ and $\{c \in k : |c| \geq 1\} \cup \{\infty\}$ in order to be able to pin down the resulting infinite Galois group.

If we only prescribe the branch points, then the collection of associated Galois extensions of $k(x)$ may be quite different from the classical case. First, consider the case that k is algebraically closed of characteristic $p > 0$. (There are such fields that are complete with respect to a nontrivial absolute value – e.g., the completion of the algebraic closure of $\mathbb{F}_p((t))$.) Then by Raynaud's solution of Abhyankar's conjecture ([Ra]), each finite group generated by its Sylow p-subgroups occurs as a Galois group over $k(x)$ with a single branch point – hence here we have "more" extensions than in the classical case. (See also the Exercise at the end of Chapter 11). If k is of characteristic 0 and not algebraically closed, then we have less.

11.4 Galois Groups over $k(x)$

11.4.1 Inductive Construction of Galois Extensions

For $c \neq 0$ in k let μ_c be the automorphism of the field $k((x))$ mapping $f(x) = \sum_{i=N}^{\infty} a_i x^i$ to $f(cx) = \sum_{i=N}^{\infty} (a_i c^i)x^i$. Note that μ_c leaves $k(x)$ invariant.

We say a finite Galois extension $L/k(x)$ is A-**normalized** if there is a basis of L over $k(x)$ that yields an A-basis for the A-span of the idempotents of $L \otimes_{k(x)} Q$.

Lemma 11.22 *Let $L/k(x)$ be a finite Galois extension contained in $k((x))$. Then there is $c \neq 0$ in k such that $L' := \mu_c(L)$ lies in Q_1, and the Galois extension $L'/k(x)$ is A-normalized.*

Proof. Choose a primitive element β for $L/k(x)$ that is integral over $k[x]$ (see Lemma 1.3). Let $\beta_1 = \beta, \ldots, \beta_m$ be the conjugates of β over $k(x)$ (where $m = [L : k(x)]$). They all lie in $k[[x]]$. For $i \neq j$ set $\lambda_{ij} = (\beta_i - \beta_j)^{-1}$, a formal Laurent series. By Section 11.1.3 there is $c \neq 0$ in k such that all β_i and all λ_{ij} converge at $x = c$. Then the $\gamma_i := \beta_i(cx)$ lie in $A_1 = k\{\{x\}\}$ (see Section 11.1.2). The $\lambda_{ij}(cx)$ do not necessarily lie in A_1, but they can be viewed as elements of A (see Section 11.2.1). Then the equality $\lambda_{ij}(cx)(\gamma_i - \gamma_j) = 1$ holds in A, hence the $\gamma_i - \gamma_j$ (with $i \neq j$) are units in A. Thus also the element $\delta = \prod_{i \neq j}(\gamma_i - \gamma_j)$ is a unit of A.

Let $\gamma := \gamma_1 = \mu_c(\beta)$. We have $L' = \mu_c(L) = k(x)(\mu_c(\beta)) = k(x)(\gamma) \subset Q_1$. The minimum polynomial of γ over $k(x)$ is $f(y) = \prod_{i=1}^{m}(y - \gamma_i)$. Hence $L'_Q := L' \otimes_{k(x)} Q \cong Q[y]/(f) \cong \bigoplus_i Q[y]/(y - \gamma_i)$ is a direct sum of copies of Q. Thus L'_Q is the Q-span of its minimal idempotents $\epsilon_1, \ldots, \epsilon_m$, and the maps $\pi_\ell : L'_Q \to Q$, $\sum_{i=1}^{m} q_i \epsilon_i \mapsto q_\ell$ are Q-algebra homomorphisms (for $\ell = 1, \ldots, m$).

View L' as naturally embedded into L'_Q. Each γ^j ($j \in \mathbb{N}$) is integral over $k[x]$, hence over A. Thus π_ℓ maps γ^j to an element of Q that is integral over A, hence lies in A since A is integrally closed (being a PID; see Theorem 11.9); It follows that all γ^j lie in $\Sigma := A\epsilon_1 + \cdots + A\epsilon_m$. We are going to show that the elements $1, \gamma, \ldots, \gamma^{m-1}$ form an A-basis of Σ. This implies that $L'/k(x)$ is A-normalized; hence the lemma.

We know that $1, \gamma, \ldots, \gamma^{m-1}$ form a Q-basis of L'_Q. Let ϵ be an idempotent of L'_Q, and write it as $\epsilon = \sum_{j=0}^{m-1} q_j \gamma^j$ with $q_j \in Q$. We have to show $q_j \in A$.

Label the elements of $G(L'/k(x))$ as $g_1, \ldots, g_m$ such that $\gamma_i = g_i(\gamma)$. The g_i extend to Q-algebra automorphisms of L'_Q, hence we get the system of equations

$$g_i(\epsilon) = \sum_{j=0}^{m-1} q_j \, \gamma_i^j, \qquad i = 1, \ldots, m.$$

The matrix $\Gamma = (\gamma_i^j)$ has determinant $\delta = \prod_{i \neq j}(\gamma_i - \gamma_j)$ (Vandermonde determinant). Since δ is a unit in A, the matrix $\Gamma^{-1} = \delta^{-1}\mathrm{adj}(\Gamma)$ has coefficients in Σ (since the γ_i^j lie in Σ). Applying $\Gamma^{-1} = (\gamma'_{ij})$ to the above system of equations yields

$$q_j \cdot 1 = \sum_{i=1}^{m} g_i(\epsilon)\gamma'_{ij}, \qquad j = 0, \ldots, m-1.$$

Since $g_i(\epsilon)$ and γ'_{ij} lie in Σ, also $q_j \cdot 1 \in \Sigma$. Finally, $q_j = \pi_\ell(q_j \cdot 1) \in \pi_\ell(\Sigma) = A$. Thus $q_j \in A$. □

Theorem 11.23 (Harbater) *Let G be a finite group generated by subgroups G_1 and G_2. Suppose for $i = 1, 2$ there exists a Galois extension $F_i/k(x)$ with group isomorphic to G_i, contained in $k((x))$. Then there exists a Galois extension $F/k(x)$ with group isomorphic to G, contained in $k((x))$.*

Proof. By the lemma we may assume that $F_2/k(x)$ is A-normalized and contained in Q_1. Now consider the automorphism of A that maps $f(x) = \sum_{i=-\infty}^{\infty} a_i x^i$ to $f(x^{-1}) = \sum_{i=-\infty}^{\infty} a_{-i} x^i$. It interchanges A_1 and A_2, hence extends to an automorphism of Q interchanging Q_1 and Q_2. Applying this automorphism and the lemma, we may assume that $F_1/k(x)$ is A-normalized and contained in Q_2.

For $i = 1, 2$ let P_i be the composite of F_i and Q_i inside Q. Then P_i is Galois over Q_i, and its Galois group $\tilde{G}_i$ embeds into $G(F_i/k(x))$ via restriction. The fixed field of $\tilde{G}_i$ in F_i equals $F_i \cap P_i^{\tilde{G}_i} \subset Q_j \cap Q_i = k(x)$, where $\{i, j\} = \{1, 2\}$. Hence $\tilde{G}_i \cong G(F_i/k(x))$ via restriction. It follows that $[P_i : Q_i] = [F_i : k(x)]$, hence $P_i \cong F_i \otimes_{k(x)} Q_i$ naturally. Thus $P_i \otimes_{Q_i} Q \cong F_i \otimes_{k(x)} Q$ naturally, and since $F_i/k(x)$ is A-normalized it follows that condition (1) of Theorem 11.17 holds. Hence that theorem yields a Galois extension $F/k(x)$ with group G. We may assume $F \subset Q$ by Remark 11.18.

Lemma 11.12 yields a k-algebra embedding $Q \to k((t))$ mapping $x - 1$ to t. The image of F under this map is a Galois extension of $k(t)$ inside $k((t))$ with group isomorphic to G. □

Remark 11.24 (Ramification data of $F/k(x)$) Here we show how the ramification behavior of $F/k(x)$ is determined by that of F_1 and F_2 (under the normalizations in the above proof). Fix some $i \in \{1, 2\}$. From Theorem 11.17 we have Q_i-algebra isomorphisms $F \otimes_{k(x)} Q_i \cong N_i = \mathrm{Ind}_{G_i}^{G}(P_i)$. Since $P_i \cong F_i \otimes_{k(x)} Q_i$, we get that $F \otimes_{k(x)} Q_i$ has a minimal ideal whose stabilizer in G is G_i, and which has a G_i-equivariant Q_i-isomorphism to $F_i \otimes_{k(x)} Q_i$. Further, $F \otimes_{k(x)} Q_i$ is the direct sum of its minimal ideals, which are all G-conjugate.

(a) Suppose k is a subfield of $\mathbb{C}$. We have F regular over k, hence $F_{\bar{k}} := F \otimes_k \bar{k}$ is a Galois extension of $\bar{k}(x)$ whose Galois group we can canonically identify with $G = G(F/k(x))$. Define the ramification type of $F/k(x))$ to equal that of $F_{\bar{k}}/\bar{k}(x)$ (see Chapter 2). We use the well-known fact (see [Jac],II, Ths. 9.8 and 9.12) that the absolute value on k extends uniquely to an absolute value on $\bar{k}$ relative to which each finite extension of k is complete.

Now the ramification type of $F/k(x)$ is given as follows by that of F_1 and F_2. The branch points of F_1 (resp., F_2) have absolute value <1 (resp., >1, or equal ∞), and together they yield exactly the branch points of F. The conjugacy class $C_{\mathbf{p}}$ of $G = G(F/k(x))$ associated with such a branch point $\mathbf{p}$ contains the corresponding class of G_1 (resp., G_2) if $|\mathbf{p}| < 1$ (resp., $|\mathbf{p}| > 1$ or $\mathbf{p} = \infty$).

A proof using the set-up of Chapter 2 runs as follows. Fix $\mathbf{p} \in \bar{k}$ of absolute value ≤ 1 (the other case is analogous). Via Taylor expansion around $\mathbf{p}$, we get a k-algebra embedding of Q_1 into $\bar{k}((t))$ mapping x to $t + \mathbf{p}$. This embedding coincides on $k(x)$ with the map $\vartheta_{\mathbf{p}} : \bar{k}(x) \to \bar{k}(t)$ from Section 2.2. Accordingly, forming tensor products we get $F \otimes_{k(x)} \bar{k}((t)) \cong (F \otimes_{k(x)} Q_1) \otimes_{Q_1} \bar{k}((t))$, a direct sum of G-conjugates of $F_1 \otimes_{k(x)} \bar{k}((t)) \cong (F_1 \otimes_{k(x)} Q_1) \otimes_{Q_1} \bar{k}((t))$, and this is compatible with the action of G_1.

From the set-up of Chapter 2 one shows that the class $C_{\mathbf{p}}$ of G is the set of all $g \in G$ with the following property: g fixes a minimal idempotent ϵ of $F \otimes_{k(x)} \bar{k}((t))$, and if s is an element in the ideal (ϵ) with $s^e = t\epsilon$ for some $e \in \mathbb{N}$ then $g(s) = \exp(2\pi\sqrt{-1}/e)s$. The claim follows.

(b) For general k, the corresponding facts can be expressed as follows, using the language of valuation theory. Fix some $i \in \{1, 2\}$, and let $\mathbf{p}$ be a place of Q_i whose valuation ring contains A_i. We can view it as a place of $k(x)$ by Remark 11.10. The completion $\Omega_{\mathbf{p}}$ of $k(x)$ at $\mathbf{p}$ can be canonically identified with the completion of Q_i at $\mathbf{p}$. Thus by the facts stated before (a), we get that $F \otimes_{k(x)} \Omega_{\mathbf{p}}$ is a direct sum of G-conjugates of $F_i \otimes_{k(x)} \Omega_{\mathbf{p}}$, and the embedding of $F_i \otimes_{k(x)} \Omega_{\mathbf{p}}$ is compatible with the action of G_i. This implies the following (see, e.g., [Ca], Ch. 7, for definitions): *The decomposition and inertia groups over* $\mathbf{p}$ *in* $G_i = G(F_i/k(x))$ *are also decomposition and inertia groups over* $\mathbf{p}$, *respectively, in* $G = G(F/k(x))$ *(via the given embedding of* G_i *in* G*).*

Corollary 11.25 *In the situation of Theorem 11.23, assume* G *is the semi-direct product of* G_1 *and* G_2, *with* G_1 *normal. If then* $\varphi : G \to G(F_2/k(x))$ *is a surjection with kernel* G_1, *then the corresponding (split) embedding problem over* $k(x)$ *is solvable. The solution field* $M/k(x)$ *embeds into* $k((x-c))$ *for some* $c \in k$.

Proof.

Case 1 F_2 is A-normalized and contained in Q_1.

As in the proof of Theorem 11.23, we may assume that F_1 is A-normalized and contained in Q_2. Let P_1, P_2 be as in that proof. Let $n = |G|$, and let $K = K_n$ be the composite of all Galois extensions of $k(x)$ inside $P_1 \cap P_2$ of degree $\leq n$. By Corollary 11.20, the group $H = G(K/k(x))$ is n-freely generated by the restriction images H_i of $\tilde{G}_i = G(P_i/Q_i)$ (for $i = 1, 2$). Since restriction $\tilde{G}_2 \to G(F_2/k(x))$ is an isomorphism (see the proof of Theorem 11.23), we also get a restriction isomorphism $\rho_2 : H_2 \to G(F_2/k(x))$.

The given map φ restricts to an isomorphism $\varphi_2 : G_2 \to G(F_2/k(x))$. Let $\lambda_2 : H_2 \to G_2$ be the isomorphism with $\varphi_2 \circ \lambda_2 = \rho_2$. Choose any isomorphism $\lambda_1 : H_1 \to G_1$. By the n-freeness property of H, the maps $\lambda_1 : H_1 \to G_1$ and $\lambda_2 : H_2 \to G_2$ extend to a homomorphism $\lambda : H \to G$. Then $\varphi \circ \lambda = \mathrm{res}_{K/F_2}$ since this holds on H_1 and H_2. Let M be the fixed field in K of $\ker(\lambda)$. Then we get an induced isomorphism $\bar{\lambda} : G(M/k(x)) \cong H/\ker(\lambda) \to G$ with $\varphi \circ \bar{\lambda} = \mathrm{res}_{M/F_2}$. This solves the given embedding problem.

The solution field M lies in $P_1 \cap P_2$, hence in Q. Thus it embeds into $k((x-1))$ by Lemma 11.12.

Case 2 The general case.

By Lemma 11.22 there is $c \in k$ such that $F_2' := \mu_c(F_2)$ satisfies the hypothesis of Case 1. Let φ' be φ composed with the isomorphism $G(F_2/k(x)) \to G(F_2'/k(x))$ induced by μ_c. By Case 1, there is $M'/k(x)$ containing F_2' and an isomorphism $\Psi' : G \to G(M'/k(x))$ such that $\varphi' = \mathrm{res}_{M'/F_2'} \circ \Psi'$. Then the isomorphism $(\mu_c)^{-1} : F_2' \to F_2$ extends to an isomorphism $M' \to M$ for some Galois extension $M/k(x)$. This M solves the original embedding problem.

Further, we know that M' embeds into $k((x-1))$. Hence by Lemma 7.8 there is a generator α' of M' over $k(x)$ satisfying $f'(x, \alpha') = 0$ for some $f'(x, y) \in k[x, y]$, monic in y, such that $f'(1, y)$ is a product of distinct monic linear factors in $k[y]$. Then M has a generator α over $k(x)$ satisfying $f'(c^{-1}x, \alpha) = 0$. The polynomial $f(x, y) := f'(c^{-1}x, y)$ is monic in y, and

$f(c, y)$ is a product of distinct monic linear factors in $k[y]$. Thus M embeds into $k((x-c))$ by Hensel's Lemma 2.1. □

11.4.2 Regular Extensions in Positive Characteristic

Let ℓ be a field, and $\mathbf{x} = (x_1, \ldots, x_m)$ a vector of independent transcendentals over ℓ. As in the case of characteristic 0, a finite Galois extension $K/\ell(\mathbf{x})$ is called **regular** over ℓ if ℓ is algebraically closed in F; and we say a group G **occurs regularly** over ℓ if it is isomorphic to such $G(K/\ell(\mathbf{x}))$. However, if K is not Galois or at least separable over $\ell(\mathbf{x})$ then one needs to modify the definition of regularity in order to preserve the usual properties; see [FJ], 9.2.

Clearly, k is algebraically closed in $k((x))$, hence the fields F and F_i from Theorem 11.23 are regular over k.

The properties of regular extensions from Section 1.1 continue to hold. We need the analogue of Lemma 1.1(ii).

Lemma 11.26 *Let $f(\mathbf{x}, y) \in \ell(\mathbf{x})[y]$ be irreducible over $\ell(\mathbf{x})$, and let $K = \ell(\mathbf{x})[y]/(f)$ be the corresponding field extension of $\ell(\mathbf{x})$. Assume K is Galois over $\ell(\mathbf{x})$. Then K is regular over ℓ if and only if f is irreducible over $\bar{\ell}(\mathbf{x})$. If this holds, then $f(\mathbf{x}, y)$ is irreducible over $\ell_1(\mathbf{x})$ for every extension field ℓ_1 of ℓ such that $x_1, \ldots, x_m, y$ are independent transcendentals over ℓ_1.*

Proof. We may assume that ℓ has positive characteristic. The proof of Lemma 1.1 carries over to this situation, with one exception. Namely, if K is regular over ℓ then the argument from Lemma 1.1 shows only that f is irreducible over $\tilde{\ell}(x)$, where $\tilde{\ell}$ is the separable closure of ℓ. It remains to show that f is irreducible over $\bar{\ell}(x)$.

Indeed, let L (resp., L_0) be the composite of K and $\bar{\ell}(x)$ (resp., $\tilde{\ell}(x)$) inside some algebraic closure of $\ell(x)$. Then L is Galois over $\bar{\ell}(x)$, say with group $\bar{G}$, since $L = \bar{\ell}(x)(\alpha_1, \ldots, \alpha_n)$, where the α_i are the roots of the separable polynomial f. Similarly, $L_0 = \tilde{\ell}(x)(\alpha_1, \ldots, \alpha_n)$ is Galois over $\tilde{\ell}(x)$; call its group $\tilde{G}$. The group $\bar{G}$ permutes the α_i, hence leaves L_0 invariant. Let M be the fixed field of $\bar{G}$ in L_0. Clearly, $M = L_0 \cap \bar{\ell}(x)$. But L_0 is separable and $\bar{\ell}(x)$ is purely inseparable over $\tilde{\ell}(x)$. (For the latter note that for each $a \in \bar{\ell}$ we have $a^q \in \tilde{\ell}$ for some power q of the characteristic of ℓ; hence for each $g(x) \in \bar{\ell}(x)$ there is such q with $g(x)^q = g^{[q]}(x^q) \in \tilde{\ell}(x)$. Here $g^{[q]}$ denotes the polynomial obtained by raising each coefficient of g to the qth power.)

It follows that $M = \tilde{\ell}(x)$. Hence the restriction map $\bar{G} \to \tilde{G}$ is surjective. It is injective because L_0 contains the α_i. Hence $\bar{G} \cong \tilde{G}$. Thus $[L : \ell(x)] = [L_0 : \tilde{\ell}(x)] = \deg(f)$. Since $L = \bar{\ell}(x)(\alpha_1)$ with $f(\alpha_1) = 0$ it follows that f remains irreducible over $\bar{\ell}(x)$. □

11.4.3 Galois Realizations of Cyclic Groups

Theorem 11.23 reduces us to the case of cyclic p-groups. Such groups can be realized over $\ell(x)$ for any field ℓ, in many ways (see [FJ], [Ha1], [Liu], [Ma1], [Se1],). We begin with the case $p \neq \text{char}(\ell)$. Our treatment of this case is close to that in [Liu].

Lemma 11.27 *Let ℓ be any field, and n a positive integer not divisible by the characteristic of ℓ. Then there is a cyclic Galois extension $F/\ell(x)$ of degree n, contained in $\ell((x))$.*

Proof. Under our assumptions, the equation $y^n - 1 = 0$ is separable over ℓ, hence the nth roots of unity generate a Galois extension L of ℓ. Let $\xi_n \in L$ be a primitive nth root of unity. For each γ in $\Gamma := G(L/\ell)$ we have $\gamma(\xi_n) = \xi_n^{\chi(\gamma)}$ for some $\chi(\gamma) \in \{1, \ldots, n-1\}$. The group Γ acts naturally on the field $L((x))$ (via action on the coefficients of a Laurent series), and this identifies Γ with the Galois group of $L((x))$ over $\ell((x))$.

Suppose $g(x) \in L[x]$ is of the form $g(x) = (x-a)\tilde{g}(x)$ with $\tilde{g}(a) \neq 0$ (and $a \in L$). Then the polynomial $y^n - g(x)$ is separable and irreducible as polynomial in y over $L(x)$ (by Eisenstein's criterion). Since L contains the nth roots of unity, we obtain a Galois extension of $L(x)$ of degree n by adjoining an element u with $u^n = g(x)$. Its Galois group is cyclic, generated by an element ω with $\omega(u) = \xi_n u$. If $g(0) \neq 0$ is an nth power in L then we can take $u \in L[[x]]$ by Hensel's Lemma 2.1 (because the polynomial $y^n - g(0)$ splits into n distinct linear factors over L).

Now let $b \neq 0$ be any generator of L over ℓ, and set

$$g(x) = \prod_{\gamma \in \Gamma} (1 + \gamma(b)x)^{\chi(\gamma^{-1})},$$

an element of $L[x]$. Let $v \in L[[x]]$ with $v^n = 1 + bx$ (exists by the above). Then

$$u = \prod_{\gamma \in \Gamma} \gamma(v)^{\chi(\gamma^{-1})}$$

lies in $L((x))$ and satisfies $u^n = g(x)$. Hence $F_L := L(x)(u)$ is Galois over $L(x)$ of degree n with group $\langle \omega \rangle$, where $\omega(u) = \xi_n u$. For each $\gamma \in \Gamma$ we have

$$\gamma(u) = u^{\chi(\gamma)} f(x), \qquad \text{with} \qquad f(x) \in L(x),$$

which is proved by a straightforward computation using $\chi(\gamma_1\gamma_2) \equiv \chi(\gamma_1)\chi(\gamma_2)$

(mod n), in particular $\chi(\gamma)\chi(\gamma^{-1}) \equiv 1 \pmod{n}$, and $v^n \in L(x)$. It follows that Γ leaves F_L invariant. For each $\gamma \in \Gamma$ we have

$$\gamma\omega(u) = \omega\gamma(u). \tag{11.2}$$

Indeed, with $m = \chi(\gamma)$ we get $\gamma\omega(u) = \gamma(\xi_n u) = \xi_n^m u^m f(x) = \omega(u)^m f(x) = \omega(u^m f(x)) = \omega\gamma(u)$.

Let Γ_0 be the group of automorphisms of F_L induced by Γ, and let Λ be the group generated by Γ_0 and ω. Clearly $(F_L)^\Lambda = \ell(x)$, hence $|\Lambda| = |\Gamma_0| \cdot |\langle\omega\rangle|$. Thus it follows from (11.2) that Λ is the direct product of Γ_0 and $\langle\omega\rangle$. Hence the field $F := (F_L)^{\Gamma_0}$ is Galois over $\ell(x)$ with Galois group $\cong \langle\omega\rangle \cong \mathbb{Z}/n\mathbb{Z}$. Clearly, $F \subset L((x))^\Gamma = \ell((x))$. □

Clearly, if $F/\ell(x)$ is contained in $\ell((x))$ then it is regular over ℓ. For abelian extensions there is a partial converse.

Lemma 11.28 *Let ℓ be an infinite field, and G a finite abelian group. If there is a Galois extension $F/\ell(x)$, regular over ℓ, with group isomorphic to G, then there is such F contained in $\ell((x))$.*

Proof. Let α be a generator of F over $\ell(x)$ with minimum polynomial $f(y) = f(x, y) \in \ell[x, y]$, monic in y. Then $f(y)$ is separable, hence by Lemma 1.6 there is $a \in \ell$ such that the polynomial $f(a, y) \in \ell[y]$ is separable. (We use here that ℓ is infinite.) By a change of coordinates we may assume $a = 0$. Then it follows from Hensel's Lemma 2.1 that we can take $\alpha \in L[[x]]$, where L is the splitting field of $f(0, y)$ over ℓ. Set $\Gamma = G(L/\ell)$, and identify Γ naturally with the Galois group of $L((x))$ over $\ell((x))$ (as in the proof of the preceding lemma).

Since F is regular over ℓ, the polynomial $f(y)$ remains irreducible over $L(x)$ (Lemma 11.26). Thus $F_L := L(x)(\alpha)$ is Galois over $L(x)$, and its Galois group G_L is isomorphic to $G(F/\ell(x))$ via restriction to F. Clearly, $F_L = L(x)F$ is Galois over $\ell(x)$ and

$$\Lambda := G(F_L/\ell(x)) = G(F_L/F) \times G_L.$$

Since $G_L \cong G$ is abelian it follows that G_L lies in the center of Λ.

Further, F_L is invariant under Γ because Γ fixes the minimum polynomial $f(y)$ of α. Let Γ_0 denote again the subgroup of Λ induced by Γ. Using the fact that G_L lies in the center of Λ, it follows as in the preceding proof that Λ is the direct product of Γ_0 and G_L. Thus $F' := (F_L)^{\Gamma_0}$ is Galois over $\ell(x)$ with Galois group $\cong G_L \cong G$. Again, $F' \subset L((x))^\Gamma = \ell((x))$. □

Now we turn to the realization of cyclic p-groups over a field of characteristic p. We reproduce the basic Lemma 24.42 from [FJ] (based on an exercise in Lang's "Algebra").

Lemma 11.29 *Let K be a field of characteristic $p > 0$. Then each cyclic Galois extension F/K of degree $p^m \geq p$ can be embedded in a cyclic extension of K of degree p^{m+1}.*

Proof. Recall the trace function $T_{F/K}(b) = \sum_{g \in G(F/K)} g(b)$. The trace function of a Galois extension does not vanish identically (see [Jac], I, p. 285), hence $T_{F/K} : F \to K$ is a nonzero K-linear map. Thus its kernel H is a K-hyperplane of F. Let σ be a generator of $G(F/K)$. Then $F \to H$, $a \mapsto \sigma(a) - a$ is a K-linear map with kernel K. By comparing dimensions, this map is surjective, that is, every element of H is of the form $\sigma(a) - a$. (This is the additive analogue of Hilbert's Theorem 90.)

Let $\mathcal{P}(y) = y^p - y \in F[y]$ (the Artin – Schreier polynomial). This polynomial is additive (i.e., $\mathcal{P}(u+v) = \mathcal{P}(u)+\mathcal{P}(v)$), and its roots are the elements of the prime field of F. Thus for any $a \in F$ the roots of the polynomial $\mathcal{P}(y) - a$ are of the form $\alpha, \alpha + 1, \ldots, \alpha + p - 1$. If $\alpha \notin F$ then $F' := F(\alpha)$ is a cyclic extension of F of degree p.

Pick $b \in F \setminus H$. Dividing b by its trace we may assume $T_{F/K}(b) = 1$. Then $T_{F/K}(\mathcal{P}(b)) = \mathcal{P}(T_{F/K}(b)) = \mathcal{P}(1) = 0$. Thus by the additive analogue of Hilbert 90, there is $a \in F$ with $\sigma(a) - a = \mathcal{P}(b)$. Fix such $a \in F$, and let α be a root of $\mathcal{P}(y) - a$. If $\alpha \in F$ then $\mathcal{P}(\sigma(\alpha) - \alpha) = \sigma\mathcal{P}(\alpha) - \mathcal{P}(\alpha) = \sigma(a) - a = \mathcal{P}(b)$; hence $\mathcal{P}(\sigma(\alpha) - \alpha - b) = 0$ which means that $\sigma(\alpha) - \alpha - b$ lies in the prime field of F and thus in H. (Note that $T_{F/K}(1) = p^m \cdot 1 = 0$ since $m > 0$.) This contradicts our assumption $b \notin H$. Hence $\alpha \notin F$. Thus $F' = F(\alpha)$ has degree p over F.

We have $\mathcal{P}(\alpha + b) = \mathcal{P}(\alpha) + \mathcal{P}(b) = a + \mathcal{P}(b) = \sigma(a)$ (by choice of a). Hence $\alpha + b$ is a root of $\mathcal{P}(y) - \sigma(a) = \sigma(\mathcal{P}(y) - a)$, which means that σ extends to an automorphism σ' of F', mapping α to $\alpha + b$. We have $(\sigma')^j(\alpha) = \alpha + \sigma^{j-1}(b) + \cdots + \sigma(b) + b$, hence $(\sigma')^{p^m}(\alpha) = \alpha + T_{F/K}(b) = \alpha + 1 \neq \alpha$. Hence σ' has order p^{m+1}, which means that F' is Galois and cyclic over K of degree p^{m+1}. □

Now we can state the main result of this subsection.

Proposition 11.30 *Let ℓ be an infinite field, and $q = p^m$ a power of a prime p. Then there is a cyclic Galois extension $F/\ell(x)$ of degree q, contained in $\ell((x))$.*

Proof. By the first lemma in this subsection, it only remains to consider the case $p = \mathrm{char}(\ell)$. We may also assume $q > 1$. View $y^p - y - x$ as polynomial in y over $\ell(x)$. This polynomial has no root in $\ell[x]$, hence by Gauss' Lemma it has no root in $\ell(x)$. Adjoining a root yields a cyclic extension $F_0/\ell(x)$ of degree p (see above).

The preceding lemma shows that there is a cyclic extension $F/\ell(x)$ of degree $q = p^m$, containing F_0. By the Galois correspondence, each proper field extension of $\ell(x)$ inside F contains F_0. It follows that if the algebraic closure L of ℓ in F is $\neq \ell$ then $F_0 \subset L(x)$. Hence $y^p - y - x$ has a root in $L(x)$, which is again impossible. It follows that F is regular over ℓ. Now the claim follows from Lemma 11.28. □

11.4.4 Galois Groups and Embedding Problems over $k(x)$

Now we return to our complete (nontrivially) valued field k. Note that each finite group is generated by cyclic subgroups of prime power order. Thus inductive use of Proposition 11.30 and Theorem 11.23 yields the main result of Chapter 11:

Theorem 11.31 (Harbater) *Let k be a field equipped with a nontrivial ultrametric absolute value, and complete with respect to this absolute value. Then each finite group G occurs regularly over k. More precisely, G is isomorphic to the Galois group of a Galois extension $F/k(x)$ contained in $k((x))$.*

In particular, this applies for $k = \mathbb{Q}_p$ (the p-adic fields) and the Laurent series fields $k = k_0((t))$ over any field k_0.

Now we are ready to generalize Theorem 7.14 to any algebraically closed field ℓ. We use the cohomological result that $\ell(x)$ has cohomological dimension 1 (see [Se2], II, 3.3), that is, all Frattini embedding problems over $\ell(x)$ are solvable (see Section 8.3.3).

Theorem 11.32 (Harbater and Pop) *If ℓ is any algebraically closed field then all (finite) embedding problems over $\ell(x)$ are solvable.*

Proof. Since $\mathrm{cdim}(\ell(x)) = 1$ it suffices to show that all split embedding problems over $\ell(x)$ are solvable (by Lemma 8.24). So consider a split embedding problem given by a finite Galois extension $M/\ell(x)$ and (split) surjection $\varphi : H \to G(M/\ell(x))$.

We see as in the proof of Lemma 11.28 that M embeds into $\ell((x - a))$ for some $a \in \ell$ (since ℓ is algebraically closed). Replacing $x - a$ by x we may assume that M embeds into $\ell((x))$. Pick a generator β of $M/\ell(x)$, and let $g(y) \in \ell(x)[y]$ be its minimum polynomial over $\ell(x)$.

Now let k be an extension field of ℓ that is complete with respect to a nontrivial ultrametric absolute value. For example, $k = \ell((t))$ works by Example 11.5. View $\ell((x))$ as a subfield of $k((x))$ in the natural way. Then $\tilde{M} = k(x, \beta)$ is a subfield of $k((x))$ that is Galois over $k(x)$, and restriction gives an isomorphism $\rho : G(\tilde{M}/k(x)) \to G(M/\ell(x))$. The latter follows from the fact that $g(y)$ remains irreducible over $k(x)$ (Lemma 11.26).

The map $\tilde{\varphi} = \rho^{-1} \circ \varphi : H \to G(\tilde{M}/k(x))$ yields a split embedding problem over $k(x)$. Its kernel is isomorphic to the group of a Galois extension of $k(x)$ inside $k((x))$ by Theorem 11.31. Hence Corollary 11.25 shows that this split embedding problem over $k(x)$ is solvable; and the solution field can be embedded into $k((x - c))$ for some $c \in k$, hence is regular over k. Thus there is a Galois extension $\tilde{K}/k(x)$, regular over k and containing $\tilde{M}$, and an isomorphism $\tilde{\psi} : H \to G(\tilde{K}/k(x))$ with $\tilde{\varphi} = \mathrm{res}_{\tilde{K}/\tilde{M}} \circ \tilde{\psi}$. In particular,

$$\tilde{\psi}(h)(\beta) = \varphi(h)(\beta) \qquad \text{for all} \qquad h \in H. \tag{11.3}$$

We proceed as in the proof of Theorem 7.9. Consider again a generator α for $\tilde{K}/k(x)$ such that $f(x, \alpha) = 0$ for some polynomial $f \in k[x, y]$ that is irreducible and monic as polynomial in y over $k(x)$. Again we obtain a finitely generated ℓ-algebra B in k such that $f \in B[x, y]$, and such that the following holds for the field of fractions $k_B \subset k$ of B: The fields $K_B = k_B(x, \alpha)$ and $M_B = k_B(x, \beta)$ are Galois over $k_B(x)$, with $M_B \subset K_B$, and restriction gives isomorphisms $G(\tilde{K}/k(x)) \to G(K_B/k_B(x))$ and $G(\tilde{M}/k(x)) \to G(M_B/k_B(x))$.

The polynomial $f(y)$ remains irreducible over $\bar{k}(x)$ since $\tilde{K}$ is regular over k (Lemma 11.26). Since $f(y) = f(x, y)$ is monic in y, it is also irreducible in $\bar{k}[x, y]$ (Lemma 1.4). Hence f is irreducible in $\bar{k}_B[x, y]$. Thus Lemma 7.6 yields $b \neq 0$ in B such that f^ω is irreducible in $\ell[x, y]$ for each homomorphism $\omega : B \to \ell$ with $\omega(b) \neq 0$. Now apply Lemma 1.5, taking for K/F the extension $K_B/k_B(x)$ and $R = B[x]$, and requiring $\alpha, \beta \in A$. We obtain $u \neq 0$ in R with the properties from that Lemma. Define $c \in B$ as product of the above element b with the nonzero coefficients of u (where $u \in R = B[x]$ is viewed as polynomial in x). The ring $B[c^{-1}]$ is a finitely generated algebra over ℓ, hence has an ℓ-algebra homomorphism to ℓ (Lemma 7.5). Its restriction to B yields a homomorphism $\omega : B \to \ell$ with $\omega(c) \neq 0$. Extend ω to a homomorphism $\omega : R = B[x] \to \ell[x]$ (that fixes x). Since $\omega(u) \neq 0$, the map ω extends to $\omega : S \to K'$, where S is a subring of K_B containing $B[x, \alpha, \beta]$, and K' is a finite Galois extension of $\ell(x)$. Since $\omega(b) \neq 0$, we have f^ω irreducible, hence there is an isomorphism $\omega' : G(K_B/k_B(x)) \to G(K'/\ell(x)), \sigma \mapsto \sigma'$, such that $\omega(\sigma(s)) = \sigma'(\omega(s))$ for all $s \in S$.

Set $\beta' = \omega(\beta)$ and $M' = \ell(x, \beta')$. Then β' satisfies the same irreducible polynomial $g(y)$ as β (since ω is an $\ell[x]$-algebra homomorphism). Hence M' is $\ell(x)$-isomorphic to M. We identify M' with M by identifying β with β'. Then the isomorphism $\omega^* = \omega' \circ \mathrm{res}_{\tilde{K}/K_B} : G(\tilde{K}/k(x)) \to G(K'/\ell(x))$ satisfies $\omega^*(\sigma)(\beta) = \sigma(\beta)$ for all $\sigma \in G(\tilde{K}/k(x))$.

Finally, $\psi := \omega^* \circ \tilde{\psi} : H \to G(K'/\ell(x))$ is an isomorphism with $\mathrm{res}_{K'/M'} \circ \psi = \varphi$. Indeed, for each $h \in H$ we have $(\mathrm{res}_{K'/M'} \circ \psi)(h)(\beta) = \psi(h)(\beta) = \omega^*(\tilde{\psi}(h))(\beta) = \tilde{\psi}(h)(\beta) = \varphi(h)(\beta)$ by (11.3). Thus ψ solves the given embedding problem over $\ell(x)$. □

Remark 11.33 The statement of the above theorem has been called the "**geometric case of Shafarevich's Conjecture.**" Indeed, Shafarevich's Conjecture (see Section 8.3.3) expects solvability of all embedding problems over the field $\mathbb{Q}_{\mathrm{ab}}$ generated over $\mathbb{Q}$ by all roots of unity. The function field analogue of $\mathbb{Q}$ is the field $\mathbb{F}_p(x)$ (with p a prime). Adjoining all roots of unity to this field we obtain the field $\bar{\mathbb{F}}_p(x)$ (to which the above theorem applies).

As indicated in Chapter 8, the solvability of all embedding problems over a countable field determines its absolute Galois group: It is free profinite of countable rank. Such a group has the property that any closed subgroup of finite index is again free (of same rank); see [FJ], Prop. 15.27. Since K as below is finite separable over $\ell(x)$ ([FJ], Lemma 9.5), we get

Corollary 11.34 *Let ℓ be countable and algebraically closed. Let K be a field that is finitely generated and of transcendence degree 1 over ℓ. Then the absolute Galois group of K is free profinite of countable rank.*

This remains true without the restriction to the countable case. See [Ha3] and [P3].

Exercise Let ℓ be an algebraically closed field of characteristic $p > 0$. Let $F/\ell(x)$ be a finite Galois extension. Extending the terminology from the characteristic zero case, we call an element $a \in \ell$ (resp., $a = \infty$) a **branch point** of $F/\ell(x)$ if F does not embed into $\ell((x-a))$ (resp., into $\ell((x^{-1}))$).

(a) Let $f(x) \in \ell[x]$ not be of the form $g(x)^p - g(x)$ with $g(x) \in \ell[x]$. Show that the field $F = \ell(x, \alpha)$, where $\alpha^p - \alpha = f(x)$, is a cyclic extension of $\ell(x)$ of degree p having ∞ as only branch point. (Use Hensel's Lemma.)

(b) Let G be a finite group generated by r elements of order p. Show that G occurs as Galois group of an extension of $\ell(x)$ with r branch points. (Use the construction in Theorem 11.23, and Remark 11.24, plus a descent from k to ℓ as for Theorem 11.32, keeping track of branch points as in the proof of Theorem 7.9.) Actually, G occurs as Galois group of an extension of $\ell(x)$ with only one branch point by Raynaud [Ra] (see the discussion at the end of Section 11.3); this does not seem to be obtainable by our elementary method.

References

[Abh1] Abhyankar, S. S. *Algebraic Geometry for Scientists and Engineers.* Providence, RI: AMS, 1990.

[Abh2] Abhyankar, S. S. Several preprints entitled "Nice equations for nice groups," "More nice equations for nice groups," "Further nice equations for nice groups," "Again nice equations for nice groups." Purdue University, 1994–95.

[RDIGP] Abhyankar, S. S., W. Feit, M. Fried, and H. Völklein, eds. *Recent Developments in the Inverse Galois Problem, Proceedings volume of the 1993 Seattle Joint AMS Summer Conference.* Contemp. Math., Vol. 186. Providence, RI: AMS, 1995.

[At] Conway, J. H., R. T. Curtis, S. P. Norton, R. A. Parker, and R. A. Wilson. *Atlas of Finite Groups.* New York: Clarendon, 1985.

[BF1] Biggers, R., and M. Fried. Moduli spaces for covers of P^1 and representations of the Hurwitz monodromy group. *J. Reine Angew. Math.* 335 (1982), 87–121.

[BF2] Biggers, R., and M. Fried, Irreducibility of moduli spaces of cyclic unramified covers of genus g curves. *Transactions AMS* 295 (1986), 59–70.

[Bir] Birman, J. S., *Braids, links, and mapping class groups.* Princeton University Press, 1974.

[Ca] Cassels, J. W. S. *Local Fields.* Cambridge University Press, 1986.

[Cleb] Clebsch, A. Zur Theorie der Riemann'schen Fläche. *Math. Annalen* 6 (1872), 216–230.

[CR] Curtis, W., and I. Reiner. *Methods of Representation Theory II.* Wiley, 1987.

[DF1] Dèbes, P., and M. Fried. Rigidity and real residue class fields. *Acta Arithm.* 56 (1990), 291–323.

[DF2] Dèbes, P., and M. Fried. Non-rigid constructions in Galois theory. *Pacific J. Math.* 163 (1994), 81–122.

[Desch] Deschamps, Bruno. Existence de points p-adiques pout tout p sur un espace de Hurwitz, in [RDIGP], pp. 239–47.

[D] Dieudonné, J. *La Géométrie des Groupes Classiques.* Berlin: Springer Verlag, 1955.

[Do] Dörge, K. Ein einfacher Beweis des Hilbertschen Irreduzibilitätssatzes. *Math. Annalen* 96 (1927), 176–82.

[E] Eichler, M. *Introduction to the Theory of Algebraic Numbers and Functions.* New York: Academic Press, 1966.

[Fo] Forster, O. *Riemannsche Flächen.* Heidelberg: Springer, 1977.

[FP] Fresnel, J., and M. van der Put. Géométrie Analytique Rigide et Applications. *Prog. in Math.* 18, Birkhäuser, 1981.

[Fr1] Fried, M. Fields of definition of function fields and Hurwitz families – groups as Galois groups. *Comm. Algebra* 5 (1977), 17–82.
[Fr2] Fried, M. On a problem of Schur. *Michigan Math. J.* 17 (1970), 41–55.
[Fr3] Fried, M. Rigidity and applications of the classification of simple groups to monodromy, Part II. Unpublished preprint, 1986.
[Fr4] Fried, M. Exposition on an Arithmetic Group-Theoretic Connection via Riemann's Existence Theorem, in *Proceedings of Symposia in Pure Math: Santa Cruz Conference on Finite Groups*. A.M.S. Publications 37 (1980), 571–601.
[Fr5] Fried, M. On the Sprindzuk–Weissauer approach to universal Hilbert subsets. *Israel J. Math.* 51 (1985), 347–63.
[Fr6] Fried, M. Alternating groups and lifting invariants. Preprint, 1989.
[Fr7] Fried, M. Extension of constants, rigidity and the Chowla–Zassenhaus conjecture. *Finite Fields and their applications* 1 (1995), 326–59.
[Fr8] Fried, M. Riemann's existence theorem and applications. Partial book manuscript.
[FGS] Fried, M., R. Guralnick, and J. Saxl. Schur covers and Carlitz's conjecture. *Israel J. Math.* 82 (1993), 157–225.
[FHV] Fried, M., D. Haran, and H. Völklein. The absolute Galois group of the totally real numbers. *Compt. Rend. Acad. Sci. Paris* 317 (1993), 95–99.
[FJ] Fried, M., and M. Jarden. *Field Arithmetic, Ergebn. Math. und Ihrer Grenzgeb.* 11, Springer Verlag, 1986.
[FJ2] Fried, M., and M. Jarden. Diophantine properties of subfields of $\bar{\mathbb{Q}}$. *Amer. J. Math.* 100 (1978), 653–66.
[FV1] Fried, M., and H. Völklein. The inverse Galois problem and rational points on moduli spaces. *Math. Annalen* 290 (1991), 771–800.
[FV2] Fried, M., and H. Völklein. The embedding problem over a Hilbertian PAC-field. *Annals of Math.* 135 (1992), 469–81.
[Fu] Fulton, W. Hurwitz schemes and irreducibility of moduli of algebraic curves. *Annals of Math.* 90 (1969), 542–75.
[Gor] Gorenstein, D. *Finite Groups*. New York and London: Harper and Row, 1968.
[GT] Guralnick, R. M., and J. G. Thompson. Finite groups of genus zero. *J. Algebra* 131 (1990), 303–41.
[Han] Hansen, V. L. *Braids and Coverings*. Cambridge University Press, 1989.
[HJ] Haran, D., and M. Jarden. Compositum of Galois extensions of hilbertian fields. *Ann. scient. Éc. Norm. Sup.* 24 (1991), 739–48.
[HV] Haran, D., and H. Völklein. Galois groups over complete valued fields. Preprint, 1994.
[Ha1] Harbater, D. Galois coverings of the arithmetic line, in: *Lecture Notes in Math.* 1240. Heidelberg: Springer, 1987, pp. 165–95.
[Ha2] Harbater, D. Abhyankar's Conjecture on Galois Groups over Curves. *Invent. Math.* 117 (1994), 1–25.
[Ha3] Harbater, D. Fundamental Groups and Embedding Problems in Characteristic *p*, in [RDIGP], pp. 353–69.
[Hsh] Hartshorne, R. *Algebraic Geometry*. Springer Graduate Texts in Math. 52 (1977).
[Hup] Huppert, B. *Endliche Gruppen I*. Springer–Grundlehren 134, 1983.
[Hur] Hurwitz, A. Über Riemann'sche Flächen mit gegebenen Verzweigungspunkten. *Math. Annalen* 39 (1891), 1–61.
[Jac] Jacobson, N. *Basic Algebra I, II*. San Francisco: Freeman, 1974, 1980.
[Ja] Jarden, M. Infinite Galois Theory, in *Encyclopedia Math.*, in press.
[Klein] Klein, F. Über binäre Formen mit linearen Transformationen in sich selbst. *Math. Annalen* 9 (1876), 183–208.

[Lam] Lamotke, K. *Regular solids and isolated singularities*. Braunschweig: Vieweg, 1986.

[La] Lang, S. *Fundamentals of Diophantine Geometry*. New York: Springer, 1983.

[Liu] Qing Liu, Tout groupe fini est un groupe de Galois sur $\mathbf{Q}_p(T)$, d'après Harbater, in [RDIGP], pp. 261–65.

[MSV] Magaard, K., K. Strambach, and H. Völklein. Finite quotients of the symplectic braid group, in preparation.

[Malle1] Malle, G. Exceptional groups of Lie type as Galois groups. *J. reine angew. Math.* 392 (1988), 70–109.

[Malle2] Malle, G. Polynome mit Galoisgruppen $PGL_2(p)$ und $PSL_2(p)$ über $\mathbf{Q}(t)$. *Comm. Algebra* 21 (1993), 511–526.

[MM] Malle, G., and B. H. Matzat. *Inverse Galois Theory*. IWR preprint series, Heidelberg, 1993–95.

[MM2] Malle, G., and B. H. Matzat. Realisierung von Gruppen $PSL_2(\mathbb{F}_p)$ als Galoisgruppen über $\mathbb{Q}$, *Mathematische Annalen* 272 (1985), 549–65.

[Mas] Massey, W. S. *Algebraic Topology: An Introduction*, GTM, Vol. 56. Springer Verlag 1978.

[Ma1] Matzat, B. H. *Konstruktive Galoistheorie*, Lecture Notes in Math., Vol. 1284. Heidelberg: Springer, 1987.

[Ma2] Matzat, B. H. Ueber das Umkehrproblem der Galoisschen Theorie, *Jber. d. Dt. Math.-Verein.* 90 (1988), 155–83.

[Ma3] Matzat, B. H. Der Kenntnisstand in der konstruktiven Galoisschen Theorie, in G.O. Michler and C. M. Ringel, eds., *Representation Theory of Finite Groups and Finite-Dimensional Algebras*. Birkhäuser-Verlag, 1991, pp. 65–98.

[Ma4] Matzat, B. H. Zum Einbettungsproblem der algebraischen Zahlentheorie mit nichtabelschem Kern. *Invent. Math.* 80 (1985), 365–74.

[Ma5] Matzat, B. H. Rationality Criteria for Galois extensions, in *Galois Groups over* $\mathbb{Q}$, Proceedings of a 1987 MSRI workshop, eds. Ihara, Ribet, and Serre. Springer Verlag, 1989.

[Mi] Mitchell, M. H. Determination of the finite quaternary linear groups, *Transactions AMS* 14 (1913), 123–42.

[Mü] Müller, P. Monodromiegruppen rationaler Funktionen und Polynome mit variablen Koeffizienten. Dissertation, Erlangen, 1994.

[P1] Pop, F. *Fields of totally Σ-adic numbers*. Preprint, Heidelberg, 1992.

[P2] Pop, F. *Hilbertian fields with a universal local-global principle*. Preprint, Heidelberg, 1994.

[P3] Pop, F. *The geometric case of a conjecture of Shafarevich*. Preprint, Heidelberg, 1993.

[P4] Pop, F. $\frac{1}{2}$ *Riemann Existence Theorem with Galois action, Algebra and Number Theory*, ed., Frey–Ritter, de Gruyter Proceedings in Math. Berlin, New York, 1994.

[Ra] Raynaud, M. Revêtements de la droite affine en caractéristique $p > 0$ et conjecture d'Abhyankar. *Invent. Math.* 116 (1994), 425–62.

[Ru] Rudin, W. *Real and complex analysis*. New York: McGraw-Hill, 1966.

[Se1] Serre, J-P. *Topics in Galois Theory*. Boston: Jones and Bartlett, 1992.

[Se2] Serre, J-P. *Cohomologie Galoisienne, Lecture Notes in Math.*, Vol. 5, Heidelberg: Springer, 1964.

[Se3] Serre, J-P. *Corps Locaux*. Paris, Hermann, 1962.

[Shatz] Shatz, S. S. *Profinite Groups, Arithmetic and Geometry*, Annals of Math. Studies, Vol. 67, Princeton University Press, 1972.

[SV1] Strambach, K., and H. Völklein. Generalized braid groups and rigidity. J. Algebra 175 (1995), 604–15.

[SV2] Strambach, K., and H. Völklein. *Galois Realizations and Symplectic Braid Groups*. Preprint, 1995.

[Th1] Thompson, J. G. Some finite groups which appear as $Gal(L|K)$, where $K \subset \mathbb{Q}(\mu_n)$. *J. Algebra* 89 (1984), 437–99.

[Th2] Thompson, J. G. $GL_n(q)$, rigidity and the braid group. *Bull. Soc. Math. Belg.* 17 (1990), 723–33.

[V1] Völklein, H. $GL_n(q)$ as Galois group over the rationals. *Math. Annalen* 293 (1992), 163–76.

[V2] Völklein, H. Braid group action via $GL_n(q)$ and $U_n(q)$, and Galois realizations. *Israel J. Math.* 82 (1993), 405–27.

[V3] Völklein, H. Moduli spaces for covers of the Riemann sphere. *Israel J. Math.* 85 (1994), 407–30.

[V4] Völklein, H. Braid group action, embedding problems and the groups $PGL(n, q)$, $PU(n, q^2)$. *Forum Math.* 6 (1994), 513–35.

[V5] Völklein, H. $PSL_2(q)$ and extensions of $\mathbb{Q}(x)$. *Bull. AMS* 24 (1991), 145–53.

[V6] Völklein, H. Cyclic covers of P^1, and Galois action on division points, in [RDIGP], pp. 91–107.

[W] Weissauer, R. Der Hilbertsche Irreduzibilitätssatz. *J. reine angew. Math.* 334 (1982), 203–20.

[Wag] Wagner, A. Collineation groups generated by homologies of order greater than 2. *Geom. Dedicata* 7 (1978), 387–98.

[Wa] Waterhouse, W. C. *Introduction to Affine Group Schemes*. Springer GTM 66, 1979.

Index